Springer Series in Biomaterials Science and Engineering

Volume 13

Series editor
Prof. Min Wang
Department of Mechanical Engineering
The University of Hong Kong
Pokfulam Road, Hong Kong
e-mail: memwang@hku.hk

Aims and scope

The Springer Series in Biomaterials Science and Engineering addresses the manufacture, structure and properties, and applications of materials that are in contact with biological systems, temporarily or permanently. It deals with many aspects of modern biomaterials, from basic science to clinical applications, as well as host responses. It covers the whole spectrum of biomaterials – polymers, metals, glasses and ceramics, and composites/hybrids – and includes both biological materials (collagen, polysaccharides, biological apatites, etc.) and synthetic materials. The materials can be in different forms: single crystals, polycrystalline materials, particles, fibers/wires, coatings, non-porous materials, porous scaffolds, etc. New and developing areas of biomaterials, such as nano-biomaterials and diagnostic and therapeutic nanodevices, are also focuses in this series. Advanced analytical techniques that are applicable in R & D and theoretical methods and analyses for biomaterials are also important topics. Frontiers in nanomedicine, regenerative medicine and other rapidly advancing areas calling for great explorations are highly relevant.

The Springer Series in Biomaterials Science and Engineering aims to provide critical reviews of important subjects in the field, publish new discoveries and significant progresses that have been made in both biomaterials development and the advancement of principles, theories and designs, and report cutting-edge research and relevant technologies. The individual volumes in the series are thematic. The goal of each volume is to give readers a comprehensive overview of an area where new knowledge has been gained and insights made. Significant topics in the area are dealt with in good depth and future directions are predicted on the basis of current developments. As a collection, the series provides authoritative works to a wide audience in academia, the research community, and industry.

More information about this series at http://www.springer.com/series/10955

Jun Fu • Zhong-Min Jin • Jin-Wu Wang
Editors

UHMWPE Biomaterials for Joint Implants

Structures, Properties and Clinical Performance

Springer

Editors
Jun Fu
Ningbo Institute of Materials
Technology & Engineering
Chinese Academy of Sciences
Ningbo, Zhejiang, China

Zhong-Min Jin
Tribology Research Institute, School of
Mechanical Engineering
Southwest Jiaotong University
Chengdu, Sichuan, China

Jin-Wu Wang
Shanghai Key Laboratory
of Orthopaedic Implants
Department of Orthopaedic Surgery
Shanghai Ninth People's Hospital
Shanghai Jiaotong University
School of Medicine
Shanghai, China

ISSN 2195-0644 ISSN 2195-0652 (electronic)
Springer Series in Biomaterials Science and Engineering
ISBN 978-981-13-6926-1 ISBN 978-981-13-6924-7 (eBook)
https://doi.org/10.1007/978-981-13-6924-7

This Springer imprint is published by the registered company Springer Nature Singapore Pte Ltd.
The registered company address is: 152 Beach Road, #21-01/04 Gateway East, Singapore 189721, Singapore

Preface

Joint arthroplasty has been the clinical choice to restore the function of malfunctioning joints. Artificial joint prosthesis is used to replace the diseased joints. In the components of contemporary joint implants, the articulating surfaces have been widely recognized critical for the in vivo performance and longevity of the prosthesis. The articulating surface configurations can be classified as metal-on-metal, ceramic-on-ceramic, metal-on-polyethylene, or ceramic-on-polyethylene. Here, the polyethylene is referred to medical-grade ultrahigh molecular weight polyethylene, or UHMWPE, with the average molecular weight over 1,500,000 g per mole. UHMWPE is advantageous over most other materials for its excellent biocompatibility, bioinertness, chemical resistance, high strength and toughness, low coefficient of friction, and low wear, which are desired for artificial joints. It has been the material of choice for both hip and knee since the early 1960s, when John Charnley first introduced UHMWPE to manufacture the articular liners for hip prosthesis.

The aim of this book is to give a general and comprehensive introduction to the fundamental of UHMWPE as biomaterials for joint implants, major challenges of UHMWPE component in history, and the most important recent progresses of high-performance UHMWPE. The chemistry, physics, and mechanisms of the strength, toughness, wear, oxidation, and stabilization are discussed in detail. Moreover, biomechanics and biotribology of artificial prosthesis are introduced. It also presents clinical diagnostic methods to analyze the in vivo performance and to diagnose or predict the failure of the prosthesis.

In history, UHMWPE implants have experienced very bad problems in clinical use. The cyclic articulating of femoral head against the UHMWPE liner produces numerous wear debris that migrate to the surrounding tissue and cause osteolysis. Moreover, early sterilization by gamma has made it even worse. The free radicals generated during radiation sterilization cause cascade oxidation of polyethylene on-shelf and in vivo, which severely deteriorates the strength and toughness of the polymer and accelerates wear and failure of implants. In the 1980s, wear-related problems are reported to account for a large amount of the revisional surgeries.

The bad demand of wear-resistant materials has stimulated extensive innovations to improve the wear resistance of UHMWPE. In the 1990s, diverse strategies have been developed to reduce the wear rate. Based on the wear mechanism that polyethylene fibers are generated due to plastic flow and fractured upon cyclic shear, cross-linking of the polymer chains is proposed to reduce wear by counteracting against the mechanism. High-dose radiation cross-linking has been successfully used to reduce the wear rate of polyethylene by 90% or more. Highly cross-linked UHMWPE has been approved for clinical applications in 1998. Since then, millions of joint implants based on highly cross-linked UHMWPE have been implanted in patients. The 20-year clinical outcomes have become very positive. Wear-related revisional surgeries have been largely reduced according to multicenter follow-up studies.

Oxidation of highly cross-linked UHMWPE remains a challenge for both in vivo service and on-shelf behavior. This book will introduce some recent progresses on the use of very potent antioxidants, i.e., natural polyphenols, to stabilize highly cross-linked UHMWPE, which show outstanding strength, toughness, oxidation resistance, and low wear rate in comparison to the first-generation highly cross-linked UHMWPE. The oxidation and antioxidation mechanisms will be analyzed and discussed in detail.

The success of UHMWPE in joint implants would not be possible without the development of biomechanics and biotribology of joints. The daily gaiting contains sophisticated biomechanics and tribology that can never be described by using a simple model. This book introduces some fundamentals and recent progress on the biomechanics and tribology of total hip and knee implants.

The clinical performance of the permanent joint prosthesis has to be regularly monitored to evaluate their in vivo status and predict any possible problems. Objective assessment methods based on medical imaging have been widely developed and clinically used. This book will introduce the use of Roentgen stereophotogrammetric analysis technique to detect the loosening of hip prosthesis.

This book aims to offer students with fundamental knowledge on UHMWPE as biomaterials. Senior biomedical engineers, polymer scientists, joint implant professionals, and physicians will find in-depth details on the research methods, experiments, and latest progress on UHMWPE materials.

It would not be possible to accomplish this book without the outstanding contributions from the coeditors, Professor Zhong-Min Jin, and Professor Jin-Wu Wang and their teams. I am very grateful to the authors for writing the important chapters. Professor Luigi Costa from Turin University is a longtime friend and always brings new ideas, particularly on the oxidation and antioxidation mechanisms. His expertise on FTIR spectroscopy definitely benefits me and my students. Professor Shuxin Qu has made great contributions to the UHMWPE biomaterials by introducing drugs into the polymer to depress osteolysis since wear is inevitable so far. I'm also indebted to Professor Ebru Oral and Professor Orhun K. Muratoglu who were my advisors during my stay in the Harris Orthopaedics Laboratory at Massachusetts General Hospital.

Finally, I would like to thank my team members who help me a lot in the chapter writing and experiments on UHMWPE biomaterials.

Ningbo, China
December 2018

Jun Fu

Contents

Chapter 1
Clinical Applications of UHMWPE in Joint Implants

Jing Chen, Guorong Gao, and Jun Fu

Abstract Artificial joint implants have been widely applied in clinical surgery of joint replacement for those patients whose natural joints suffer from trauma, disease, or overuse. Typical artificial joints consist of a metallic component integrated with bone and a polymer component that facilitates movement. The requirements for the polymer include biocompatibility, toughness, wear resistance, lubrication, etc. Over the past 50 years, ultrahigh-molecular-weight polyethylene (UHMWPE) has been used to fabricate artificial hips, knees, shoulders, and other joints. In this chapter, a historical development of artificial joint implants for arthroplasty is briefly reviewed. The fundamental physicochemical properties and processing of UHMWPE are summarized to demonstrate its superiority in the application of artificial joint implants. In view of clinical outcomes, main challenges of conventional UHMWPE joint implants are finally discussed, such as oxidation degradation, sterilization, wear, and debris.

Keywords Ultrahigh-molecular-weight polyethylene · Artificial joints · Joint implants · Clinical application · Wear-resisting · Anti-oxidation

1.1 The History of Artificial Joint Implants for Arthroplasty

When natural joints fail due to trauma, disease, or overuse, replacement with artificial ones is needed to maintain the joint function and relieve the pain of patients. Joint replacement, also called joint arthroplasty, is a surgery that removes parts of an arthritic or damaged joint and replaces it with artificial implants or prosthesis comprised of metal, plastic, or ceramic devices. Prosthesis is designed to replicate

J. Chen · G. Gao · J. Fu (✉)
Ningbo Institute of Materials Technology and Engineering, Chinese Academy of Sciences, Ningbo, China
e-mail: fujun@nimte.ac.cn

J. Fu et al. (eds.), *UHMWPE Biomaterials for Joint Implants*, Springer Series in Biomaterials Science and Engineering 13,
https://doi.org/10.1007/978-981-13-6924-7_1

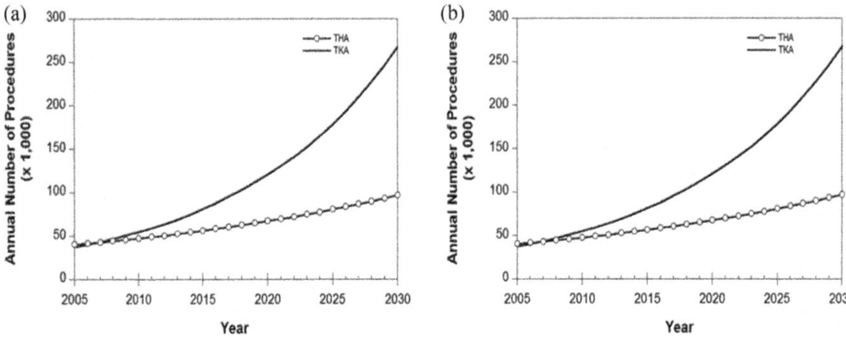

Fig. 1.1 The projected number of primary (**a**) and revision (**b**) total hip arthroplasty (THA) and total knee arthroplasty (TKA) procedures in the United States from 2005 to 2030. (Reprinted from [3] with permission from Wolters Kluwer Health, Inc.)

the movement function of a normal and healthy joint. Nowadays, roughly 570,000 primary total hip and knee replacements are performed in the United States each year [1, 2]. According to the American Academy of Orthopedic Surgeons, the procedures performed in the United States are expected to increase to at least 750,000 per year by 2030 (Fig. 1.1) [3].

Despite the long-lasting demands for artificial joints, it has taken about two centuries for human beings to establish the modern joint implant designs as well as optimal materials for joint implants. The earliest attempt of hip replacement occurred in 1891. Gluck presented the first use of ivory to replace femoral heads for a patient whose hip joints had been destroyed [4]. Since then, worldwide, surgeons carried out the interpositional arthroplasty from the late nineteenth to early twentieth century, which involved placing a variety of tissues between articulating surfaces of a remodeled joint, such as the fascia lata, skin, and pig bladder submucosa [5]. The first mold arthroplasty out of glass was performed by an American surgeon Marius Smith-Petersen in 1925. This mold consisted of a hollow hemisphere that can fit the femoral head and give rise to a new smooth surface required by the joint movement. However, the glass failed to endure great forces through the hip joint during movements. A British surgeon, George McKee, was the first to use a metal-on-metal prosthesis on a regular basis in 1953. He used the modified Thompson stem, which was a cemented hemiarthroplasty used for the neck of femur fracture. That implant showed a 28-year survival rate of 74% [6]. Unfortunately, that prosthesis was more and more unpopular in the 1970s because the local effect of metal particles was found during the revision surgery for prosthetic failures [7]. An orthopedic surgeon, Sir John Charnley, proposed low-friction arthroplasty in the early 1960s and thus created an era of modern total joint arthroplasty. This design is still used today as a golden standard. It is comprised of three parts: (1) a metal femoral stem, (2) a polyethylene acetabular component, and (3) an acrylic bone cement. Thus, a small femoral head is used to reduce wear due to its small surface area [8, 9].

Currently, a growing number of successful joint replacement operations have been achieved every year using different artificial joint materials. Scientists and engineers have been struggling to find the best combination that yields the fewest complications and excellent long-term survival, as highly motivated by the clinical applications and challenges of joint implants.

Metal-on-metal (MOM) prostheses are usually fabricated from cobalt, chromium, and molybdenum (CoCrMo) alloys. In the MOM total hip implants, a large metal femoral head on a standard femoral stem articulates against a metal acetabular cup or liner. Metal articulating surfaces could be engineered to be extremely smooth, which is beneficial to reduce the wear rate. A low wear rate is expected to prolong the lifetime of the implant and thus delay the need for any revision surgery. Thus, MOM hip replacements have traditionally been used for young patients. These alloys exhibit excellent hardness and strength but may raise concerns about long-term metal ion release [10]. Besides, during the long-term use, MOM hip implants have been estimated to release many wear particles because the particles tend to be smaller and smaller [11]. The CoCrMo particles may be more easily engulfed by cells and migrate into body fluids. It was found that patients with MOM hip implants often have an elevated level of chromium and cobalt in their blood, urine, and hair [12]. The long-term effect of elevated metal ions residual in vivo on the human beings' health remains unknown.

Since the first introduction by Pierre Boutin, a French surgeon, in 1970, ceramic heads have accounted for half of the hip arthroplasties in Europe. To address the friction and wear that frequently emerged in other materials, the ceramic materials used in orthopedics include either alumina or zirconia. Compared with metals, ceramics usually induce reduced osteolysis, which is favorable for joints or joint surface materials. Moreover, their facile processing and forming and superior mechanical properties endow some ceramics as bone substitute materials. The ceramic-on-ceramic (COC) (Fig. 1.2) prostheses exhibit a high level of hardness, scratch resistance, inert nature, good biocompatibility, and stability in physiological conditions [13]. Besides, the hydrophilicity of these prostheses improves the lubrication and thus leads to a low coefficient of friction and excellent wear resistance [14]. However, the cost of ceramics is often too high, and the risk of fracture should be taken into account because ceramics are brittle materials. Once the fracture occurs, it is a complex task to revise because ceramic femoral heads typically break into multiple fragments that are hard to clean from surrounding tissues [15]. Squeaking is another issue for COC joint prosthesis. It is defined as an audible sound of 20–20,000 Hz during movement of the hard-on-hard bearings. The squeaking rate in MoM and CoC articulation has been reported to be 0.3% to 24.6%. Unfortunately, the origination of squeaking behavior is not fully understood yet. It has been widely accepted that patient, implant, and surgical factors play a role in the squeaking phenomenon [16].

Since Charnley first utilized ultrahigh-molecular-weight polyethylene (UHMWPE) in 1962, the most widely accepted implant configuration contains a metal component articulating against a polymeric component prepared from UHMWPE. Nowadays, such metal-on-UHMWPE joint implants have become an

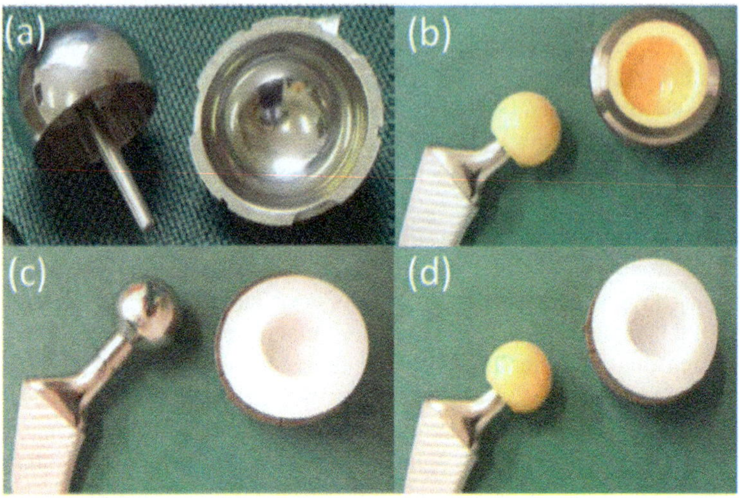

Fig. 1.2 Representative modern joint prosthesis. (**a**) Metal-on-metal, (**b**) ceramic-on-ceramic, (**c**) metal-on-polyethylene, and (**d**) ceramic-on-polyethylene. (Reprinted from [17] with open access)

international standard for curing degenerative joint disorders. Total hip replacement for both femoral and acetabular surfaces became more and more popular during the 1980s and then grew steadily thereafter. These joints initially had a UHMWPE cup cemented into the acetabulum articulating against a metal head. The metal head was then attached to a stem inserted into the femur. Over the past 20 years, cement-free designs have been adopted, which involves a polymer component fixed to the pelvis by a metal shell and articulating against a metal or ceramic femoral head. On average, more than 90% of all total joint replacements that use UHMWPE-bearing surfaces survive for more than 10 years, according to the Swedish hip registry [18]. Such joints tend to be revised at a rate of about 1% per year in the first decade after implantation. Data show that implant survivorship reduces after 10 years, especially for patients who are younger than 55 years [18].

This chapter presents a general introduction to the clinical applications of UHMWPE in joint implants, which consists of four sections including (1) fundamentals to UHMWPE, (2) conventional UHMWPE for joint implants, (3) toughened UHMWPE for joint implants, and (4) clinical outcomes of conventional UHMWPE joint implants.

1.2 Fundamentals to UHMWPE

The medical grade UHMWPE used in total joint implants is a kind of linear homopolymer. Before giving the definition of UHMWPE, it is necessary to have a general concept about polymer. A *polymer* is defined as a big molecule constructed

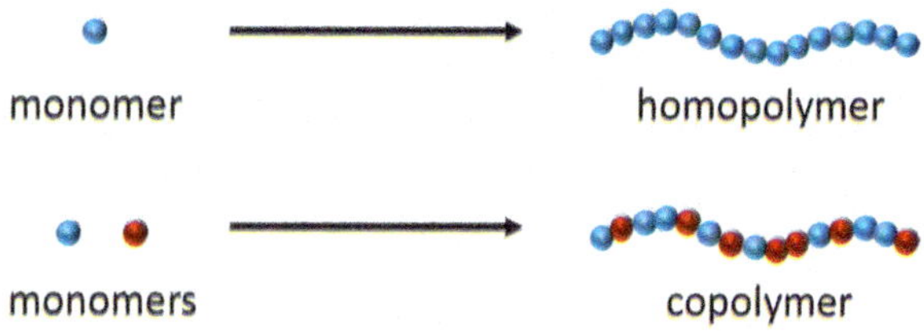

Fig. 1.3 Schematic presentation of structures of homopolymer and copolymer

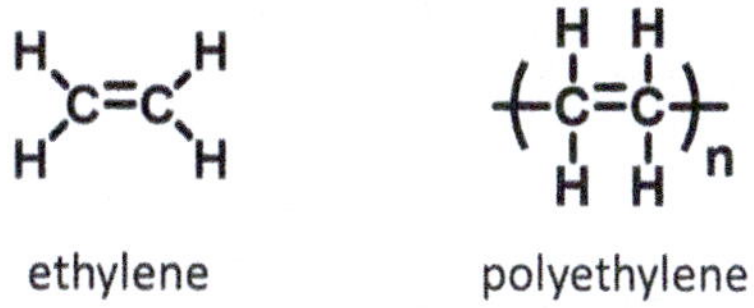

Fig. 1.4 Chemical structures of ethylene and polyethylene

by numerous small molecules via covalent bonds. The small molecules used as the basic building blocks for these large molecules are known as *monomers*. The monomer segments can all be the same to generate a *homopolymer*. If there are more than two monomers involved, a *copolymer* can be obtained. The illustration of homopolymer and copolymer is shown in Fig. 1.3. The UHMWPE used for joint implants are homopolymers. Here, we focus on the synthesis, structure, and property of UHMWPE.

Polyethylene (PE), the largest tonnage plastic material, has one of the simplest molecular structures ($[CH_2CH_2]_n$) formed from ethylene (Fig. 1.4). It was first commercially produced in 1939 for the use in electric insulation. There are four different industrial routes to produce PE including (1) high-pressure processes, (2) Ziegler processes, (3) the Phillips process, and (4) the Standard Oil (Indiana) process. A pressure of 1000–3000 atm and a temperature of 80–300 °C are often used for the high-pressure process. Free-radical initiators, such as benzoyl peroxide or oxygen, are generally used, and conditions need to be carefully controlled to prevent a runaway reaction. Generally, the high-pressure process yields low-density PE, typically in the range 0.915–0.945 g cm^{-3}, with relatively low molar mass. Ziegler processes are based on coordination reactions catalyzed by metal alkyl systems. A typical Ziegler–Natta catalyst is the complex prepared from titanium tetrachloride and triethylaluminum. It is fed into the reaction vessel first, and subsequently ethylene is added. The reaction is carried out at low pressures and low temperatures, typically no more than 70 °C, with rigorous exclusion of air and moisture. The PE products are of intermediate density of about 0.945 g cm^{-3} with molecular weights ranging from 5 to 10 million. In the Phillips process, ethylene monomers are dissolved in cyclohexane and then polymerized by a supported metal oxide (mainly CrO_3) catalyst at 130–160 °C and 1.4–3.5 MPa. The resulting commercial products have a melt flow index of only 0.2–5 and the highest density (0.96 g cm^{-3}). The process invented by Standard Oil (Indiana) is based on the use of a supported transition metal oxide combined with a promoter. The reaction temperature is in the range of 230–270 °C and the pressures are 40–80 atm.

The size of a polymer molecule can be defined either by its molecular weight (M_w) or by the number of repeat units in the molecule (degree of polymerization, DP). The relative molar mass of the polymer is thus the product of the relative MW of the repeat unit and the DP. Due to the methods used to prepare polymers, the distribution of the MW is heterogeneous for polymers rather than the exact value of MW for small molecules. In principle, there are several ways to calculate such average MW. The most straightforward is the simple arithmetic mean, usually called the number average molar mass, M_n, which is defined by Eq. 1.1:

$$M_n = \frac{\sum N_i M_i}{\sum N_i} \tag{1.1}$$

where M_i is the molar mass of the molecular species i and N_i is the number of molecules of i in the sample.

Another definition is the weight-average molar mass (M_w), which is scaled to M_i^2 by Eq. 1.2:

$$M_w = \frac{\sum N_i M_i^2}{\sum N_i M_i} \tag{1.2}$$

For a polymer consisting of molecules of the same molar mass $M_n = M_w$, but in all other cases, M_w is greater than M_n. We can thus use the ratio of M_w to M_n as an indication of molecular dispersity in a particular polymer sample. This ratio is called the polydispersity index (PDI), where the sample with PDI = 1 is said to be homo- or monodispersed.

There are various methods to determine the M_w of polymers including end-group analysis, measurement of colligative properties, light scattering, ultracentrifugation, and measurement of dilute solution viscosity. In principle, absolute values of MW can be obtained by those methods except viscosity measurement. The viscosity method relies on prior calibration using standard polymers with known molar mass.

UHMWPE is a linear homopolymer of ethylene. The term "ultrahigh" refers to an average MW larger than 1.5 million g/mol. UHMWPE is a unique polymer with excellent physical and mechanical properties. Moreover, UHMWPE exhibits the best sliding wear resistance and impact toughness and thus has been used to replace steel in many applications requiring wear resistance [19]. Since 1962, UHMWPE has been used in orthopedics as a bearing material for joint implants. Nowadays, around 1.4 million joint replacement procedures are performed worldwide. More than 90% of all total hip implants worldwide in the past decade have used UHMWPE. Despite the great success of restorative procedures, UHMWPE implants have a finite lifetime due to the wear and damage.

UHMWPE was first synthesized by Ziegler in the early 1950s and then commercially produced in 1955. A highly reactive organotitanate catalyst was used to polymerize the ethylene gas. The resultant polymer chains are remarkably linear with measured branching of less than 3%.

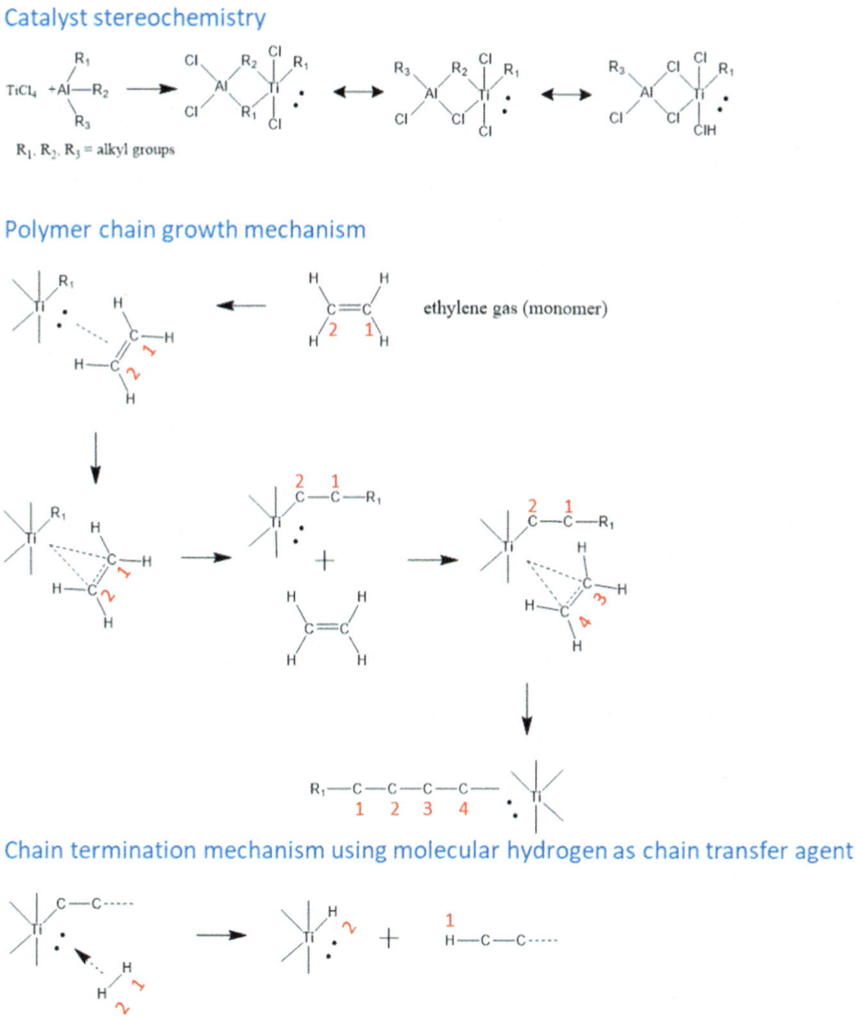

Fig. 1.5 Ziegler–Natta mechanism for the polymerization of ethylene. (Reprinted from [19] with permission from Taylor & Francis)

As shown in Fig. 1.5, the Ziegler–Natta mechanism is usually used to depict the polymerization of ethylene. Generally, the catalyst consists of titanium and aluminum alkyls. During the reaction of catalyst stereochemistry, the aluminum alkyls form a titanium alkyl bond. Ethylene is then inserted into this chemical bond to generate a titanium ethylene alkyl. The chain grows by inserting more ethylene molecules into this titanium carbon bond. A chain transfer agent such as molecular hydrogen is typically employed to control molecular weight because it reacts cleanly, leaves no residue, and is low cost. A chain transfer agent terminates a growing chain without deactivating the catalyst.

Several parameters such as reaction temperature, activation of the catalyst, and concentration of alkyls affect the polymerization of UHMWPE. Ziegler–Natta polymerization does not easily allow for secondary reactions. In situ modification of UHMWPE is not practiced. The Ziegler–Natta reaction can be seen as an inherent "clean" strategy, and the product requires no purification beyond removal of the slurry diluent by staged dryers. The crude polymer is normally white and opaque due to the scattering by the polyethylene crystallites. The average particle diameters of the commercial product usually range from 50 to 250 microns.

The melt viscosity of UHMWPE is very high due to the ultrahigh molecular weight. It cannot flow like low MW polyethylene when the temperature is above its melting temperature. No processing aids or additives are allowed for medical-grade UHMWPE for clinical applications. Therefore, conventional thermoplastic processing techniques such as injection molding, screw extrusion, and blow molding are not suitable for UHMWPE. Semi-manufactures of UHMWPE are typically used by compression molding and ram extrusion under high temperatures and pressures.

A typical compression molding press installed at Poly Hi Solidur Meditech in Germany was designed by Hoechst in 1970s for the production of CHIRULEN. There are two sheets (1 m × 2 m) molded in a single press cycle: one is pressed between the upper and middle platens, and the second is produced between the middle and lower ones. In this way, the heating and loading systems are totally controlled by computer. The entire press is performed in a clean room to prevent the introduction of extraneous matter into the sheet.

The ram extrusion of UHMWPE was developed by converters in the United States in 1970s. Similar to the compression molding facilities, a medical-grade extruder is typically used in a clean room to reduce and avoid extraneous matters into the UHMWPE. The schematic illustration of ram extrusion is shown in Fig. 1.6, where UHMWPE powder is first fed into an extruder comprised of a hopper that allows the powder to enter into a receiving chamber, a horizontal reciprocating ram, a heated die, and an outlet successively. The UHMWPE is maintained under

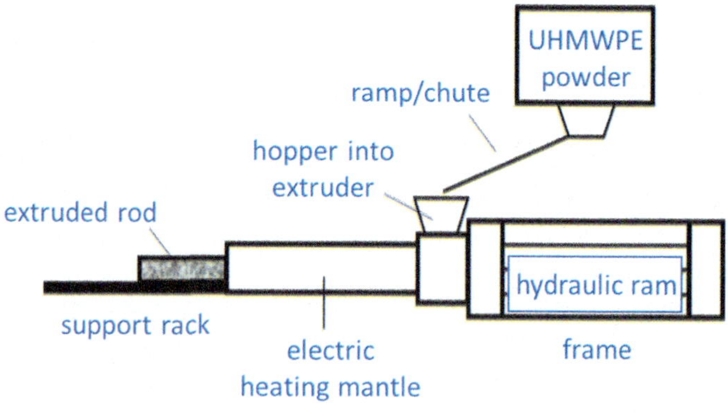

Fig. 1.6 Ram extrusion of UHMWPE

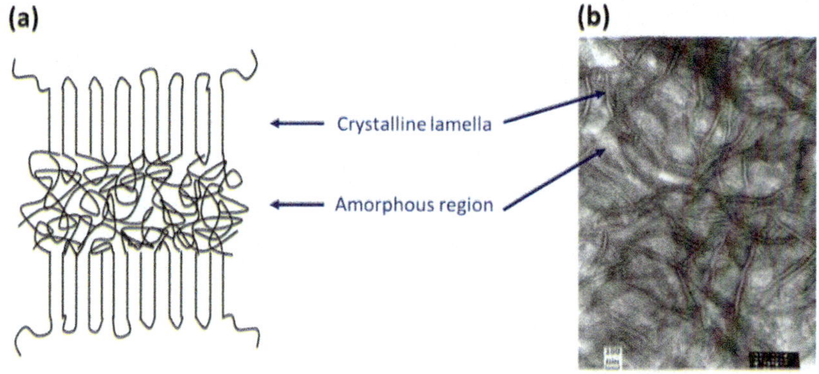

Fig. 1.7 Crystalline morphology of UHMWPE (**a**) and TEM image of UHMWPE (**b**)

pressure by the ram and the back pressure of the melting UHMWPE. Beyond the outlet, the rod of UHMWPE is slowly cooled in a series of electric heating mantles.

Medical-grade UHMWPE is a semicrystalline polymer that can be described as many ordered regions (crystalline lamellae) embedded in a disordered amorphous matrix (Fig. 1.7a). The degree of crystalline is very important for the properties of UHMWPE. With a high crystallinity, the polymer shows a high elastic modulus, superior yield strength, improved resistance to creep deformation, and enhanced fatigue resistance, which are critical for joint implants. The degree and orientation of crystalline regions within a polyethylene strongly depend on various factors such as MW, processing conditions, and external loading.

The crystalline morphology of UHMWPE can be visualized by using transmission electron microscopy (TEM) on ultrathin slices. The amorphous phase can be selectively stained and turn gray in the micrograph, while the lamellae appear as light domains with gray outlines (Fig. 1.7b). The composite structures of UHMWPE are featured as an interconnected network of amorphous and crystalline regions.

UHMWPE shows outstanding tensile properties. Tensile testing is commonly conducted to characterize the mechanical properties of UHMWPE. A dumbbell-shaped specimen is punched out and stretched at a specified crosshead speed in a tensile load frame. The load (F) and displacement (L) are converted to stress and strain. Moreover, Young's modulus, yield stress, ultimate tensile stress, and elongation at break are all determined from the stress–strain curve (Fig. 1.8). Upon loading, the spherulites of UHMWPE start to deform, showing linear elastic deformation at low strain. Once the load becomes adequate to break the spherulites, the spherulites rupture and the polymer undergoes plastic deformation, that is, the polymer yields. After yielding, the stress decreases and then gradually increases when the lamellae start to deform along the elongation, with a gradual orientation. During this procedure, the amorphous chains and lamellae are gradually stretched, resulting in strain hardening. At very high strain, the stress abruptly increases until the sample fails. Thus, the ultimate stress, yield strength, elongation at break, and fracture toughness are determined.

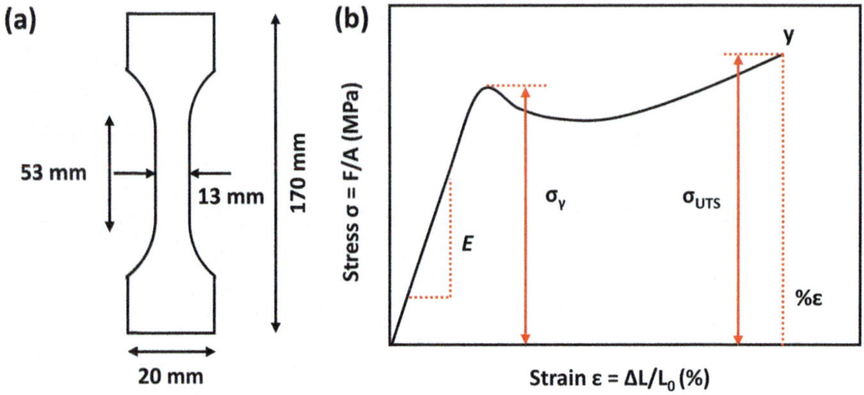

Fig. 1.8 Typical "dogbone" tensile specimen (**a**) and a typical engineering tensile stress–strain curve of UHMWPE (**b**). (Reprinted from [20] with permission from Elsevier)

Fatigue testing is a method to determine the crack propagation in a specimen subjected to an oscillating tensile load. The numbers of oscillations (N) used in these tests are typically several million cycles. The purpose of the test is to monitor the crack propagation resistance under cyclic loadings. In this test, a compact specimen, as used in the J-integral tests (Fig. 1.9a), is placed in a hydraulic load frame and then oscillated at either a fixed displacement or between fixed loads. During the test, the crack length (Δa) is periodically measured. The stress intensity factor (K), which depends on the load range used, sample geometry, and crack length, is computed from the raw data (a vs N), which are then used to prepare a curve of log da/dN versus log ΔK (Fig. 1.9b).

1.3 Conventional UHMWPE for Joint Implants

In the late 1950s, Dr. John Charnley, a joint replacement pioneer, found that natural joints have a low friction coefficient due to the synovial fluid between the contacting surfaces as it is compressed. The pressurized fluid film protects the cartilage from wear. Arthritis and other joint diseases can cause the cartilage to lose its ability to lubricate joints. Therefore, the design of artificial joints aims to mimic such lubrication, which is dependent on the boundary lubrication of the materials in contact with each other. Polytetrafluoroethylene (PTFE) was Charnley's earliest polymeric-bearing material for artificial hip joints in 1958. The earliest design had a PTFE acetabular element articulating against a PTFE femoral component.

Although PTFE implants initially restored pain-free mobility to those suffering from joint disease, they were worn out rapidly. Many debris from PTFE wear led to inflammation and pain, and thus there was an urgent need to revise nearly all of the

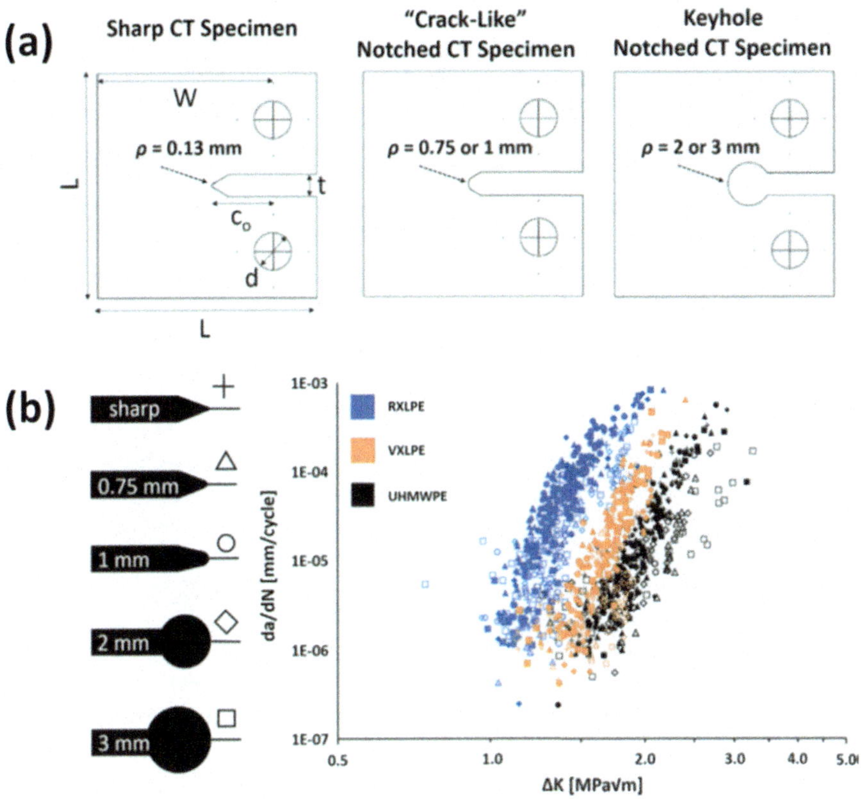

Fig. 1.9 Specimens with various notch geometries used for fatigue analysis (**a**) and fatigue crack propagation rate (da/dN) data as a function of the stress intensity range (ΔK) for virgin UHMWPE (UHMWPE, black), cross-linked (75 kGy) and remelted UHMWPE (RXLPE (RXLPE, blue) and vitamin E (0.1 wt%) blended and cross-linked (75 kGy) UHMWPE (VXLPLE, orange). (Reprinted from [21] with permission from Elsevier)

implanted joints just 2–3 years after implantation. In 1962, Charnley used UHMWPE to increase joint longevity. Initially, it was available only to a limited number of surgeons trained personally by Charnley. Since then, it had become more widely used during the 1970s. Total hip replacement for both femoral and acetabular surfaces experienced explosive popularity during the 1980s and then grew steadily thereafter. These joints initially had a UHMWPE cup cemented into the acetabulum articulating against a metal head. The metal head was then attached to a stem inserted into the femur.

From the standpoint of medical application, UHMWPE orthopedic components should be sterilized before clinical use. High-energy radiation was the most commonly used sterilization technique. The source of γ radiation is the decay of an unstable ^{60}Co nucleus, while electron beams are generated from a thermally excited tungsten filament accelerated by electric fields. The electron beam is easier to control and requires a very short period of treatment (usually in seconds). Usually, UHMWPE components are stored on the shelf for long durations prior to implantation (6 months or longer). In addition, UHMWPE inserts of total joint replacements have historically been packaged in air and thereafter sterilized by γ radiation. However, it is well established that such ionizing irradiation can result in chain scission and degradation of PE. A large number of radicals are thus formed to cause unexpected cross-linking, which have detrimental effects on the morphology, mechanical properties, and wear of UHMWPE [22–24].

Once a polymer material is exposed to strong energy environments, it is likely to form free radicals due to the bond scission, and thus the mechanical properties of polymer reduces, as induced by the chain fragmentation. In orthopedics, this degradation is mainly associated with the radiation and electron beams commonly used during sterilization mentioned above. If oxygen is present when the degradation occurs, it is called oxidation. The oxidation process cannot be interrupted if it has been initiated, and the oxidation rate increases continuously with a series of reactions that involve free radicals and oxygen. The extent of oxidation is strongly dependent on the amount of oxygen. The origin of oxygen is very complicated including (1) from atmosphere present at sterilization, (2) penetrated by diffusion into the polymer during processing and storage, and (3) being used in vivo. Therefore, the oxidation can continue during storage and in vivo after implantation [22].

In order to solve these oxidation issues, some manufacturers sterilize UHMWPE using non-radiation-based methods, such as ethylene oxide (EtO) or gas plasma sterilization. EtO is widely used for sterilization of UHMWPE components sealed in gas permeable packages. UHMWPE sterilized with EtO does not undergo any variation in chemical or physical structures. Gas plasma (GP) is a surface sterilization strategy based on the action of ionized gas (i.e., hydrogen peroxide or peracetic acid), which deactivates biological organisms. Commercially available GP sterilization methods are usually performed at low temperatures (< 50 °C) and have no significant effects on the physical, chemical, and mechanical properties of UHMWPE. In response to long-term postirradiation aging and oxidation, some manufacturers further developed the sterilization system with high-energy radiation performed in vacuum or under inert gases (N_2 or Ar).

Wear is the process of removing parts of a material from the surface during reciprocal movement along another surface with greater hardness. In artificial joint components, the UHMWPE is removed because it is soft and relatively weak compared with the metal or ceramic materials in femoral head and femoral knee components. During the wear process, the removed PE particles can migrate to the prosthetic tissues and thus induce aseptic loosening, through a mechanism involving the formation of reactive tissue and consequent osteolysis, which has been recognized as the main reason for the failure of implants based on conventional UHMWPE [25]. Unfortunately, the exact mechanism of immune reaction that occurs in periprosthetic osteolysis of joint replacements is still unclear. Up to now, there are several types of immune processes that are believed to be relevant to osteolysis. A foreign-body and granulomatous response to UHMWPE wear particles may denote a nonspecific chronic inflammatory reaction involving activated mononucleated macrophages and fibroblasts but few T lymphocytes [26]. The activation of macrophages is strongly related to the size, shape, volume, and number of radiation-sterilized UHMWPE debris particles [27]. Especially those 0.3–10 μm in size are phagocytable and are therefore the most biologically active [28]. In addition, the influence of the chemical composition of the UHMWPE particles has recently been suggested. The reactivity might be related to the composition of the surfaces of the particles themselves and in particular to the level of oxidation of the UHMWPE [29, 30]. It is found that oxidized particles from γ-irradiated UHMWPE would be more effective in activating the macrophages than the non-oxidized ones from EtO-sterilized UHMWPE. The properties of the absorbed molecules such as the hydrophilic/hydrophobic feature and the release of radicals also affect the surface reactivity of the particles [30]. In fact, although catastrophic failures due to supreme wear and heavy oxidation are quite infrequent, wear is also dependent on time. The abrasion and the production of abrasive particles therefore remain a severe issue for young and active patients with long life expectancies. Cross-linked UHMWPE appears to be the solution to the wear issue.

1.4 Toughened UHMWPE for Joint Implants: Lessons Learned

In order to strengthen and toughen UHMWPE, composites can be engineered by blending UHMWPE powders with micro- or nano-objectives (particles or fibers) before consolidation. In the 1970s, carbon fiber-reinforced (CFR) UHMWPE composites were first used for orthopedic implants and were commercially produced as Poly II in Zimmer, Inc., Warsaw, USA. However, the short-term clinical failures eventually resulted in the abandonment of Poly II [31]. The CFR UHMWPE was reinforced by randomly oriented carbon fibers during the direct compression molding of UHMWE matrix [32]. Initially, the incorporated carbon fibers were deemed to be the reason for reinforcement [33]. However, further investigation confirmed that the poor fiber–matrix interface decreased the ductility and crack resistance [34]. In practice, there are many difficulties in the blending process, resulting in incomplete consolidation of the UHMWPE powders and carbon fibers. Therefore, severe issues such as wear, fracture, and extensive delamination made the composites hazardous for clinical applications. Moreover, carbon fibers with very high modulus, once exposed from the UHMWPE matrix, may severely damage the femoral head.

Alternative self-reinforced composites, also called homocomposites, are another strategy to reinforce UHMWPE. In this way, the matrix and reinforced component all come from the same material (i.e., UHMWPE). In the 1990s, self-reinforced UHMWPE composites were widely studied, which were developed both by sintering oriented fibers together and by reinforcing a polymer matrix with UHMWPE fibers. Within the composites, UHMWPE fibers provide much

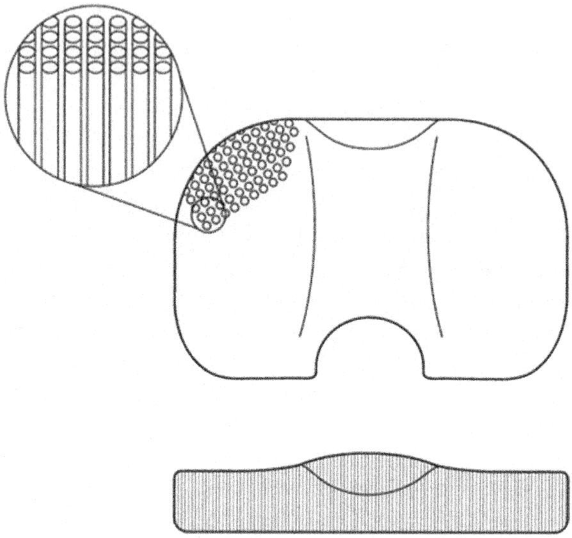

Fig. 1.10 Schematic presentation of self-reinforced UHMWPE composites. (Reprinted from [39] with permission from Elsevier)

higher tensile modulus (around 113 GPa) and ultimate tensile strength (2–4 GPa) compared with bulk UHMWPE consolidated from powders [35, 36]. Moreover, the melting temperature of UHMWPE fibers is 10 °C higher than the powders, which could be compression molded either by themselves or by combining with powders during a narrow temperature range to keep the initial orientation [37]. This self-reinforcement strategy was first used to fabricate oriented UHMWPE composites for orthopedic bearings by sintering compressed continuous Spectra 1000 fibers at Zimmer [38]. The schematic presentation is shown in Fig. 1.10. The sintered UHMWPE fibers are oriented orthogonal to the bearing surface. The obtained UHMWPE composites exhibit dramatically increased modulus and strength by an order of magnitude when tested along the direction of fiber orientation. However, self-reinforced UHMWPE homocomposites have never been commercialized for total joint replacement applications due to the main challenge of the processing compared with radiation-processed materials.

1.5 Clinical Outcomes of Conventional UHMWPE Joint Implants

Joint replacement surgeries have witnessed a rapid development over the past decades. An obvious trend has been observed that people aged 45–60 years are increasingly opting for joint replacement surgeries. Technological advancements have resulted in better prosthetics, longer lifespan, and more comfort for the patients. Increasing geriatric population in several countries, such as the United States, Canada, and Japan, is expected to drive the demand for medical UHMWPE over the long term. According to a new report by Grand View Research, Inc. [40], the global medical-grade UHMWPE market demand was 60.9 kilotons in 2015 and is expected to reach 204.8 kilotons by 2024, growing at a CAGR of 15% from 2016 to 2024. The market value is expected to reach 1.36 billion USD by 2024.

Although UHMWPE has achieved a success in total joint arthroplasty for over 30 years, we are still facing many challenges in clinical applications due to the material failures which are typically caused by the gamma irradiation-induced oxidative degradation and the high cyclic stress environment of the joints. With early joint replacement components, gamma irradiation in air at a dose level of approximately 25 kGy was used as a major sterilization method due to its low cost. However, oxidative degradation can take place when UHMWPE is exposed to gamma radiation in the presence of air. More seriously, such oxidative degradation can continue, and the number of oxidation products will increase without any further irradiation once there is an oxygen source [41]. Clinical investigations have shown that the oxidative degradation has a negative impact on the performance of acetabular hip components [42–44] and tibial knee components [45, 46] in terms of wear and fracture resistance.

To minimize the oxidation and its subsequent effects on wear and mechanical properties of UHMWPE, orthopedic implant manufacturers have turned to modified sterilization protocols, such as gamma radiation sterilization in vacuum packaging or inert gas packaging with reduced amount of oxygen. These strategies can greatly reduce or eliminate the potential of oxidation due to the lack of oxygen source. But the free radicals still remain in vivo. The oxidation of UHMPE components sterilized under low oxygen conditions was, in fact, investigated, and the highest oxidation indices tend to occur at stress concentrations inherent in the UHMWPE component design [47, 48].

The gross fracture and component cracking of conventional UHMWPE total joint replacement components has been found [42, 49, 50]. All of these design features are due to inherent stress concentrators. Oxidation of these regions may further increase the susceptibility to fracture. For example, a recent study of retrieved conventional UHMWPE acetabular liners indicated that the percentage of retrieved liners that showed cracking increased with increasing the oxidation level [49] (Fig. 1.11).

Another significant issue in the clinic is the abrasion wear (Fig. 1.12). It frequently results in the removal of a partial component of one material from the surface during reciprocal movement along another surface with greater hardness. In orthopedic joint components, the UHMWPE is easily removed because the interactions of its chains are relatively weak compared to those between the metal and ceramic materials. Such removed PE particles induce aseptic loosening through a mechanism involving the formation of reactive tissue and consequently osteolysis [25]. The surface wear mechanism is able to produce relatively small particles with the size less than 1 μm typically, while the fatigue wear mechanism usually generates larger particles. The smaller particles can cause an osteolytic response more seriously than the larger ones [51]. The existence of such particles could cause the immune reaction that occurs in periprosthetic osteolysis of joint replacements, but the mechanism is still unclear. It could be related to some types of immune processes. For

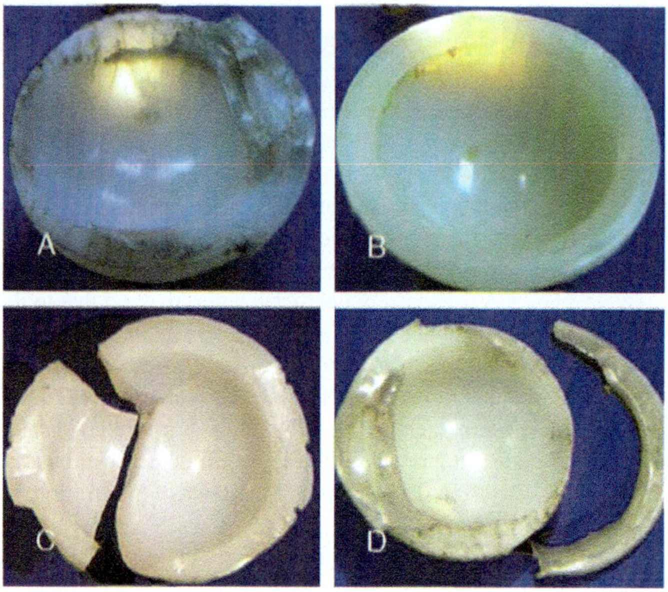

Fig. 1.11 (**a**) Highly oxidized liners with extensive rim cracks extending onto articular surface. (**b**) Liner with multiple subsurface rim cracks demonstrated by transillumination. (**c**) Liner revised after complete mechanical failure secondary to fragmentation. (**d**) Liner with extensive oxidation and impingement damage leading to extensive cracking and separation of the rim. (Reprinted from [49] with permission from Elsevier)

example, the granulomatous response to UHMWPE particles is a nonspecific chronic inflammatory reaction involving activated mononucleated macrophages and fibroblasts but few T lymphocytes [26].

1.6 Conclusions

To date, UHMWPE has been a popular joint implant material because of its low coefficient of friction, low wear over the long term, and stability and biocompatibility in the body. Since it has been placed in the knee and hip joints in the 1960s, UHMWPE bearings have also been used in shoulder, elbow, wrist, ankle, and great toe replacements. The development of UHMWPE material continuously breaks the bottleneck in surgical outcome. It is expected to see a significant decrease in revisions because of UHMWPE-related failures such as particle-induced osteolysis and delamination-related instability. Radiation cross-linking has appeared as a standard bearing surface to achieve enhanced wear resistance of UHMWPE and is likely to decrease the incidence of osteolysis. With the development of polyethylene technologies including synthesis, processing, and fabrication, it is expected to offer

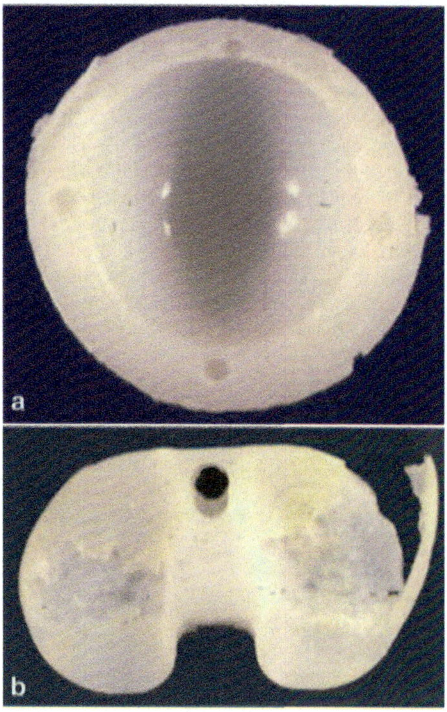

Fig. 1.12 (**a**) A clinically retrieved total hip acetabular component. (**b**) A clinically retrieved total knee tibial insert. (Reprinted from [52] with permission from Wolters Kluwer Health, Inc.)

more flexibility in implant design such that more anatomical reconstruction of joints can be performed with the least amount of invasive interference to improve the life quality of joint implant patients.

References

1. Kurtz S, Mowat F, Ong K, Chan N, Lau E, Halpern M (2005) Prevalence of primary and revision total hip and knee arthroplasty in the United States from 1990 through 2002. J Bone Joint Surg Am 87A(7):1487–1497
2. Maradit KH, Larson DR, Crowson CS, Kremers WK, Washington RE, Steiner CA, Jiranek WA, Berry DJ (2015) Prevalence of total hip and knee replacement in the United States. J Bone Joint Surg Am 97(17):1386–1397
3. Kurtz S, Ong K, Lau E, Mowat F, Halpern M (2007) Projections of primary and revision hip and knee arthroplasty in the United States from 2005 to 2030. J Bone Joint Surg Am 89(4):780–785
4. Speed JS, Smith H (1940) Arthroplasty – a review of the past ten years. Surg Gynecol Obstet 70:224–230
5. Learmonth ID, Young C, Rorabeck C (2007) The operation of the century: total hip replacement. Lancet 370(9597):1508–1519
6. Brown SR, Davies WA, DeHeer DH, Swanson AB (2002) Long-term survival of McKee-Farrar total hip prostheses. Clin Orthop Relat Res 402:157–163

7. McKellop H, Park SH, Chiesa R, Doorn P, Lu B, Normand P, Grigoris P, Amstutz H (1996) In vivo wear of 3 types of metal on metal hip prostheses during 2 decades of use. Clin Orthop Relat Res 329:S128–S140
8. Charnley J (1961) Arthroplasty of the hip: a new operation. Lancet 277(7187):1129–1132
9. Charnley J (1973) Arthroplasty of the hip: a new operation. Clin Orthop Relat Res 95:4–8
10. MacDonald SJ, Brodner W, Jacobs JJ (2004) A consensus paper on metal ions in metal-on-metal hip arthroplasties. J Arthroplast 19(8):12–16
11. Firkins PJ, Tipper JL, Saadatzadeh MR, Ingham E, Stone MH, Farrar R, Fisher J (2001) Quantitative analysis of wear and wear debris from metal-on-metal hip prostheses tested in a physiological hip joint simulator. Biomed Mater Eng 11(2):143–157
12. Hallab NJ, Anderson S, Caicedo M, Skipor A, Campbell P, Jacobs JJ (2004) Immune responses correlate with serum-metal in metal-on-metal hip arthroplasty. J Arthroplast 19(8):88–93
13. Katti KS (2004) Biomaterials in total joint replacement. Colloids Surf B Biointerfaces 39 (3):133–142
14. Christel PS (1992) Biocompatibility of surgical-grade dense polycrystalline alumina. Clin Orthop Relat Res 282:10–18
15. Hannouche D, Nich C, Bizot P, Meunier A, Nizard RM, Sedel L (2003) Fractures of ceramic bearings – history and present status. Clin Orthop Relat Res 417:19–26
16. Mai K, Verioti C, Ezzet KA, Copp SN, Walker RH, Colwell CW Jr (2010) Incidence of 'squeaking' after ceramic-on-ceramic total hip arthroplasty. Clin Orthop Relat Res 468 (2):413–417
17. Schwartsmann CR, Boschin LC, Gonçalves RZ, Yépez AK, de Freitas Spinelli L (2012) New bearing surfaces in total hip replacement. Rev Bras Ortop 47(2):154–159
18. Malchau H (2002) Prognosis of total hip replacement: update of results and risk-ratio analysis for revision and re-revision from the Swedish National Hip Arthroplasty Register 1979–2000. In: 69th annual meeting of American Academy of Orthopedic Surgeons, Dallas, Texas
19. Kelly JM (2002) Ultra-high molecular weight polyethylene. J Macromol Sci C Polym Rev C42 (3):355–371
20. Spiegelberg S (2009) Chapter 24 – Characterization of physical, chemical, and mechanical properties of UHMWPE A2 – Kurtz, Steven M. In: Kurtz S (ed) UHMWPE biomaterials handbook, 2nd edn. Academic, Boston, pp 355–368
21. Ansari F, Gludovatz B, Kozak A, Ritchie RO, Pruitt LA (2016) Notch fatigue of ultrahigh molecular weight polyethylene (UHMWPE) used in total joint replacements. J Mech Behav Biomed Mater 60:267–279
22. Bracco P, del Prever EMB, Cannas M, Luda MP, Costa L (2006) Oxidation behaviour in prosthetic UHMWPE components sterilised with high energy radiation in a low-oxygen environment. Polym Degrad Stab 91(9):2030–2038
23. Birkinshaw C, Buggy M, Daly S, Oneill M (1989) The effect of gamma-radiation on the physical structure and mechanical-properties of ultrahigh molecular-weight polyethylene. J Appl Polym Sci 38(11):1967–1973
24. Bostrom MP, Bennett AP, Rimnac CM, Wright TM (1994) The natural-history of ultra-high-molecular-weight polyethylene. Clin Orthop Relat Res 309:20–28
25. Harris WH (2001) Wear and periprosthetic osteolysis – the problem. Clin Orthop Relat Res 393:66–70
26. Goodman SB (2007) Wear particles, periprosthetic osteolysis and the immune system. Biomaterials 28(34):5044–5048
27. Ingham E, Fisher J (2005) The role of macrophages in osteolysis of total joint replacement. Biomaterials 26(11):1271–1286
28. Chiba J, Schwendeman LJ, Booth RE, Crossett LS, Rubash HE (1994) A biochemical, histologic, and immunohistologic analysis of membranes obtained from failed cemented and cementless total knee arthroplasty. Clin Orthop Relat Res 299:114–124
29. Fubini B (1997) Surface reactivity in the pathogenic response to particulates. Environ Health Perspect 105:1013–1020

30. Brach del Prever EM, Bistolfi A, Costa L, Bracco P, Linari A, Botto Micca F, Crova M, Gallinaro P (2003) The biological reaction to polyethylene wear debris can be related with oxidation of the UHMWPE cups. Chir Organi Mov 88:291–303
31. Wright TM, Astion DJ, Bansal M, Rimnac CM, Green T, Insall JN, Robinson RP (1988) Failure of carbon fiber-reinforced polyethylene total knee-replacement components. A report of two cases. J Bone Joint Surg Am 70(6):926–932
32. Farling GM (1976) Human body implant of graphitic carbon fiber reinforced ultra-high molecular weight polyethylene. USA Patent, 4055862
33. Ainsworth RFG, Bardos D (1977) An improved bearing material for joint replacement prostheses: carbon fiber-reinfored UHMW polyethylene. Trans Third Soc Biomater 3:119
34. Connelly GM, Rimnac CM, Wright TM, Hertzberg RW, Manson JA (1984) Fatigue crack propagation behavior of ultrahigh molecular weight polyethylene. J Orthop Res 2(2):119–125
35. Deng M, Shalaby SW (1997) Properties of self-reinforced ultra-high-molecular-weight polyethylene composites. Biomaterials 18(9):645–655
36. Suh NP, Mosleh M, Arinez J (1998) Tribology of polyethylene homocomposites. Wear 214 (2):231–236
37. Mosleh M (1998) An UHMWPE homocomposite for joint prostheses. In: Jacobs JJ, Craig TL (eds) Alternative bearing surfaces in total joint replacement. American Society for Testing and Materials, West Conshohoken, pp 256–265
38. Price H, Lin S, Hawkins M, Parr J (1997) Reinforced polyethylene for articular surfaces. USA Patent, 5609638
39. Siskey R, Smelt H, Boon-Ceelen K, Persson M (2016) UHMWPE homocomposites and fibers. In: Kurtz S (ed) UHMWPE biomaterials handbook, 3rd edn. William Andrew Publishing, Oxford, pp p398–p411
40. Grand View Research (2016) Medical grade ultra high molecular weight polyethylene (uhmwpe) market analysis by application (total hip replacement, knee replacement, shoulder replacement, ankle replacement, small joints) and segment forecast to 2024
41. Costa L, Bracco P (2004) Mechanisms of crosslinking, oxidative degradation and stabilization of UHMWPE. In: Kurtz S (ed) UHMWPE biomaterials handbook. Elsevier Academic Press, Boston, pp 309–323
42. Kurtz S, Rimnac C, Hozack W, Turner J, Marcolongo M, Goldberg V, Kraay M, Edidin A (2005) In vivo degradation of polyethylene liners after gamma sterilization in air. J Bone Joint Surg 87(4):815–823
43. McKellop H, Shen F, Lu B, Campbell P, Salovey R (2000) Effect of sterilization method and other modifications on the wear resistance of acetabular cups made of ultra-high molecular weight polyethylene. J Bone Joint Surg 82A(12):1708–1725
44. Sutula LC, Collier JP, Saum KA, Currier BH, Currier JH, Sanford WM, Mayor MB, Wooding RE, Sperling DK, Williams IR, Kasprzak DJ, Surprenant VA (1995) Impact of gamma-sterilization on clinical-performance of polyethylene in the hip. Clin Orthop Relat Res 319:28–40
45. Won C, Rohatgi S, Kraay M, Goldberg V, Rimnac C (2000) Effect of resin type and manufacturing method on wear of polyethylene tibial components. Clin Orthop Relat Res 376:161–171
46. Williams IR, Mayor MB, Collier JP (1998) The impact of sterilization method on wear in knee arthroplasty. Clin Orthop Relat Res 356:170–180
47. Kurtz S, Rimnac C, Rimnac C, Hozack W, Turner J, Marcolongo M, Goldberg V, Kraay M, Edidin A (2005) In vivo degradation of polyethylene liners after gamma sterilization in air. J Bone Joint Surg 87(4):815–823
48. Currier B, Currier J, Mayor M, Lyford K, Van Citters D, Collier J (2007) In vivo oxidation of gammabarrier-sterilized ultra-high-molecular-weight polyethylene bearings. J Arthroplast 22 (5):721–731
49. Birman M, Noble P, Conditt M, Li S, Mathis K (2005) Cracking and impingement in ultra-high-molecular-weight polyethylene acetabular liners. J Arthroplast 20(Suppl. 3):87–92

50. Chiu Y-S, Chen W-M, Huang C-K, Chiang C-C, Chen T-H (2004) Fracture of the polyethylene tibial post in a NexGen posterior-stabilized knee prosthesis. J Arthroplast 19(8):1045–1049
51. Green T, Fisher J, Matthews J, Stone M, Ingham E (2000) Effect of size and close on bone resorption activity of macrophages by in vitro clinically relevant ultra high molecular weight polyethylene particles. J Biomed Mater Res 53(5):490–497
52. Greenwald A, Bauer T, Ries M (2001) New polys for old: contribution or caveat? J Bone Joint Surg 83:27–31

Chapter 2
Highly Crosslinked UHMWPE for Joint Implants

Guorong Gao and Jun Fu

Abstract Wear is a major obstacle limiting the longevity of implanted conventional UHMWPE components. Efforts to solve the wear problem in UHMWPE have spurred enormous studies in highly crosslinked UHMWPE (HXPE). HXPE-bearing couples have been clinically used in total joint arthroplasties for two decades, and the follow-up studies confirmed their effectiveness in reducing in vivo wear and related disease. This chapter provides a comprehensive review of crosslinking, oxidation and thermal stabilization, structure and property relationship, and in vitro and in vivo wear performance of HXPE for total joint replacements. The first part of this chapter describes the fundamental of UHMWPE crosslinking, including chemical reactions and the formation of crosslinked structures induced by high-energy radiations, detection of macro-free radicals by ESR, and calculation of crosslink density based on Flory theory. The second part outlines the oxidation due to irradiation-induced residual free radicals in HXPE and thermal treatments to eliminate the free radicals. The third part summarizes the effect of crosslinking on crystalline structures and mechanical properties, including fatigue, tensile, and impact properties of HXPE. The fourth part focuses on the wear mechanism and in vivo wear properties of HXPE. Finally, the fifth part reviews the follow-up studies of HXPE used in total hip and total knee arthroplasties.

Keywords Ultrahigh molecular weight polyethylene · Joint arthroplasty · Radiation crosslinking · Free radicals · Oxidation stability · Mechanical properties · Fatigue resistance · Wear rate

---

G. Gao · J. Fu (✉)
Ningbo Institute of Materials Technology and Engineering, Chinese Academy of Sciences, Ningbo, China
e-mail: fujun@nimte.ac.cn

J. Fu et al. (eds.), *UHMWPE Biomaterials for Joint Implants*, Springer Series in Biomaterials Science and Engineering 13,
https://doi.org/10.1007/978-981-13-6924-7_2

2.1 Introduction

Ultrahigh molecular weight polyethylene, also known as UHMWPE, or polyethylene, has been an orthopedic-bearing material for total joint arthroplasties since the 1960s [1]. Not only the polyethylene biomaterials themselves but also their clinical applications have expanded since the 1990s. Although total joint arthroplasty using UHMWPE-bearing couple has been one of the most successful operations of the last century, long-term bone resorption around the implants (osteolysis) was observed, limiting the longevity of the arthroplasty [2–4]. One major cause of this phenomenon is the immune reactions associated with the wear debris primarily generated from the UHMWPE component [5, 6]. Over the past two decades, researchers all over the world confirmed the finding that elevated doses of radiation crosslinking would reduce wear rates by an order of magnitude during in vitro hip simulator testing [7, 8]. Nowadays, highly crosslinked UHMWPE (HXPE) is the "gold standard" for joint prosthesis.

Before HXPE were used, gamma radiation-sterilized UHMWPE was widely promoted in arthroplasty, which was usually mentioned as conventional UHMWPE. The crosslinking was imparted by a single sterilization dose (25–40 kGy) of gamma radiation. Compared to virgin UHMWPE, gamma radiation substantially improved the wear resistance of conventional UHMWPE [9, 10]. Studies in hip simulators [11] and in radiographic wear [12] showed higher wear rates for the non-crosslinked, gas-sterilized components when compared with gamma-sterilized controls. However, significant wear debris were still generated from conventional UHMWPE components over time [13, 14], which may lead to osteolysis and the need for revision.

Gamma sterilization also had a drawback that it rendered the polymer vulnerable to oxidation before implantation [13, 15] and in vivo [16–19]. The oxidation, in turn, contributed to mechanical degradation and embrittlement of UHMWPE, thus ultimately leading to mechanical failure of the implants [18, 20, 21].

HXPE were clinically introduced starting in 1998 to reduce wear and the incidence of revision resulting from osteolysis [22]. A second motivation for developing highly crosslinked polyethylene was to reduce oxidation, which had been associated with clinical failures after gamma sterilization and long-term shelf aging. The strategies to improve oxidation resistance after irradiation varied among orthopedic implant producers. One approach, known as annealing, involved a single thermal treatment below the crystalline melt transition in polyethylene [23]. A second approach to polyethylene formulation, referred to as remelting, involved thermal treatment above the melt transition [8].

This chapter introduces (1) the fundamentals, (2) the oxidation and thermal stabilization of crosslinked UHMWPE, (3) the structure and property relationship, (4) in vitro wear performance, and (5) in vivo wear performance of radiation crosslinked UHMWPE.

2.2 Fundamental of Radiation Crosslinking of UHMWPE

2.2.1 Chemical Reactions During and After Exposure to High-Energy Radiations

UHMWPE crosslinks when exposed to ionizing radiation through free radical recombination reactions. Ionizing radiation cleaves the C–H and C–C bonds, forming free radicals, some of which recombine to form crosslinks. The cleavage of C–C bonds reduces the molecular weight of the polymer. The reaction of free radicals generated by the cleavage of a C–H and C–C bond results in long-chain branching or Y-linkages. The recombination of two free radicals formed by the cleavage of C–H bonds results in crosslinking or H-linkage (Fig. 2.1). The formation rate of H-linkages increases with increasing irradiation temperature [24, 25].

Both gamma (γ) and electron beam (e-beam) irradiation can be used to crosslink UHMWPE. While the cascade formation of crosslinks and residual free radicals is similar for both methods, important differences include the penetration and radiation dose rate. The γ irradiation sources are commonly based on the artificial isotope of cobalt (^{60}Co) that generates γ photons. While the penetration of γ source into UHMWPE has no practical limitations, the activity level of the γ source limits the radiation dose rate. The penetration of e-beam radiation is limited by its kinetic energy, measured in million electron volts (MeV). With a 10 MeV electron beam incident on the UHMWPE surface, the radiation penetrates about 4–4.5 cm. The radiation dose rate that a commercial e-beam accelerator provides is about two orders of magnitude larger than that from a commercial γ source. The e-beam produces adequate number of radicals in a few seconds.

During e-beam irradiation, adiabatic heating of UHMWPE is common. The adiabatic heat generated increases with increasing radiation dose rate [8]. E-beam radiation offers a greater control of dose rate, which presented a mean to control the temperature rise in UHMWPE during irradiation throughout the development of highly crosslinked UHMWPEs. This also meant that the polymer could be melted during irradiation rather than after irradiation.

Radiation dose is energy per unit mass of the material being irradiated, and it is measured in kGy units, or kJ/g. That is, a 100-kGy radiation dose corresponds to 100 kJ/g of energy that will be converted to heat, resulting in a temperature rise, ΔT, in the polymer per the following equation:

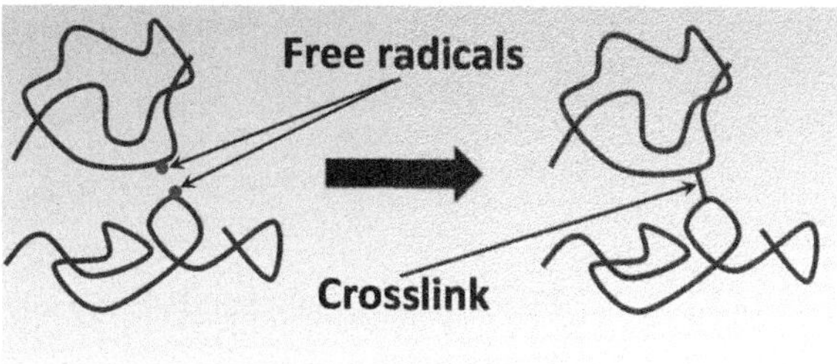

Fig. 2.1 During irradiation, carbon–hydrogen bonds are broken, forming free radicals along the backbone of the UHMWPE molecule. The reaction of two free radicals in two separate chains results in the formation of a crosslink

$$\text{Dose} = c_p \Delta T + \Delta H_m \tag{2.1}$$

where c_p (= 2.5 kJ/g/°C) is the specific heat, and ΔH_m is the heat of fusion of the polymer. Assuming that the polymer is about 50% crystalline, the heat of fusion would be about 150 kJ/g, and about 430 kGy of radiation dose would be needed to completely melt the UHMWPE from room temperature. This is a rather high-dose level and would adversely affect the mechanical properties. The optimum radiation dose level for UHMWPE for total hips is about 100 kGy [7, 26]. With a radiation dose of 100 kGy, UHMWPE needs to be heated to at least 120 °C to achieve near-complete melting of the crystalline domains. The dose rate has to be high, and the polymer needs to be insulated to minimize heat loss to the surroundings to achieve near-adiabatic conditions. The efforts on using e-beam to melt the UHMWPE led to the development of warm-irradiated UHMWPE, now used in the fabrication of total hip and total knee components [8].

2.2.2 Formation of Free Macroradicals and Crosslinked Structures in Irradiated UHMWPE

2.2.2.1 Formation of Macroradicals During Irradiation

E-beam and γ rays, employed for both sterilization and crosslinking processes, have a mean energy of some orders of magnitude higher than that of chemical bonds, which leads to the scission of C–C (Reaction 1, Scheme 2.1) and C–H bonds (Reaction 3, Scheme 2.1), generating H• radicals and primary and secondary alkyl macroradicals, respectively [28]. These alkyl macroradicals are dispersed throughout both the amorphous phase of the polymer and likely in the crystalline phase.

In previous studies, researchers have not detected primary alkyl macroradicals in irradiated UHMWPE [29–31]. In addition, NMR and FTIR studies have not revealed an appreciable increase in the concentration of terminal methyl units, which would be associated with radiolytic cleavage of the polymer chains. Because breaking of

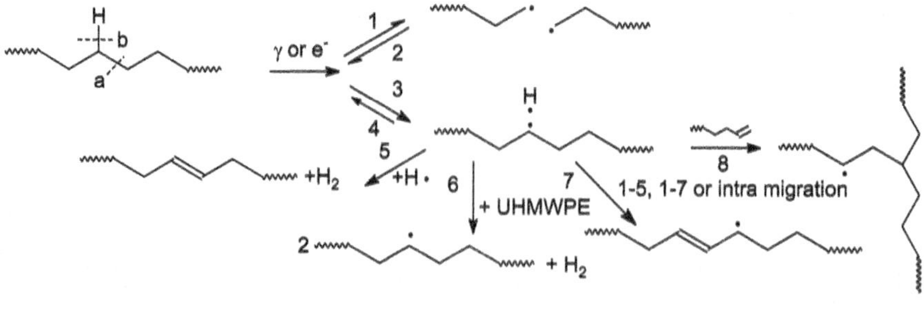

Scheme 2.1 Reaction pathways. (Reprinted from Ref. [27], Copyright 2008, with permission from Elsevier)

C–C is a random (stochastic) process, it can be assumed that the primary alkyl macroradicals resulting from Reaction 1 (Scheme 2.1) undergo recombination, in both amorphous and crystalline phases, giving back a C–C bond, with dissipation of energy in the polymer bulk (Scheme 2.1, Reaction 2).

Orthopedic UHMWPE has a weight average molecular mass of 2×10^{6} amu. The polymer has a high viscosity, even in the molten state. Thus, macroradicals have very low mobility, either in the molten or in the solid state, whereas the H• radical, which has a diameter smaller than 1 Å, can diffuse in the polymer mass, even in the crystalline phase, where distances between C atoms are in the order of 4 Å.

Reaction 5 (Scheme 2.1) is extremely favored, being exothermal ($\Delta H = -288$ kJ/mol), with a very low entropy variation. The intramolecular process (Reaction 5) is extremely fast, and the secondary macroradicals decay, giving vinylene double bonds and molecular hydrogen (a gaseous product), which in turn can diffuse through the polymer mass. Among the vinylene double bonds, the *trans*-bonds are thermodynamically more stable. However, NMR studies have reported the presence of both *cis*- and *trans*-vinylene in the amorphous phase and *trans*-vinylene only in the crystalline phase [32, 33], formed according to a pseudo-zero-order kinetic [34].

H• radicals resulting from Reaction 3 (Scheme 2.1) are very mobile, and they can extract other H atoms intramolecularly, producing hydrogen. Intermolecular extraction (Reaction 6, Scheme 2.1) is possible, being exothermal ($\Delta H = -30$ kJ/mol) and associated with a very low entropy variation.

Reaction 6 (Scheme 2.1) leads to an increase of isolated macro-alkyl radicals in the polymer mass. Secondary alkyl radicals can then migrate along the polymer chain via H transfer. When the hydrogen transfer results in setting of a radical in α to a vinylene double bond (Reaction 7, Scheme 2.1), the radicals structure changes from secondary to allyl macroradicals, which are more stable and can survive even after a few years on the shelf at room temperature. The H transfer reaction in Reaction 3 is thermodynamically favored because it leads to the formation of more stable allyl macroradicals. The activation energy for this process is only 40 kJ/mol, and thus the reaction can occur even at room temperature at a high rate.

The concentration of surviving macroradicals decreases with time after irradiation [35]. Their living time is on the order of 24 h in the amorphous phase, but macro-alkyl radicals can be found in the polymer bulk even after years, probably trapped in the crystalline phase [36]. The mechanism of decay of macro-alkyl radicals is still unknown.

These radical species are responsible for radical reactions (crosslinking, initiation, and propagation of the oxidative cascade).

2.2.2.2 Formation of Y-Crosslinking

UHMWPE is not just a simple sequence of methylene, (CH_2), but it also contains small but measurable concentrations of vinyl double bonds, tertiary carbons, and methyl groups. These short- and long-chain branches, as well as the residues of the catalyst, are incorporated in the amorphous phase, not in the crystalline lamellae.

The amounts of structural irregularities, as well as the degree of crystallinity, depend on the polymer synthesis method and on the subsequent processing conditions of the resin.

Double bonds can react with alkyl macroradicals, resulting in the formation of branching and increasing of the molecular mass. When a terminal vinyl double bond reacts with the alkyl macroradical on an adjacent polymer chain (Reaction 8, Scheme 2.1), a Y-crosslink is formed [37]. Reaction 8 is exothermal ($\Delta H = -88$ kJ/mol), but it is controlled by inductive and steric effects. In UHMWPE, Reaction 8 occurs only with the terminal vinyl groups at the end of the polymer chains [28]. There is experimental evidence from NMR studies to support the formation of Y-crosslinks from terminal vinyl species, as shown in Scheme 2.1. Costa et al. [27] observed the FTIR absorbance of UHMWPE films irradiated at 30, 60, and 100 kGy in vacuum. In those spectra, the following changes in the IR absorption due to irradiation can be observed: a decrease in the absorption at 908 cm^{-1} (vinyl double bonds) and an increase in the absorption at 965 cm^{-1} (*trans*-vinylene double bonds) and at 982 cm^{-1} (conjugated double bonds). The decrease in vinyl absorption suggests the crosslinking degree of material. The *trans*-vinylene and conjugated double bonds absorption are proportional to the radiation dose.

Due to steric hindrance, Reaction 8 shown in Scheme 2.1 is not observed with the vinylene double bonds. In low-density polyethylene, this reaction can occur with vinylidene species [28]. However, these types of vinylidene bonds are rarely observed in UHMWPE and are mentioned nonetheless for completeness.

2.2.2.3 Formation of H-Crosslinking

Another crosslinking mechanism leading to the formation of H-crosslinks is shown in Scheme 2.2. Secondary alkyl macroradicals can decay, moving along the polymer chain, via deprotonation (Reaction 1, Scheme 2.2) or coupling (Reaction 2). Both reactions are exothermal ($\Delta H = -260$ kJ/mol and -313 kJ/mol for Reactions 1 and 2, respectively).

The formation of H-crosslinking is blocked by steric hindrance when UHMWPE is in the solid state [38]. Radicals in the crystalline phase behave differently from

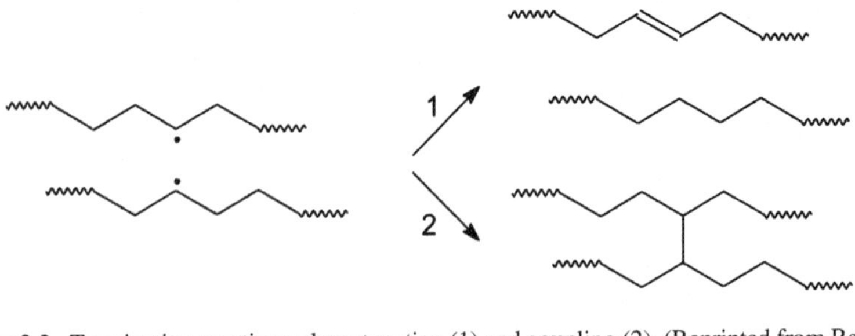

Scheme 2.2 Termination reactions: deprotonation (1) and coupling (2). (Reprinted from Ref. [38], Copyright 2016, with permission from Elsevier)

those in the amorphous phase. Models of the crystalline phase in the solid state show that the minimum distance between two C atoms is 4 Å, much higher than the mean C–C bond length (1.54 Å). Therefore, the formation of H-crosslinks (Reaction 2, Scheme 2.2) is sterically hindered in the crystalline phase. The presence of H-crosslinks in polyethylene irradiated in solid state has been ruled out in previous NMR studies. In addition, given the high rate of transfer of the secondary alkyl macroradical, the encounter of the two species involved in Reaction 2 would be statistically favored. Then, if the reaction were easily feasible, it would be a termination pathway for the radical species, which would not be found for long durations after irradiation [32].

Moreover, experimental evidence shows that the level of crosslinking increases with the radiation dose, but it reaches a plateau for doses higher than 100 kGy, when the majority of the vinyl double bonds have been consumed [28]. A different behavior can take place in the molten state, where the mobility of radicals is higher. Under these conditions, NMR studies confirm the presence of some H-crosslinks [32].

2.2.3 Macroradicals in UHMWPE Detected by ESR

The free radicals, including alkyl and allyl radicals, and oxygen-induced radicals in UHMWPE could be directly detected by electron paramagnetic resonance (EPR or ESR).

ESR spectroscopy is based on the fundamental properties of electron. An unpaired electron in atoms or molecules (e.g., hydrogen atoms, transition metal ions, free radicals) possesses both magnetic moment and angular momentum. The ratio between magnetic moment and the angular momentum is known as gyromagnetic ratio (γ). The net magnetic moment of a material (solid, liquid, or gas) can be zero because of the random nature of the spinning electrons in a material. In an external magnetic field (H), the spin magnetic moment aligns parallel or antiparallel to the field, and the spinning electrons are split or divided into high (+)- and low (−)-energy states. The energy separation ΔE between the states is known as Zeeman energy, and is given by $\Delta E = g\beta H$, where H is the external steady magnetic field, β is the electron Bohr magneton ($\beta = 9.27 \times 10^{-24}$ J/T), and g is the spectral splitting factor, commonly known as the g value. Transitions between these states can be stimulated by an oscillating magnetic field of frequency f when and only when its energy equals ΔE; that is, $hf = \Delta E = g\beta H$, where h is Planck's constant ($h = 6.63 \times 10^{-34}$ Js). The condition $hf = g\beta H_r$ is known as resonance condition, where H_r is the external magnetic field at resonance and is directly proportional to microwave frequency. An ESR spectrometer requires a resonator (a microwave cavity) that operates at a fixed frequency, and it records the resonance absorption signal as a function of external magnetic field.

An unpaired electron formed as a result of a broken covalent bond in a molecule, for example, creating a radical (by definition), and the ESR will most likely produce

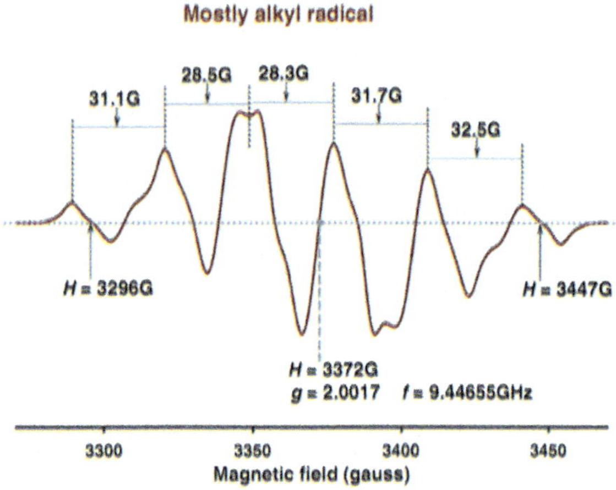

Fig. 2.2 ESR spectrum of an alkyl radical —CH_2—•CH—CH_2—. (Reprinted from Ref. [39], Copyright 2016, with permission from Elsevier)

a single line, or a singlet, provided there is no net nuclear magnetic moment coupled to the unpaired spin. If a proton is situated near the unpaired spin (at the broken site of a molecule), each electron Zeeman energy state will split into two (for nuclear spin $I = 1/2$) or three (for nuclear spin $I = 1$) states. For each nuclear spin, electron spin energy level splits into $2I + 1$ levels. The magnetic field positions of the resonance lines for $I = 1/2$ are given by $H = H_r \pm a/2$, where H_r is the resonance field position for the radical and "a" is the hyperfine coupling constant, or hyperfine constant.

For a radical containing five equivalent protons (alkyl radical in UHMWPE, —$CH_{2,\beta}$—•CH_α—$CH_{2,\beta}$—) in its neighborhood, each of $I = 1/2$, six resonance lines (Fig. 2.2) are observed in its ESR spectrum ($2 \times 5/2 + 1 = 6$). One α-proton and four β-protons in the alkyl radical effectively act like five equivalent protons because of random orientation of the molecule in a heterogeneous polymer matrix.

For allyl radicals, the clean ESR spectrum is a seven-line fingerprint, as there are four β-hydrogen and two α-hydrogen atoms in its neighborhood (Fig. 2.3a). Figure 2.3b shows an ESR spectrum of the radical with hyperfine coupling constant $a_{\text{allyl}} = 13.6$ G and the g value, $g = 2.0016$, approximately.

In 2005, Jahan et al. reported the detection of the peroxy radicals in irradiated UHMWPE [39]. They used GUR 1050 UHMWPE sheets ($9 \times 3 \times 0.2$ mm^3) irradiated by X-ray in vacuum. Immediately after irradiation, the sample was transferred to the ESR cavity with continuous nitrogen purge. In conjunction with microwave power saturation technique (PST), ESR spectra were recorded at and below room temperature. At different time intervals, the flow of nitrogen was stopped to allow oxygen (air) to react with the radicals. The resonance line in Fig. 2.4a is due to a mixture of alkyl and allyl radicals. When the microwave power was increased from 1 mW to 32 mW, the resonance lines due to alkyl and allyl were saturated (Figure 2.4b). When the sample was exposed to air for 1 min, a broad asymmetric signal was formed (Fig. 2.4c). When the temperature was decreased to 118 K, by purging the flow system with cold N_2 gas, the same broad single line split into characteristic lines of a peroxy radical with $g_{\parallel} = 2.032$ and $g_{\perp} = 2.003$ (Fig. 2.4d).

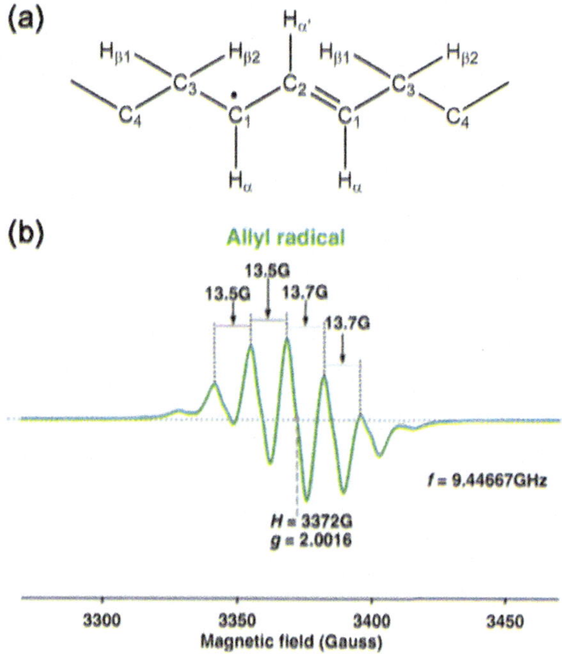

Fig. 2.3 (**a**) Molecular structure of the allyl radical showing approximate orientation of the α- and β-hydrogen atoms. (**b**) ESR spectrum due to allyl radical. (Reprinted from Ref. [39], Copyright 2016, with permission from Elsevier)

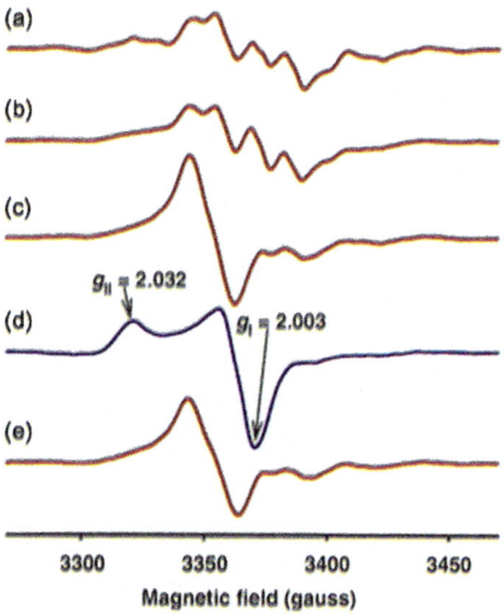

Fig. 2.4 Characteristic spectrum of peroxy radical (d) was isolated from the complex signal by using appropriate environment, microwave power, and temperature. They are presented, respectively, as follows: (**a**) N_2, 8 mW, 293 K; (**b**) N_2, 32 mW, 293 K; (**c**) air, 32 mW, 293 K; (**d**) cold N_2, 32 mW, 118 K; (**e**) air, 32 mW, 293 K following (**d**). (Reprinted from Ref. [39], Copyright 2016, with permission from Elsevier)

Resonance line in Fig. 2.4e was recorded after the sample was brought back to room temperature (293 K) again. Because of the unique orientation of the oxygen molecule with unpaired spin (—O_2•—) with respect to the PE molecular axis, the peroxy radical produces (theoretically) asymmetric lines in the ESR spectrum with two

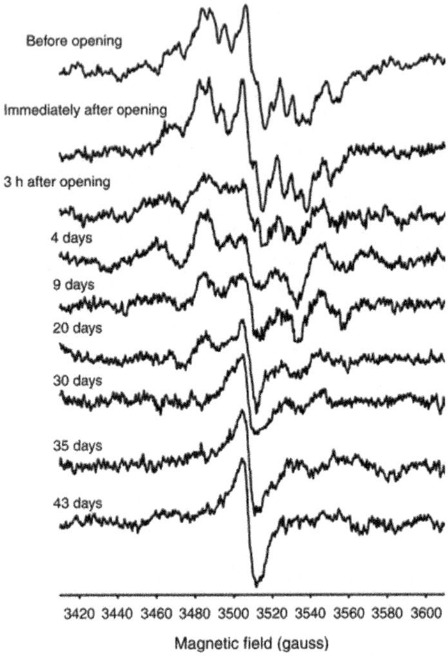

Fig. 2.5 ESR spectrum recorded as a function of time at room temperature for GUR 4120 γ irradiated in an evacuated quartz tube and then stored for 6 years at 75 °C in vacuum. (Reprinted from Ref. [39], Copyright 2016, with permission from Elsevier)

g values, $g_{\parallel}$ and $g_{\perp}$. The peroxy radical was found to be very short-lived (half-life less than 1 h) [39].

When UHMWPE samples were irradiated and stored in inert environments and 75 °C, radicals were still significantly present 15 years later [39]. When γ-irradiated GUR 4120 sample was aged at 75 °C for 6 years in vacuum, oxidation-induced radicals can still be detected once the sample is exposed to air for oxidation to occur (Fig. 2.5). To avoid oxidation, especially in vivo oxygen permeation-induced long-time oxidation, appropriate process was needed to eliminate or catch the residual primary radicals.

2.2.4 *Crosslink Density (Flory Theory)*

The crosslink density of irradiated UHMWPE was determined from swelling experiments [40]. For swelling experiments, about $3 \times 3 \times 2$ mm^3 cubes were used. The equilibrium volume–swelling ratio, q_{eq}, was determined using a Perkin-Elmer TMA/DMA 7 ($n = 3$ for each homologous series). Each test sample was placed in a quartz basket-probe assembly and lowered into a xylene/antioxidant bath at room temperature. Iraganox 1010 (Ciba Additives, Ciba-Geigy Co.) was added in xylene at a concentration of 0.2 *wt%*. The xylene was then heated to 130 °C at 5 °C/min and held at 130 °C for 2 h. The swelling was then recorded with the upward motion of the probe until the equilibrium swelling was achieved (Fig. 2.6).

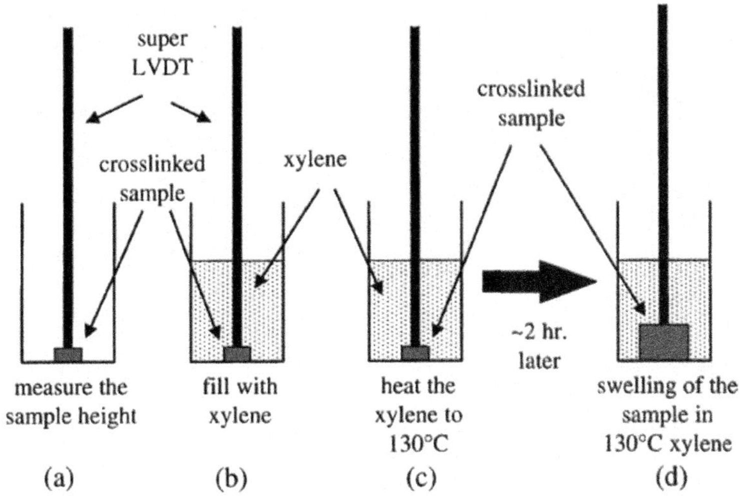

Fig. 2.6 The schematic illustration to the swelling experiments by using a Perkin-Elmer TMA/DMA 7. (Reprinted from Ref. [26], Copyright 1999, with permission from Elsevier)

The experiments in three orthogonal directions confirmed the isotropy of swelling. From this one-dimensional change in height, the apparent volume–swelling ratio (T_{eq}) was calculated:

$$T_{eq} = \left(\frac{h_f}{h_i}\right)^3 \tag{2.2}$$

where h_i and h_f denote the height of the test specimen before and after the equilibrium swelling is reached, respectively. As there are contributions from volumetric expansion due to heating and melting to T_{eq}, the following correction was used to obtain the actual equilibrium volume–swelling ratio, q_{eq}:

$$q_{eq} = \left[T_{eq} - \alpha\Delta T - \left(\frac{\rho_s}{\rho_m} - 1\right)\right] \tag{2.3}$$

where α is the volumetric coefficient of expansion ($= 1.3 \times 10^{-5}$ K^{-1} [40]) of the test sample and ΔT is the temperature difference between room temperature and temperature at which the experiment was carried out (i.e., 130 °C). ρ_s ($= 0.92$ g/cm^3) is the density of crosslinked test sample at room temperature, and ρ_m ($= 0.855$ g/cm^3 [40]) is the theoretical density of the amorphous phase. Using the values listed above, the correction can be calculated:

$$-\alpha\Delta T - \left(\frac{\rho_s}{\rho_m} - 1\right) = -1.3 \times 10^{-5} \times 110 - \left(\frac{0.92}{0.855} - 1\right) = -0.077 \tag{2.4}$$

In practical examples, T_{eq} is usually used as q_{eq} without the correction [26].

The measured q_{eq} values were then used to determine the crosslink density, d_x, and the molecular weight between crosslinks, M_c, as a function of radiation dose using the following equations [41–43]:

$$d_x(\text{mol/cm}^3) = -\frac{\ln\left(1 - q_{eq}^{-1}\right) - q_{eq}^{-1} + Xq_{eq}^{-2}}{V_1 q_{eq}^{-1/3}} \tag{2.5}$$

$$M_e(\text{g/mol}) = \frac{\rho_{solid}}{d_x} \tag{2.6}$$

where

$$V_1 = 136\ \text{cm}^3/\text{mol} \tag{2.7}$$

$$X = 0.33 + \frac{0.55}{q_{eq}} \tag{2.8}$$

In the above equations, V_1 is the molar volume of the solvent at 130 °C, and X is the polymer solvent interaction parameter at 130 °C.

The q_{eq} in Eq. (2.3) was deduced from the following steps.

The final volume (following the swelling in the hot xylene), V_f, of the test sample is expressed as follows:

$$V_f = V_i + \Delta V_t + \Delta V_m + \Delta V_s \tag{2.9}$$

where V_i is the initial volume, ΔV_t is the change in volume due to the thermal expansion, ΔV_m is the change in volume due to melt transition, and ΔV_s is the change in volume due to swelling. The change in volume due to the thermal expansion can be expressed as follows:

$$\Delta V_t = \alpha \Delta T V_i \tag{2.10}$$

Knowing the density of melt, ρ_m, and density of solid, ρ_s, the change in volume due to melting can be calculated by the difference between the volume of the melt, V_m, and initial volume:

$$\Delta V_m = V_f - V_i = \left(\frac{\rho_s}{\rho_m} - 1\right) V_i \tag{2.11}$$

Finally, the change in volume due to swelling can be expressed in terms of equilibrium volume–swelling ratio, q_{eq}, which is the ratio of the final volume of the test sample due to swelling only and its initial volume:

Fig. 2.7 Marked measurement direction before and after swelling

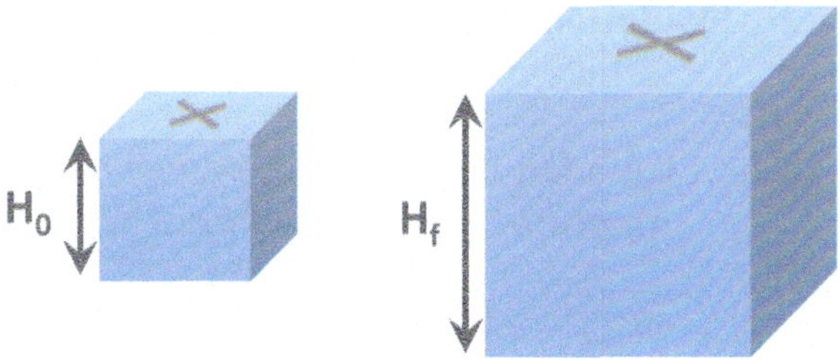

$$\Delta V_s = V_i\left(q_{eq} - 1\right) \tag{2.12}$$

Inserting Eqs. (2.9)–(2.11) in Eq. (2.8) yields Eq. (2.3).

Similar method for determination of crosslink density is speculated in ASTM F 2214. In this method, a cubic specimen is placed in a dry chamber, and then xylene mixed with a little of antioxidant is introduced into the chamber. The chamber is heated to 130 °C. The sample height is monitored as a function of time until steady state (equilibrium) is achieved. The swelling ratio, q_s, is computed from height change (Fig. 2.7) of swelled samples:

$$q_s = (V_f/V_0) = (H_f/H_0)^3 \tag{2.13}$$

where V_f = final volume, V_0 = initial volume, H_f = final height, and H_0 = initial height.

According to Flory's network theory [44], which explains the swelling ratio of a polymer–solvent system as a competition between elastic forces and forces derived from the free energy of mixing, the following expression is derived for the crosslink density, v_d, as a function of the steady-state swelling ratio (q_s), the Flory interaction parameter (χ_1), and the molar volume of the solvent (φ_1):

$$v_d = -\frac{\ln\left(1 - q_s^{-1}\right) + q_s^{-1} + \chi_1 q_s^{-2}}{\varphi_1\left(q_s^{-1/3} - q_s^{-1}/2\right)} \tag{2.14}$$

The expression in Eq. (2.14) can be reduced to calculate the molecular weight between crosslinks, M_c, where $\bar{v}$ is the specific volume of the polymer:

$$M_c = \left(\bar{v}v_d\right)^{-1} \tag{2.15}$$

The parameters used for UHMWPE swelling in xylene at 130 °C is $\chi_1 = 0.33 + 0.55/q_s$, $\varphi_1 = 136\ \mathrm{cm^3/mol}$, $\bar{v}^{-1} = 920\ \mathrm{g/dm^3}$.

2.3 Oxidation and Thermal Treatments of Radiation Crosslinked UHMWPE

2.3.1 Oxidation Due to the Presence of Free Radicals Trapped in Crystalline Phase [45]

There is evidence that crosslinks are not formed abundantly in the crystalline regions of polyethylene [46, 47]. This is presumably because the free radicals formed on the chains in the lamellae do not possess enough mobility to recombine with each other. Another possible reason is that the lattice spacing in the crystallites is larger than the carbon–carbon bond length needed for a crosslink to form, requiring kinking of the chains, which is not energetically favored. The entropy for these free radicals to move out of the crystallites toward the crystalline–amorphous interface is higher than that for movement between the chains. The resultant is an abundance of free radicals, called "residual free radicals," which are trapped for prolonged periods of time in the crystalline lamellae [36, 48].

If the irradiated UHMWPE is not treated to stabilize these residual free radicals and allow them to escape from the crystallites and recombine to form crosslinks, diffused oxygen reacts with these free radicals, leading to hydroperoxides, the breakdown of which is accompanied by chain scission and the deterioration of mechanical properties [49]. This oxidation was manifested in the form of embrittlement in UHMWPE components that were terminally γ sterilized [17, 50].

Thermal treatment of radiation-crosslinked UHMWPE was proposed to avoid oxidation. Three main forms of thermal treatment are used: (1) melting after irradiation, (2) annealing below the melting point of the irradiated polymer, and (3) sequentially irradiated and annealing below the melting point of the irradiated polymer.

2.3.2 Remelting

In order to remove the free radicals trapped in the crystalline domains, the irradiated polymer was heated to above its melting point. This procedure does not only eliminate the crystalline domains or the lamellae to release the trapped radicals but also impart them with enough mobility to find and recombine with each other. Thus, the residual free radicals are quenched, and oxidative stability of the irradiated UHMWPE is improved. Melting of irradiated UHMWPE is carried out in an air convection oven. Typically, UHMWPE irradiated in its bar stock form is machined into finished implants after the melting step to remove any surface oxidation prior to the machining of final components.

2.3.3 Annealing

Heating above the melting temperature of UHMWPE effectively quenches free radicals but changes the microstructure. Free radical quenching below the melting temperature (annealing) preserves the microstructure but is not as effective in quenching free radicals.

Annealing treatments have experienced three development courses, including Durasul®, Crossfire®, and X3®.

Durasul® was derived from sterilization. It was the first annealed crosslinked material. Ram-extruded GUR 4150 was machined into components (hip or knee), packaged in nitrogen, and gamma sterilized to 30 kGy [7]. Following sterilization, the packaged components were heated for 144 h at 50 °C (stabilization). Crosslinks form during sterilization. Excessive free radicals are eliminated during the stabilization step. In 1998, ram-extruded GUR 1050 replaced GUR 4150 as the starting UHMWPE.

In the Crossfire® process, a GUR 1050 rod is irradiated to 75 kGy in air. The rod is then annealed at 130 °C for 8 h to eliminate free radicals. The outer surface of the rod is machined away, and components are machined from the rod. Components are packaged in nitrogen and γ irradiated to 30 kGy (N_2/Vac). The total radiation dose is 105 kGy. Crossfire® acetabular inserts were clinically introduced in late 1998 for use in the hip, but not used for knee components.

2.3.4 X3

X3 is a highly crosslinked UHMWPE produced by a sequential irradiation and annealing process. The sequential process results in a material with a free radical content of 1% that of conventional UHMWPE γ sterilized in nitrogen, resulting in an oxidation resistance similar to that of virgin UHMWPE [51].

In 2002, Dumbleton et al. hypothesized that it might be possible to create a UHMWPE with little to no free radicals by integrating a sequential irradiation and annealing process in which a low dose of radiation was used followed by annealing. On average, the crosslinks would be far enough apart that there would be sufficient mobility to extinguish the free radicals. The process could then be repeated [51, 52].

By 2002, the experience with crosslinked UHMWPEs showed that a radiation dose of 50–100 kGy, and preferably closer to 100 kGy, would be needed to provide sufficient crosslinking for greater wear reduction. It was believed that annealing to remove free radicals had the advantage of retaining the original UHMWPE microstructure and properties, unlike the remelting process. However, the annealing process needed to be made more effective to remove sufficient free radicals to provide oxidation resistance. The sequential process appeared to be the answer rather than a single irradiation and annealing cycle.

Compression-molded GUR 1020 was chosen over GUR 1050 as the starting material due to its higher ductility and impact strength. Experimental work identified the following sequential irradiation and annealing process as the best balance between cycle time and properties:

Cycle 1: 30-kGy irradiation followed by annealing for 8 h at 130 °C
Cycle 2: 30-kGy irradiation followed by annealing for 8 h at 130 °C
Cycle 3: 30-kGy irradiation followed by annealing for 8 h at 130 °C

The cumulative radiation dose is 90 kGy. Following the sequential irradiation and annealing, the outer 1–3 mm of the material was removed, and components were machined and packaged for gas plasma sterilization. The material produced by the sequential irradiation and annealing process was named X3 and introduced for acetabular components in May 2005. Over the following years, X3 tibial and patella components were introduced for cruciate retaining (CR) and posterior-stabilized (PS) TKR. In 2008, total stabilizer (TS) components in X3 became available.

2.4 Crystal Structures and Mechanical Properties of Crosslinked UHMWPE

UHMWPE is comprised of a network of crystalline lamellae embedded in an amorphous matrix (Fig. 2.8). The polymer chains crystallize during consolidation of UHMWPE powder under isostatic pressure. The glass transition temperature of the amorphous UHMWPE is well below room temperature. Therefore, the mechanical properties and fatigue strength of UHMWPE depend directly on the content of the crystalline domains [45].

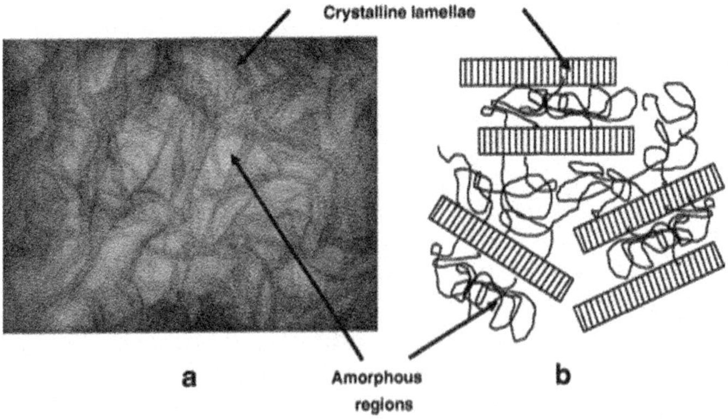

Fig. 2.8 A transmission electron micrograph (**a**) and schematic description (**b**) of the semicrystalline morphology of UHMWPE. (Reprinted from Ref. [26], Copyright 2007, with permission from Elsevier)

The crosslinking process induced changes on the microstructure and mechanical properties of UHMWPE. Crosslinking results in a restriction of chain mobility in the amorphous region and, as a result, reduces the overall plasticity of the polymer.

2.4.1 Effect of Radiation Crosslinking and Thermal Treatments on Crystallinity and Melting Point of UHMWPE

The crystallinity of the UHMWPE resin is very high, i.e., 80% [53, 54], and it decreases a lot to about 50% after consolidation.

A most commonly used method to measure the melting point and crystallinity of UHMWPE is differential scanning calorimetry (DSC). The crystallinity was calculated by normalizing the heat of fusion to that of 100% crystalline polyethylene (290 J/g [55], or 291 J/g [8], or 289.3 J/g [55], or 287.3 J/g [51]). The enthalpy peak from 20 to 160 °C [45, 56], or 60 to 155 °C [8], or 60 to 160 °C [26], or 80 to 160 [55], or 95 to 149 °C [57] is usually chosen to calculate the crystallinity.

Besides, FTIR is another effective method to determine the crystallinity of UHMWPE. In this method, the degree of crystallinity was determined by relating the crystalline band at 1896 cm^{-1} with the amorphous band at 1305 cm^{-1}. The crystalline percentage is given by [53, 54, 57]:

$$\%\text{Crystallinity} = \frac{A1896/A1305}{A1896/A1305 + 0.25} \times 100\% \tag{2.16}$$

The microscopic FTIR was found very useful for measuring the crystallinity of the inhomogeneous samples, where a high spatial resolution was required. It was not possible to use normal DSC since it requires a fairly large amount of material and thus cannot give the same spatial resolution. Using this method, Costa et al. plotted the crystallinity map of retrieved hip prosthesis [53].

In 1993, Zhao et al. [58] investigated the structure changes of compression-molded UHMWPE ($M_w = 1.5 \times 10^6$) after irradiation up to 40 Mrad (1 Mrad = 10 kGy) at 0.67 Mrad/h. At low irradiation doses (up to 10 Mrad), the apparent heat of fusion (ΔH_m) increases markedly; as the irradiation dose further increased to 40 Mrad, ΔH_m slowly dropped (Fig. 2.9a). Small-angle X-ray scattering (SAXS) results (Fig. 2.9b) presented an increase in the scattering intensity of irradiated samples with an increasing irradiation dose below 5 Mrad, and subsequent decreases with an increase at irradiation dose above 10 Mrad. Calculations according to the SAXS profiles suggest an increase in lamella thickness followed by a decrease with increasing irradiation dose (Fig. 2.9c). The melting points remain almost unchanged with varied irradiation doses (Fig. 2.9d).

At low irradiation dose, polymer chain scission occurs upon irradiation, which facilitates folding and crystallization. Further crystallization occurs both in

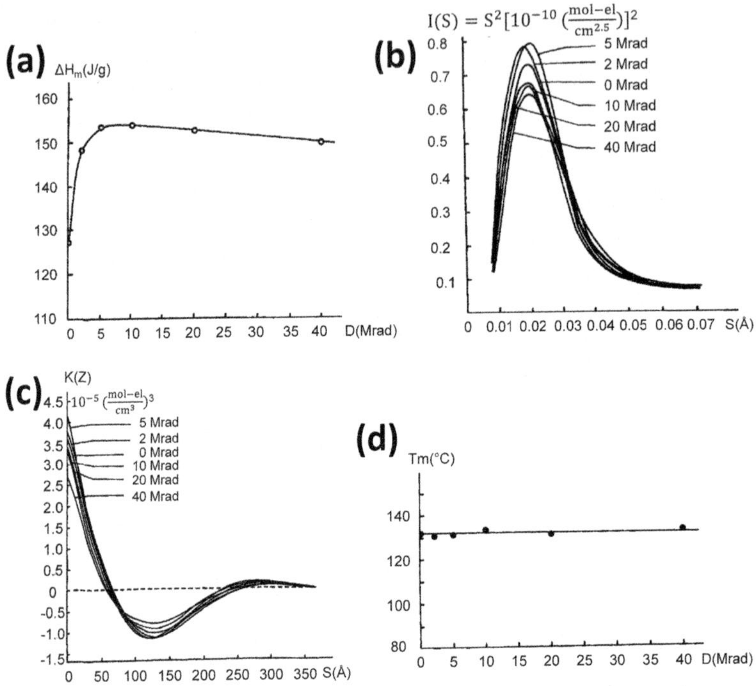

Fig. 2.9 (**a**) Apparent heat of fusion as a function of irradiation dose. (**b**) SAXS curves of UHMWPE, irradiated with various doses. (**c**) Thickness of lamellae as a function of irradiation dose. (**d**) Melting temperature as a function of irradiation dose. (Reproduced from Ref. [58] by permission of John Wiley & Sons Ltd)

crystalline and amorphous regions, while the former leads to the increase of lamella thickness. As the irradiation dose further increases, the degree of crystallinity can no longer increase, because the formation of new crystals is hindered by the formation of crosslinks in the amorphous region.

Oral et al. irradiated consolidated GUR 1050 UHMWPE blocks to 25, 50, 75, and 100 kGy in air at room temperature, which were labeled as CI-25, CI-50, CI-75, and CI-100 with CI for cold irradiation. An unirradiated control was referred to as CI-0. Some of the irradiated blocks were heated to 170 °C until they melted completely and were held in the melt for at least 2 h and then cooled down. These samples were labeled CISM-25, CISM-50, CISM-75, and CISM-100 for cold irradiation and subsequent remelting. The crystallinity of CI UHMWPE increased slightly compared to that of CI-0 UHMWPE, but there was no significant difference between the crystallinity of UHMWPE irradiated at different radiation doses ($p > 0.1$). The crystallinity of CISM UHMWPE was significantly lower than that of CI UHMWPE at all doses ($p < 0.005$, Table 2.1). On average, 13% of the crystallinity of the irradiated UHMWPE was lost due to postirradiation melting (Table 2.1).

Crystallinity increased as a result of irradiation but remained unchanged at different radiation dose levels (Table 2.1). One possible explanation for this is the

Table 2.1 Crystalline content of cold-irradiated (CI) and cold-irradiated and subsequently melted (CISM) UHMWPE as a function of radiation dose

Dose (kGy)	Crystallinity (%) of cold irradiated (CI)	Crystallinity (%) of cold-irradiated and subsequently melted (CISM)
0	63 ± 2	63 ± 1
25	67 ± 1	58 ± 2
50	66 ± 1	59 ± 1
75	66 ± 2	59 ± 0
100	69 ± 1	58 ± 1

rearrangement and crystallization of short chains formed by chain scission. Another explanation could be the potential increase in the surface energy of the crystals resulting from the decrease in the number of available low-energy conformations to tie molecules between crystallites and the formation of "taut-tie molecules." The increase in the surface energy of the crystals could increase the crystallinity as measured by DSC.

The decrease in the crystallinity of the irradiated samples upon melting (Table 2.1) may be explained by the relaxation of the taut-tie molecules, reducing the surface energy of the crystals and the decreased mobility of the crosslinked chains, slowing down the crystallization kinetics. The crosslinked chains lack the mobility to fold into lamellae.

Irradiation and subsequent remelting significantly reduce the crystallinity. The γ-irradiated and remelted GUR 1050 rod showed crystallinity decreasing from 50.1% ± 0.5% (control) to 45.6% ± 0.7% (50 kGy), 46.3% ± 0.8% (100 kGy), and 47.1% ± 0.4% (200 kGy) [59].

In comparison to the 61.3% ± 0.8% crystallinity of conventional UHMWPE, X3 [51] showed a value of 61.7% ± 0.6%. The sequential irradiation (30 kGy/pass) and annealing favor a maintenance of high crystallinity.

On the other hand, γ irradiation and subsequent annealing slightly decreased the crystallinity. For example, the 100 kGy irradiated and annealed (110 °C) GUR 1050 rod showed a crystallinity of 60.8% ± 0.9% [20], much higher than that of irradiated and remelted GUR 1050 (45.7% ± 0.3%) [60, 61].

Muratoglu et al. [8] presented a novel method of *adiabatic heating* to increase the crosslink density of UHMWPE. UHMWPE is irradiated in air at elevated temperature with a high-dose rate e-beam and subsequently melt-annealed. For example, UHMWPE preheated to 125 °C was irradiated at 125 °C using e-beam at 9.5 Mrad per pass. This high-dose rate led to *adiabatic heating* and partial melting of the polymer. After irradiation, the disks were remelted at 150 °C and then cooled to room temperature. This method is referred to as *WIAM* for *w*arm *i*rradiation with *a*diabatic heating and subsequent *m*elting. The crystallinity of the *WIAM* UHMWPE is 45% ± 0.4%, while that of conventional UHMWPE is 58% ± 0.6%. The DSC thermograms of *WIAM* UHMWPE exhibited two melting peaks at 118 °C and 139 °C, indicating the presence of a biphasic structure, in contrast to the single melting peak at 137 °C for virgin UHMWPE.

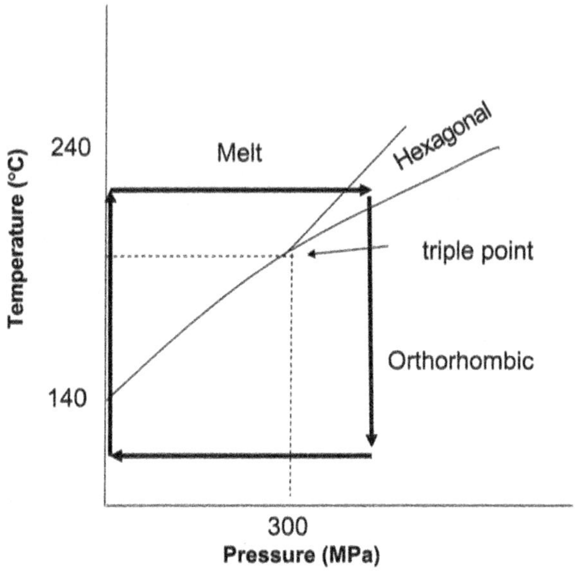

Fig. 2.10 Schematic phase diagram of UHMWPE. (Reproduced from Ref. [63] by permission of John Wiley & Sons Ltd)

The crystalline structures of irradiated UHMWPE could be manipulated by subsequent high-pressure crystallization (HPC). UHMWPE exhibits a hexagonal phase at high temperature and high pressure (Fig. 2.10), where the crystal stems are randomly rotated and the chain mobility is increased. As a result, the increased mobility allows the growth of "extended chain crystals," which are larger in size than the "folded chain crystals" in UHMWPE crystallized from the melt at ambient pressure. In addition, the overall crystallinity is increased by HPC, thereby increasing its yield strength and ultimate tensile strength [62, 63].

Pruitt et al. investigated the high-pressure treatment on the microstructure of both crosslinked and non-crosslinked UHMWPE [64]. Four UHMWPE material groups were examined: non-crosslinked control (PE); 50-kGy crosslinked control (XPE); non-crosslinked and high pressure crystallized at 180 °C, 300 MPa (HPC-PE); and 50-kGy crosslinked and high pressure crystallized at 240 °C, 500 MPa (HPC-XPE). For PE, HPC-PE, XPE, and HPC-XPE, the crystallinity was 50.2%, 70.9%, 50%, and 75%, respectively; the lamellae thickness was 28.1 nm, 131.2 nm, 23.1 nm, and 50.6 nm, respectively. The high-pressure treatment was accompanied by an increase in crystallinity and lamellae thickness.

Oral et al. [63] high pressure crystallized 25, 65, and 100-kGy irradiated UHMWPE at 200 °C and 380 MPa. The crystallinity of all HPC-irradiated UHMWPEs was higher compared to corresponding samples with the same radiation dose. The average crystallinity of 25, 65, and 100-kGy irradiated UHMWPE is 56%, 57%, and 57%, while high-pressure treatments, respectively, elevated these values to 72%, 63%, and 61%.

2.4.2 *Effect of Irradiation and Thermal Treatment on Mechanical Properties of HXPE*

Radiation crosslinking decreases chain mobility and chain stretch; as a result, it decreases the ductility of the polymer. This is manifested in reductions in the ultimate tensile strength (UTS), elongation at break (EAB) values in tensile tests, and impact toughness in Izod impact tests. Furthermore, the fatigue crack propagation resistance also decreased by radiation dose. Postirradiation melting further adversely affects the mechanical properties, while postirradiation annealing is less detrimental to the mechanical properties.

2.4.2.1 Fatigue Resistance

Due to the cyclic nature of the loads applied to the artificial joints, the fatigue crack propagation (FCP) resistance of UHMWPE materials is of great significance. FCP studies are usually conducted by compact tension tests. The detailed testing methods could be found in ASTM E 647. Briefly, a pre-gapped specimen was cyclically loaded at specific frequency and stress ratio, and the crack length was monitored at fixed periods. The crack propagation rates, *da*/*d*N, are plotted against stress intensity (ΔK). Usually, the stress intensity necessary to generate crack growth at a rate of 10^{-6} mm/cycle was measured to define $\Delta K_{\mathrm{incep}}$.

Pruitt et al. [59] investigated the fatigue initiation and propagation resistance of radiation crosslinked orthopedic-grade UHMWPE at varying irradiation doses. The samples used to study were extruded GUR 1050 rod receiving gamma radiation at 50, 100, and 200 kGy. After irradiation, the samples were melted at 170 °C for 4 h with additional 125 °C for 2 days. The controls included both non-irradiated (control) and non-irradiated but thermal treated rod stock (annealed).

The ΔK continued to decrease as radiation dose was increased (Fig. 2.11a). Also, it is apparent that radiation crosslinking resulted in a decreased crack propagation inception value, $\Delta K_{\mathrm{incep}}$ (Fig. 2.11b). Representative SEM micrograph of fractured surface showed crisscross patterns associated with ductile tearing in pristine UHMWPE, while there was distinct reduction of ductile features with increasing degree of crosslinking. These results indicate that the degree of crosslinking ($\geq$ 100 kGy) is extremely detrimental to fatigue propagation resistance.

The limited ductility in highly crosslinked UHMWPE limits the polymer's ability to accommodate plasticity at the crack tip. The decreased plasticity at the crack tip enables more of the crack driving force to be utilized in crack advance rather than dissipated through plastic work.

Gencur et al. [60] evaluated the effect of gamma crosslinking with two different postirradiation thermal treatments on the FCP resistance of UHMWPE. Three treatment groups, two highly crosslinked (100 kGy) and one virgin, were examined in this study. The two highly crosslinked groups were postirradiation thermal treated either above the melt temperature (remelted at 150 °C) or below (annealed at 110 °C) the melt temperature.

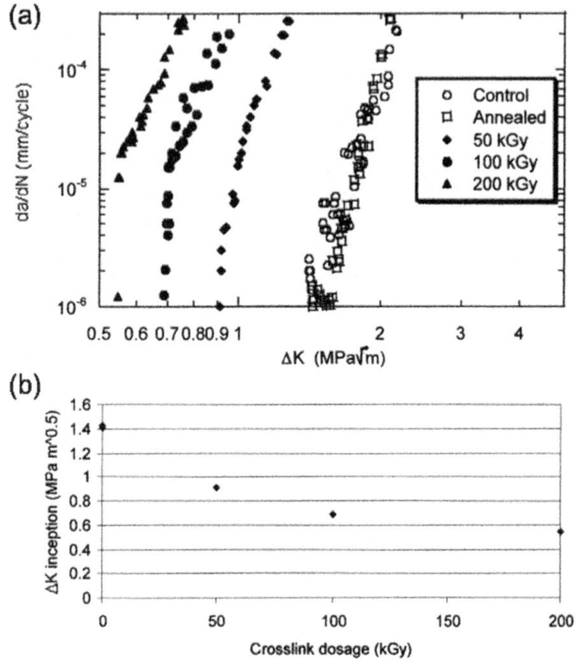

Fig. 2.11 (**a**) Fatigue crack propagation results for five UHMWPE groups and (**b**) ΔK_{incep}, value plotted as a function of radiation dose. (Reproduced from Ref. [59] by permission of John Wiley & Sons Ltd)

The highly crosslinked treatment groups demonstrated an overall reduced FCP resistance when compared to the virgin treatment group (Fig. 2.12), while the annealed UHMWPE was not significantly different from the remelted UHMWPE. The virgin treatment group showed the largest estimated ΔK_{incep} (2.0 MPa√m), followed by the annealed (1.1 MPa√m) and the remelted (0.9 MPa√m). Loss of chain mobility would be expected to affect the ability of the polymer to plastically deform [65] and, thus, would be reflected in reduced ΔK associated with crack inception. The annealed UHMWPE had an overall increased cyclic stress intensity for a given *da*/*d*N than the remelted UHMWPE. It is possible that annealing as a postirradiation treatment may be better for the preservation of the FCP resistance of UHMWPE than remelting.

The fracture surfaces of virgin and crosslinked samples all showed characteristic chevron markings, with the major axis of each chevron parallel to the direction of crack growth (Fig. 2.13a–c). The virgin group showed a fracture surface with more chevron markings than the highly crosslinked materials (Fig. 2.13a–c). In addition, all groups also showed characteristic bands perpendicular to the direction of crack growth, spaced approximately 3–6 mm apart, depending on the treatments (Fig. 2.13d–f). The perpendicular folds were less pronounced and were on the lower end of the spacing range (~3 mm) in the highly crosslinked groups when compared to the virgin group (~6 mm). The crosslinks restrict the motion of chains in the amorphous regions via chain slip [66].

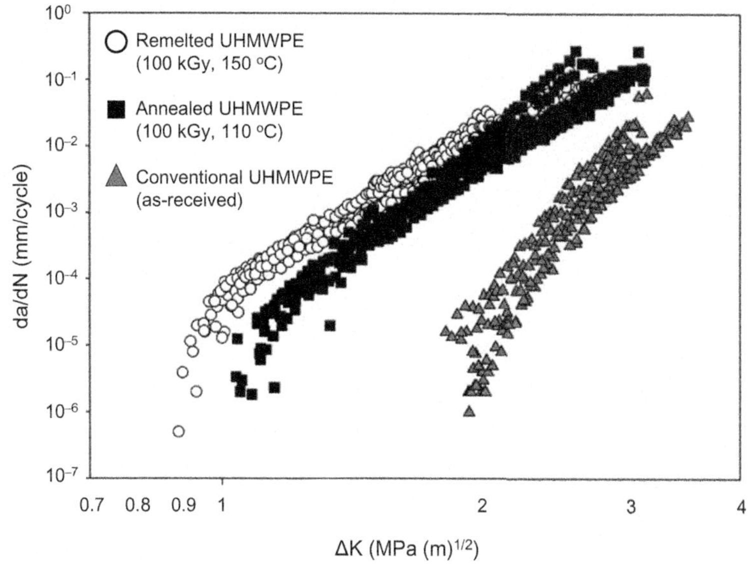

Fig. 2.12 ΔK vs. da/dN for highly crosslinked remelted, annealed, and virgin UHMWPE. (Reprinted from Ref. [60], Copyright 2006, with permission from Elsevier)

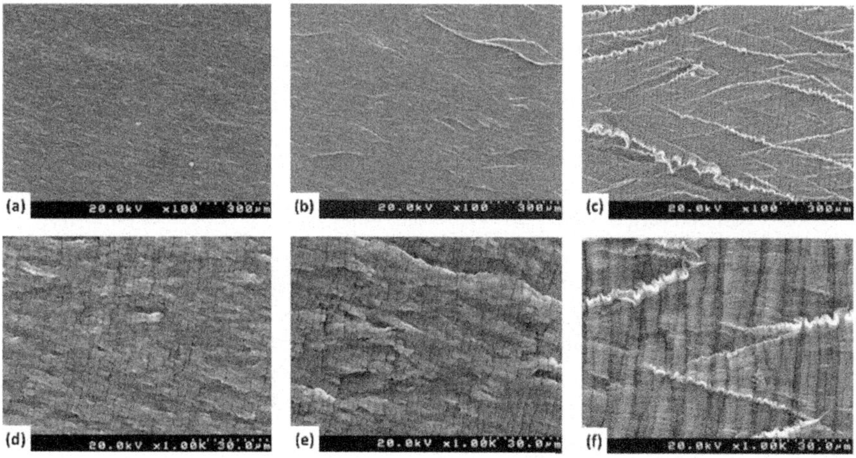

Fig. 2.13 Low-magnification, representative fracture surfaces for (**a**) remelted, (**b**) annealed, and (**c**) virgin, treatments of UHMWPE. High-magnification, representative fracture surfaces for (**d**) remelted, (**e**) annealed, and (**f**) virgin, treatments of UHMWPE. (Reprinted from Ref. [60], Copyright 2006, with permission from Elsevier)

Puértolas et al. [55] confirmed that annealing preserves the fatigue resistance well, with respect to remelting. The crack propagation resistance, ΔK_{incep}, suffers a monotonic decrease after irradiation from 1.53 ± 0.06 to 1.16 ± 0.01 for 50 and

Table 2.2 Results of ΔK_{incep} obtained from crack propagation experiments [60, 67]

Samples	ΔKincep
50 kGy	1.53 ± 0.06
100 kGy	1.37 ± 0.05
150 kGy	1.16 ± 0.01
50 kGy, remelted	1.58 ± 0.05
100 kGy, remelted	1.36 ± 0.05
150 kGy, remelted	1.03 ± 0.07
50 kGy, annealed	1.64 ± 0.05
100 kGy, annealed	1.49 ± 0.06
150 kGy, annealed	1.29 ± 0.04

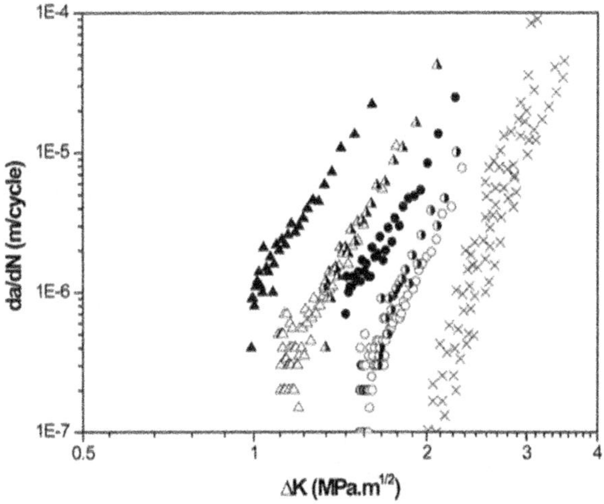

Fig. 2.14 Fatigue crack propagation curves for 50 kGy, (○); 50 kGy, remelt, (●); 50 kGy, anneal (◐); 150 kGy, (△); 150 kGy, remelt, (▲); 150 kGy, anneal, (◭) materials; and ram-extruded virgin GUR 1050 UHMWPE, (×). (Reproduced from Ref. [55] by permission of John Wiley & Sons Ltd)

150 kGy, respectively (Table 2.2). Remelting has a detrimental effect on ΔK_{incep} as well, provoking 10% decrease at the highest irradiation dose (Table 2.2). In contrast, the annealing imparted an improvement between 7% and 11% in fatigue resistance, with the ΔK_{incep} values ranging from 1.64 ± 0.05 to 1.29 ± 0.04 for 50- and 150-kGy irradiated and annealed materials (Table 2.2). Figure 2.14 displays the fatigue crack propagation plots of the as-irradiated, annealed, and remelted materials. The annealed materials exhibited the same *da*/*d*N values as corresponding irradiated materials at a given ΔK. The increased crosslink density introduced by irradiation prevents UHMWPE from exhibiting ductile behavior, and hence the ΔK_{incep} lowers [66, 68].

Oral et al. [45] investigated the separate effects of irradiation and subsequent melting on the fatigue resistance of UHMWPE. From their results, ΔK_{incep} monotonously decreased by radiation dose (Fig. 2.15). Postirradiation melting further decreased the fatigue propagation resistance with a range between 14% and 30% (Fig. 2.15). The morphology of fracture surfaces revealed the loss of ductility with increase in radiation dose and more detrimental effects on ductility at lower radiation doses after postirradiation melting. The authors speculated that loss of crystalline

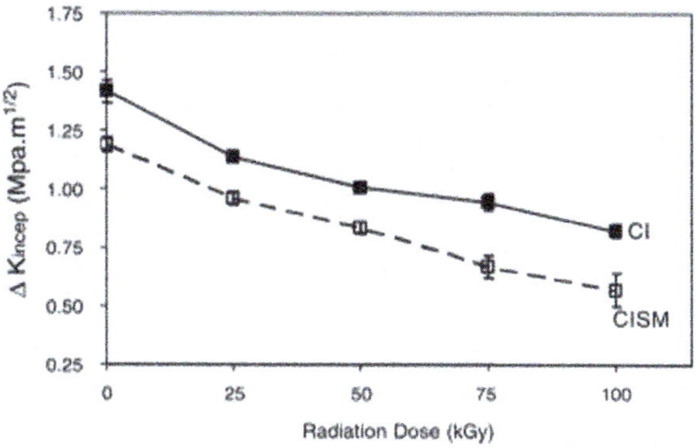

Fig. 2.15 Stress factor range at crack inception as a measure of fatigue crack propagation resistance of cold-irradiated (CI) and cold-irradiated and subsequently melted (CISM) UHMWPE as a function of radiation dose. (Reprinted from Ref. [45], Copyright 2006, with permission from Elsevier)

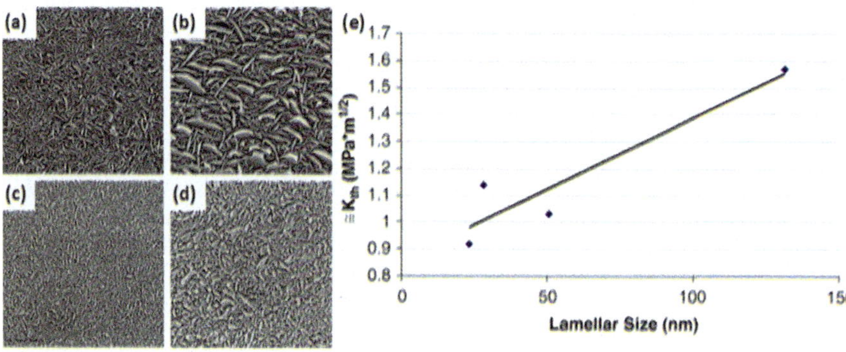

Fig. 2.16 SEM images (10,000×) of the etched surfaces of (**a**) PE, (**b**) HP-PE, (**c**) XPE, and (**d**) HP-XPE. (**e**) Fatigue threshold versus lamellae size for UHMWPE. Reprinted from Ref. [64], Copyright 2006, with permission from Elsevier)

content after postirradiation melting resulted in loss of plastic deformation ability of these materials. Therefore, postirradiation melting results in significant loss of FCP resistance of UHMWPE.

Pruitt et al. [64] revealed that a high crystallinity was beneficial to fatigue crack propagation resistance. They compared the microstructure and FCP resistance of four groups of UHMWPE: non-crosslinked control (PE); 50-kGy crosslinked control (XPE); non-crosslinked and high pressure crystallized at 180 °C, 300 MPa (HPC-PE); and 50-kGy crosslinked and high pressure crystallized at 240 °C, 500 MPa (HPC-XPE). The high-pressure treatment resulted in significantly enlarged lamellae (Fig. 2.16). For the non-crosslinked group shown in Fig. 2.16a, b, the smaller lamellae were on the order of 0.1 and 0.2 μm for PE and HPC-PE, respectively, while the larger lamellae were on the order of 0.2 and 0.5 μm for PE and HPC-PE, respectively. For the crosslinked group, the XPE have lamellae on the order of 0.05 μm (Fig. 2.16c), and the HPC-XPE have lamellae on the order of 0.1 μm (Fig. 2.16d). The ΔK_{th} and lamellar size showed a linear correlation with $R^2 = 0.87$ (Fig. 2.16e). These results suggest that the crystalline lamellae may play a critical role in crack inception and that larger crystals may serve to deflect an initiating crack, thereby reducing the effective stress intensity acting at the crack tip.

2.4.2.2 Tensile Properties

Tensile, one of the static mechanical properties of UHMWPE, is one of the major factors that limits the lifetime of UHMWPE in joint prostheses. In the case of metal backed patellar components with low conformity, the contact stresses can exceed the yield stress of UHMWPE, leading to cold flow (plastic deformation) of the component, eventually resulting in catastrophic failure [69]. Radiation crosslinking decreased several tensile mechanical properties of UHMWPE, such as UTS and EAB. On the other hand, the two postirradiation thermal treatments, remelting and annealing, have different effects on the polymer's microstructure and thus have different effects on the polymer's tensile property. The representative tensile properties of crosslinked UHMWPE from literatures are summarized in Table 2.3.

According to Table 2.3, there is a definite trend that the EAB decreased significantly as a function of radiation dose. Thus, the crosslinked UHMWPE undergoes relatively less plastic deformation and strain hardening prior to fracture compared to non-crosslinked UHMWPE [57]. Crosslinking occurs in the amorphous regions and impedes the relative motion of polymer chains, leading to reduced ductility, but only between the crystalline lamellae [20]. The EAB remained constant when irradiated materials undergo thermal treatments [55, 67]. Only the increased density of crosslinking and the decreased population of tie molecules after irradiation prevented the ductile behavior [67].

The tensile yield strength did not differ significantly between non-crosslinked and crosslinked UHMWPE [45, 57]. However, it is noted that the yield strength decreases after thermal treatments [55, 59, 63], especially after remelting, compared to that for as-irradiated materials [55]. A preliminary structural interpretation of the yield strength would be related to the degree of crystallinity present in the material [70], independent of treatment method. The crystalline deformation modes play a major role at elastic deformation regime of the crystals before yielding [67]. Galeski [71] and Brooks [72] et al. further confirm that the lamellar thickness controls the yield strength.

Kurtz et al. [61] observed significant relationships between the crystallinity, the absorbed dose, and the yielding and ultimate behavior of the UHMWPE. For each tensile rate, a nonlinear (quadratic) relationship was fit between crystallinity and the true yield strength of the polymers. The irradiation dose was significantly related to the true ultimate strength and strain of the UHMWPE.

2.4.2.3 Impact Strength

The Izod impact test has been widely used to determine the impact strength of UHMWPE. Radiation crosslinking significantly lowers the toughness of

Table 2.3 Tensile and impact properties of crosslinked UHMWPE

Resin and fabrication	Crosslinking		Post-crosslinking treatment	UTS (MPa)	YS (MPa)	E (MPa)	EAB (%)	Izod impact strength (kJ/m^2)	References
Compression-molded GUR 1050	E-beam radiation (kGy)	0	None	37.5 ± 4.4	19.1 ± 1.3	748 ± 19	360 ± 70	121 ± 7	[23]
		50		36.9 ± 0.1	20.7 ± 0.2	803 ± 13	350 ± 70	99 ± 7	
		100		42.1 ± 2.0	20.6 ± 0.1	820 ± 5	280 ± 50	66 ± 2	
		150		38.6 ± 4.1	20.5 ± 0.2	841 ± 15	220 ± 40	53 ± 3	
		50	Remelted at 150 °C for 2 h	33.1 ± 4.2	16.7 ± 0.2	642 ± 18	370 ± 60	91 ± 5	
		100		32.9 ± 4.0	15.8 ± 0.6	650 ± 25	320 ± 50	70 ± 5	
		150		27.2 ± 4.5	15.3 ± 1.3	658 ± 14	220 ± 60	57 ± 8	
		50	Annealed at 130 °C for 2 h	31.9 ± 1.3	19.2 ± 0.2	764 ± 50	350 ± 10	96 ± 4	
		100		29.5 ± 2.6	18.9 ± 0.3	811 ± 14	260 ± 20	61 ± 1	
		150		29.6 ± 2.0	19.3 ± 0.3	868 ± 50	220 ± 20	49 ± 3	
Extruded GUR 1050	γ radiation (kGy)	0	170 °C for 4 h and 125 °C for 2 days	315.5 ± 31.6[a]	20.2 ± 1.0	495 + 56	1.82 ± 0.01[b]		[57, 59]
		25		284.8 ± 18.0[a]	19.0 ± 0.4	433 ± 14	1.74 ± 0.03[b]		
		50		237.6 ± 12.3[a]	19.9 ± 0.8	412 ± 50	1.59 ± 0.01[b]		
		100		185.7 ± 7.5[a]	18.9 ± 0.7	386 ± 23	1.50 ± 0.02[b]		
		200		126.0 ± 14.0[a]	20.2 ± 1.0	266 ± 30	1.37 ± 0.06[b]		
Extruded GUR 4150	E-beam radiation (kGy)	0	Remelted at 150 °C for 2 h	46 ± 3.0	22.0 ± 0.4				[26]
		25		37 ± 0.6	19.6 ± 0.5				
		40		38 ± 1.9	18.8 ± 0.6				
		50		37 ± 4.9	19.6 ± 0.5				
		75		37 ± 4.0	19.9 ± 0.1				
		100		35 ± 5.1	20.2 ± 0.1				
		150		28 ± 2.0	19.6 ± 0.2				
		200		29 ± 1.6	19.6 ± 0.4				
		300		27 ± 2.5	20.0 ± 0.1				
		0.70		18.9 ± 0.2	37.1 ± 1.4				

(continued)

Table 2.3 (continued)

Resin and fabrication	Crosslinking		Post-crosslinking treatment	UTS (MPa)	YS (MPa)	E (MPa)	EAB (%)	Izod impact strength (kJ/m^2)	References
Extruded GUR 1050	γ sterilized 2.5–4.0 Mrad		None	55 ± 4	24 ± 0		410 ± 9	72 ± 5.1	[8]
	E-beam irradiated at 125 °C to 9.5 Mrad		Remelted at 150 °C	30 ± 1	20 ± 0		330 ± 25	62 ± 2.5	
Extruded GUR 1050	γ radiation (kGy)	0	None	231.1 ± 10.1	21.3 ± 0.19	495 ± 14.0	375 ± 7.1		[64]
		0	180 °C, 300 MPa	78.8 ± 10.0	23.4 ± 0.28	675 ± 19.8	230 ± 21.6		
		50	None	157.7 ± 11.2	19.0 ± 0.17	334 ± 7.8	317 ± 31.5		
		50	240 °C, 500 MPa	167.8 ± 20.8	20.6 ± 0.42	695 ± 5.1	325 ± 23.8		
Compression-molded GUR 1050	E-beam radiation (kGy)	0	None	50 ± 6	23 ± 0	353 ± 5	481 ± 40		[45]
		25		53 ± 3	23 ± 1	324 ± 15	428 ± 47		
		50		40 ± 2	21 ± 2	311 ± 19	300 ± 11		
		75		33 ± 3	20 ± 2	330 ± 18	285 ± 23		
		100		33 ± 1	21 ± 2	341 ± 10	214 ± 7		
		0	Remelted at 170 °C	51 ± 3	21 ± 1	311 ± 12	484 ± 29		
		25		31 ± 3	17 ± 2	243 ± 17	336 ± 24		
		50		34 ± 3	20 ± 1	256 ± 4	336 ± 19		
		75		29 ± 2	19 ± 1	280 ± 6	266 ± 6		
		100		28 ± 2	17 ± 0	235 ± 6	233 ± 16		
Extruded GUR 1050	None		None	50.2 ± 1.6	23.5 ± 0.30		421 ± 11		[61]
	γ sterilized 30 kGy		None	47.1 ± 4.2	24.11 ± 0.14		373 ± 8		
	γ radiation (kGy)	100	Annealed at 110 °C	46.4 ± 3.4	24.79 ± 0.12		248 ± 11		
		100	Remelted at 150 °C	37.1 ± 3.2	21.36 ± 0.13		232 ± 8		
Extruded GUR 1050	γ sterilized 30-kGy N_2-Vac™		None	60.0 ± 4.2	24.5 ± 0.4		370 ± 10		[23]
	γ radiation 105-kGy Crossfire™		Annealed	61.1 ± 5.0	25.4 ± 0.4		281 ± 21		

Compression-molded GUR 1020	γ sterilized 30 kGy		None	54.8 ± 2.5	23.2 ± 0.4	862 ± 16	363 ± 10		[51]
	γ radiation 90 kGy		X3: sequentially annealed	56.7 ± 2.1	23.5 ± 0.3	876 ± 8	267 ± 7		
Compression-molded GUR 1050	γ radiation (kGy)	25	None	54[c]	23[c]	1370 ± 340	358[c]		[63]
		65	None	49[c]	22[c]	1780 ± 300	256[c]		
		100	None	51[c]	23[c]	1630 ± 420	255[c]		
		25	200 °C in vacuum	44[c]	19[c]	1030 ± 70	92,356[c]		
		65	200 °C in vacuum	38[c]	20[c]	2070 ± 800	286[c]		
		100	200 °C in vacuum	41[c]	22[c]	1410 ± 500	223[c]		
		25	200 °C, 380 MPa in water	60[c]	22[c]	940 ± 530	320[c]		
		65	200 °C, 380 MPa in water	48[c]	23[c]	Not available	246[c]		
		100	200 °C, 380 MPa in water	38[c]	22[c]	4200 ± 320	190[c]		
Extruded GUR 4150	γ radiation (Mrad)	0	None					101.5 ± 9.1	[7]
		4.5						87.1 ± 3.2	
		9.5						59.9 ± 1.8	
		14.5						50.6 ± 2.6	
		100						17.3 ± 1.7	
		4.5	Remelted at 150 °C for 5 h					89.3 ± 2.2	
		9.5						68.7 ± 5.0	
		14.5						58.7 ± 4.8	
Compression-molded GUR 1050	γ sterilized 37 kGy		None	48.3 ± 3.2	20.9 ± 0.3	785 ± 94	346 ± 28	86.7 ± 2.5	[73]
	E-beam radiation 100 ± 10 kGy		Remelted	19.9 ± 1.5	19.0 ± 0.3	642 ± 110	210 ± 13	56.1 ± 4.6	

[a]True tensile stress
[b]True tensile strain
[c]Mean value

polyethylene (Table 2.3) [7, 8, 55, 73]. The crosslink density was responsible via a diminution in the ductile capability that imparted a reduced fracture resistance. Consistently, fracture strain results confirmed the loss of ductility with higher crosslink density, which probably reduces the amorphous deformation modes of the polymer [45, 57]. The impact strength of radiation crosslinked UHMWPE was not sensitive to the thermal treatments (either remelting or annealing, Table 2.3) [55, 60]. However, others declare that remelting caused a slight increase in the impact strength compared with that of the corresponding material that was not remelted (Table 2.3) [7].

2.5 Wear of Radiation Crosslinked UHMWPE

2.5.1 Wear Mechanism

In the late 1980s and early 1990s, understanding the wear mechanism of UHMWPE became central to the aim of improving the wear resistance of this polymer. The analysis of acetabular components retrieved during an autopsy or a revision operation showed that the highly worn areas were polished to a glassy finish on gross examination [74]. By using a scanning electron microscope, numerous multidirectional scratches along with fine, drawn-out fibrils with a diameter of 1 micrometer or less oriented parallel to each other became apparent on the worn articular surfaces. These fibrils were thought to be the most likely source of the wear particles observed in periprosthetic osteolytic tissue [75]. Thus, wear appeared to occur mostly through the breakup of fibrils formed by large strain plastic deformation and orientation of the surface.

The fibrils were weakened in the transverse direction and subsequently ruptured during multidirectional motion of the hip. These in vivo findings were later reproduced in the laboratory by Bragdon et al. [75] and Wang et al. [76, 77] by using wear testing devices. These researchers showed that the wear rate of UHMWPE increased markedly when the articulation is switched from a unidirectional to a multidirectional motion. The underlying mechanism is the orientation of the surface fibrils in the direction of motion. This orientation increases the strength of the fibrils in this direction and weakens them in the transverse direction. Hence, debris generation occurs easily when the direction of motion is altered. For example, when the articulation comprises crossing motions, as in the hip, the particulate debris is generated by the rupture of the surface fibrils.

2.5.2 Effect of Crosslink Density on Wear

Crosslinking of UHMWPE molecules would decrease the chain mobility needed for the large-scale plastic deformation and slow down the surface fibril formation and hence improve the wear resistance.

Muratoglu et al. [26] investigated the wear of three types of crosslinked UHMWPE by using a pin-on-disk (POD) wear tester at a bi-directional mode.

The results showed no unified correlation for the wear behavior of crosslinked UHMWPE with the crystallinity, UTS, and yield strength (YS). However, the POD wear rate showed an identical linear dependence on molecular weight between crosslinks (M_c, inversely proportional to crosslink density (d_x)). Figure 2.17 is a plot of the molecular weight between crosslinks (M_c) and wear rate of the crosslinked UHMWPEs. Irrespective of the crosslinking method used, the combined behavior of the three-homologous series studied showed a linear correlation between the M_c of the polymer and its wear rate. The wear of UHMWPE decreased to undetectable levels at around an M_c of 4800 g/mol.

This observation emphasizes the importance of the crosslinking of UHMWPE in determining the wear behavior. Crosslinking inhibits the important mechanism responsible for wear, i.e., orientation and reorientation. As a result, the wear behavior of UHMWPE changes drastically following crosslinking.

In another work [8], Muratoglu et al. further present a novel *warm irradiation* with *adiabatic* heating and subsequent *melting* (WIAM) process to crosslinking UHMWPE and eliminate residual free radicals. The absorbed dose level was 9.5 Mrad. The WIAM UHMWPE was machined into acetabular liners and sterilized with EtO. The wear property of these liners was tested on the Boston 12-station hip simulator (AMTI, Watertown, MA) [75, 78] and compared with the control liners machined from GUR 1050 UHMWPE ram-extruded bar stock and gamma sterilized in nitrogen (2.5–4.0 Mrad). The results showed that, over 20 million cycles in the hip simulator, the WIAM polyethylene showed no detectable wear (Fig. 2.18a). These liners exhibited a slight weight increase because of slight fluid absorption, while the control liners for 22- and 28-mm femoral head sizes exhibit significant wear. At 20 million cycles, the total weight loss averaged -410 ± 30 mg for the control liners with 28-mm inner diameter and -360 ± 45 mg for control liners with 22-mm inner diameter [8].

Figure 2.18b–e are representative photographs of the articulating surfaces of the control and WIAM liners after 5 and 20 million cycles. The control liners were highly polished with no remnants of the original machining marks at 5 (Fig. 2.18b)

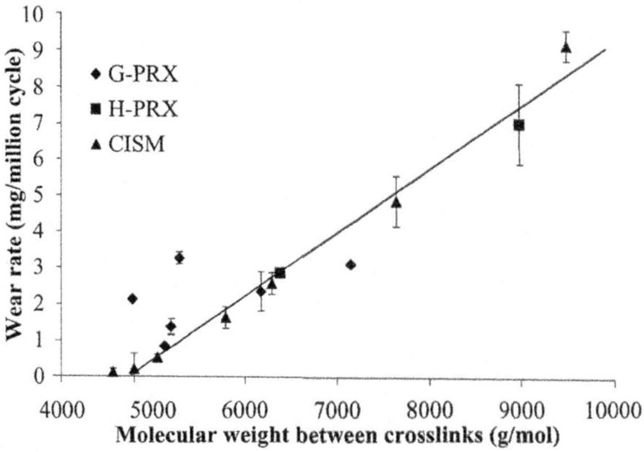

Fig. 2.17 The variation of wear rate as a function of the molecular weight between crosslinks of radiation crosslinked GUR 4150 (CISM) and peroxide crosslinked GUR 1050 (H-PRX) and Himont 1900 (G-PRX) resins. (Reprinted from Ref. [26], Copyright 2006, with permission from Elsevier)

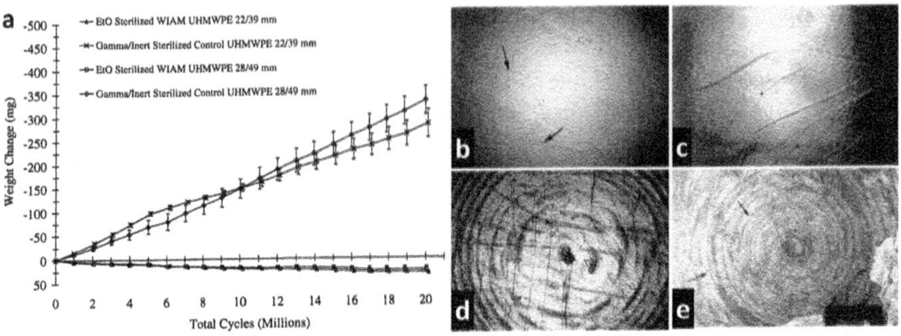

Fig. 2.18 (**a**) The weight of WIAM and control UHMWPE liners studied in the hip simulator in Sulzer Interop shells with 28-mm inner and 49-mm outer diameter (28/49) and 22-mm inner and 39-mm outer diameter (22/39) liners after correction. Representative optical micrographs of the articulating surfaces of control UHMWPE after (**b**) 5 and (**c**) 20 million cycles of simulated gait and those of the articulating surfaces of WIAM UHMWPE after (**d**) 5 and (**e**) 20 million cycles of simulated gait. (Reprinted from Ref. [8], Copyright 2001, with permission from Elsevier)

and 20 (Fig. 2.18c) million cycles. The loss of the machining marks and the highly polished surfaces is indicative of wear. The articulating surfaces of the WIAM liners retained the original machining marks after 5 (Fig. 2.18d) and 20 (Fig. 2.18e) million cycles. Considering that the height of the machining marks is about 10–50 μm, the cumulative linear wear in WIAM liners was less than 10–50 μm in 20 million cycles, or 2 μm per million cycles. After 20 million cycles, the original machining marks on the articulating surface of the WIAM liners were still present but were difficult to visualize because of the accumulated scratching on the surface (Fig. 2.18e). These results indicate a high wear resistance of WIAM UHMWPE.

Wang et al. [77] investigated the interfacial friction, crosslink density, and cross-shear angle on wear of UHMWPE and established a wear model for UHMWPE under lubricated multidirectional sliding conditions based on the effective frictional work concept. According to the theory, the wear factor k is related to the coefficient of friction μ, the crosslink density X_c, and the maximum cross-shear angle α by the following equation:

$$k = k' \frac{d(\mu - \mu_0)}{2\gamma_c} \left(\frac{1}{X_c} - \frac{1}{X_0} \right) \times \left(1 - \frac{\sin 2\alpha}{2\alpha} \right) \tag{2.17}$$

where k' is a constant, d the diameter of the fibrils, m_0 the critical coefficient of friction for initiating surface failure, and X_0 the critical crosslink density for suppressing surface rupture. The wear factor k (unit: mm^3/Nm) can be calculated from the volumetric wear rate by the following equation:

$$k = \frac{\Delta V}{PL} \tag{2.18}$$

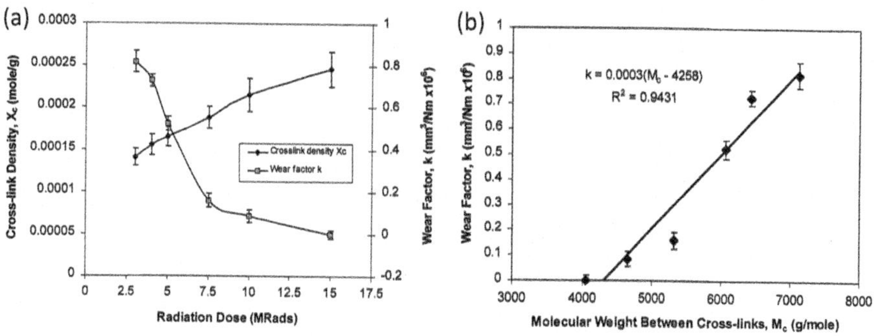

Fig. 2.19 (**a**) Crosslink density and wear factor vs. radiation dose. (**b**) Wear factor vs. molecular weight between crosslinks. (Reprinted from Ref. [77], Copyright 2001, with permission from Elsevier)

To derive Eq. (2.17), the effect of crosslink density on friction factor of GUR 4050 was carefully investigated. The UHMWPE irradiated with γ rays at 3.0, 4.0, 5.0, 7.5, 10.0, and 15.0 Mrads was annealed at 130 °C. Subsequently, the UHMWPE bars were machined into hemispherical sockets for tests in hip simulator.

Figure 2.19a shows the wear factor k and the crosslink density X_c as a function of the radiation dose. While k decreased significantly with increasing radiation dose, X_c increased as the dose increased. Figure 2.19b shows k as a function of $1/X_c$ (or M_c). A linear relationship between k and M_c was obtained. However, the best-fit line of the data ($R^2 = 0.94$) does not go through the origin as M_c approaches zero. The line intersects with the horizontal axis at $M = 1/X_c = 1/X_0 = 4250$ g/mole. These results imply that the wear resistance of UHMWPE increased by increasing radiation dose when treated by γ irradiation; furthermore, there exists a critical value of the crosslink density above which rupture between fibrils cannot occur.

2.5.3 *Joint Simulator Studies*

Crosslinked polyethylene has demonstrated superior wear resistance during in vitro testing by using hip or knee simulators.

Essner et al. [79] compared the hip joint simulator wear performance of HXPE bearings to conventional PE (CPE) bearings. CPE acetabular cups with i.d. 28 mm were machined from untreated virgin GUR 1050 material and subsequently sterilized in nitrogen by gamma irradiation at 30 kGy (Stryker Orthopaedics, Mahwah, NJ). HXPE acetabular cups with i.d. 36 mm or 32 mm were machined from irradiated (75 kGy) and annealed (below 130 °C) GUR 1050 bar stock, subsequently packaged in nitrogen and sterilized by γ radiation at 30 kGy. These cups were matched with alumina ceramic femoral heads. A 12-station hip joint simulator was used (MTS, Eden Prairie, MN).

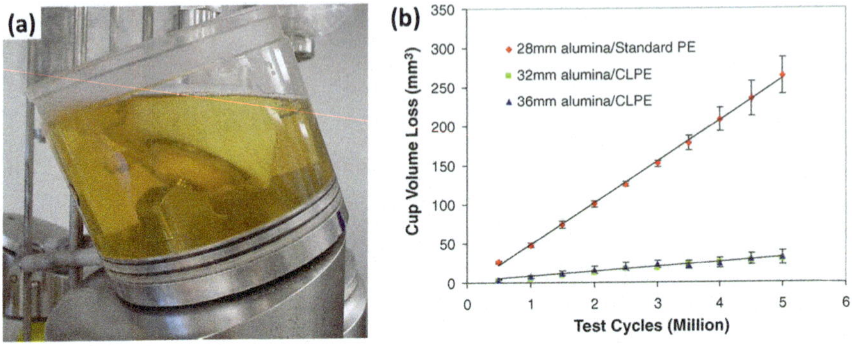

Fig. 2.20 Anatomical positioning of components in the hip simulator. (Reprinted from Ref. [79], Copyright 2005, with permission from Elsevier)

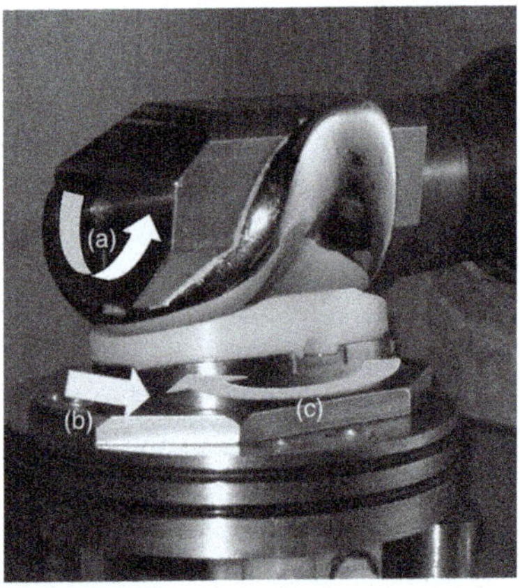

Fig. 2.21 One station of the 6-station AMTI knee simulator. The arrows indicate the motions of femoral flexion (**a**), femoral translation (**b**), and tibial rotation (**c**). Reprinted from Ref. [80], Copyright 2007, with permission from Elsevier)

The acetabular components were mounted in the anatomical position against matching femoral heads (Fig. 2.20a). All liners showed a linear relationship between wear and test duration (Fig. 2.20b). Both the 32- and 36-mm HXPE liners had a wear rate 87% lower than the 28-mm CPE liners. There was no difference in the wear rate for the HXPE liners when used against 32- and 36-mm alumina heads [79].

Muratoglu et al. [80] compared the resistance to adhesive/abrasive wear of HXPE and CPE tibial inserts of identical design on both articular and backside surfaces. HXPE inserts were machined from a 65-kGy e-beam irradiated and melted GUR 1050 compression-molded UHMWPE, which had been subsequently sterilized with gas plasma. CPE group was machined from GUR 1050 compression-molded UHMWPE bar stock and sterilized with γ irradiation (25–40 kGy) in nitrogen. The inserts were tested on a 6-station AMTI knee simulator. The anatomic location of the components was set as per the optimum recommended surgical placement (Fig. 2.21). Undiluted bovine calf serum with added 20 mmol of EDTA and 0.2 wt % of sodium azide was used as the lubricant.

Using a coordinate measuring machine (CMM), the difference between the digitized surfaces before and after wear test was collected and represented the volume of the articular surface scar. The linear regression of the slope of the changes in scar volume subsequent to 2 million cycles was used to calculate a volumetric wear rate, by assuming that bedding-in processes including creep are completed by 2 million cycles. The ordinate of this regression line at origin was taken to represent the total volumetric creep. Gravimetric assessment of wear was calculated from 5 to 7 million cycles. The weight change of insert was corrected for fluid uptake. Gravimetric analysis was not performed during the first 5 million cycles to avoid any global deformation that could have adversely affected the accuracy of coordinate measuring.

The average gravimetric wear rate of the CPE (23 $\pm$ 6 mm^3/MC) was significantly higher than that of the HXPE (4.6 $\pm$ 2 mm^3/MC) ($p = 0.034$). The CMM method also showed a significantly lower wear rate with HXPE (4.8 $\pm$ 2.3 mm^3/MC) than CPE (15.1 $\pm$ 4.2 mm^3/MC) ($p = 0.034$). The articular surface creep volume on the lateral plateaus of HXPE (28.2 $\pm$ 9.1 mm^3) was significantly higher than CPE (8.8 $\pm$ 5.1 mm^3). In this test, the backside wear mostly depends on the extent of micro motion present between tibial insert and tibial baseplate.

Wang et al. [51] compared the in vitro wear performance of contemporary cruciate retaining (CR) and posterior-stabilized (PS) knee inserts manufactured from X3 to that from CPE. All tibial inserts were articulated against cobalt–chromium alloy femoral components. The wear testing of X3 PS tibial inserts was conducted on a machine similar to that used for CR knee wear evaluations; stair climbing kinematics were applied to ensure stress transfer to the polyethylene post [80].

Five million cycles were conducted for each insert. The wear rates were 17.7 $\pm$ 2.2 and 5.7 $\pm$ 1.5 mm^3/MC for CR-type CPE and X3 inserts, respectively, representing a 68% decrease compared to CPE ($p = 0.002$). No surface or subsurface cracks, pits, or delamination were observed for either X3 or conventional inserts.

For PS arthroplasties, the wear rate for CPE and X3 was 4.1 $\pm$ 0.7 and 1.5 $\pm$ 0.5 mm^3/MC, respectively, representing a 64% decrease compared to CPE ($p = 0.008$). No surface or subsurface cracks, pits, or delamination were observed on articular surfaces or posts for either material.

They also compared the wear performance of accelerated aged X3 and CPE inserts, which were sealed in chambers at 70 °C in 5 atm of oxygen for 14 days. The wear rate is 1.5 $\pm$ 0.5 mm^3/MC before aging and 1.0 $\pm$ 0.2 mm^3/MC after aging for X3 inserts. There was no surface appearance difference between aged and unaged X3 components after testing. The wear rate is 4.1 $\pm$ 0.7 mm^3/MC before aging and 25.1 $\pm$ 7.8 mm^3/MC after aging for CPE components. Condylar surfaces and posts for the X3 components showed burnishing, deformation, striations, and scratching; no cracks or delamination were observed. Aged CPE components exhibited cracking and severe delamination. The excellent wear properties of X3, combined with oxidation resistance equivalent to that of virgin polyethylene, suggest that this material is suitable for knee joint arthroplasty components.

Bragdon et al. [81] investigated the wear performance of HXPE in the presence of third-body particulate debris in a hip simulator. Cement debris, corrosion products from the metal tapers, metal fragments from other fixation devices, bone particles, and metal beads or fibers from porous coatings are common third-body contaminants in the hip joint found at revision surgery. These hard particles can accelerate polyethylene wear through ploughing of the surface. In addition, these particles can damage the femoral head, which can further increase polyethylene wear.

The CPE acetabular liners were machined from ram-extruded GUR 1050 bar and γ sterilized in nitrogen to a dose level of 25–40 kGy. The HXPE liners with the same design were machined from highly crosslinked GUR 1050 disks, which were e-beam irradiated at 125 °C to a total dose of 95 kGy and melted at 150 °C for 2 h and were sterilized with ethylene oxide.

The wear testing was performed on the 12-station Boston hip simulator (AMTI, Watertown, MA). Bovine calf serum was used as lubricant. Two different types of third-body particles were separately added to represent severe and mild abrasive wear. Aluminum oxide particles with a mean size of 1 μm were used to investigate severe abrasive wear. Bone cement particles containing barium sulfate with 30 μm were used to investigate mild abrasive wear. The concentration of particles was 0.15 mg/cm^3.

Before the introduction of third-body particles, the wear loss of CPE liners after first million cycles was 9.7 ± 0.6 mg. In contrast, the HXPE liners showed no weight loss during the first million cycles. In the presence of aluminum oxide particles, the average weight loss of CPE liners after 5 million cycles was 400 ± 250 mg, while the average weight loss of HXPE components was 64 ± 50 mg. In the presence of bone cement particles, the weight loss of CPE liners after 5 million cycles was 38.1 ± 14.4 mg, while the HXPE components showed a markedly reduced weight loss of 1.9 ± 2.2 mg. In third-body wear, there was an increase in the average incremental wear rate of both the CPE and HXPE liners, as the femoral heads had been progressively damaged.

In another work, the wear of HXPE acetabular liners against CoCr femoral heads with 28, 38, and 46 mm diameter in the presence of third-body poly(methyl-methacrylate) was compared to the wear performance of CPE liners against 46-mm diameter head [82]. The larger heads afford larger range of motion of the reconstructed hip as well as greater stability, thereby reducing the risk of dislocation [83, 84]. Acetabular liners based on HXPE and CPE tested on a Boston hip simulator over 5 million cycles in the presence of bone-cement third-body particles showed that the average wear rate is 29.3 ± 3.0 mg/MC for the 46-mm CPE liners, while no weight loss was found for the 46-mm and 38-mm HXPE liners. The 28-mm HXPE liners showed an average wear rate of 3.7 ± 4.2 mg/MC. The wear rates of each set of HXPE components were all significantly lower than that of the CPE group ($p \leq 0.05$). The HXPE material is far more resistant to the third-body wear than CPE, even when very large diameter femoral heads were used.

Saikko et al. [85] investigated the wear of CPE and HXPE acetabular cups against polished and roughed CoCr femoral heads in hip simulator. The roughened femoral heads ($R_a = 0.14$–0.18 μm) represented the type of roughening observed in roughest explanted femoral heads [86–88]. The CPE and HXPE (95 kGy and 150 °C remelted) cups were tested against 28-mm CoCr heads. The mean wear rate and standard deviation of CPE cups against polished and against roughened heads were 11.6 ± 0.07 and 64.4 ± 10.1 mg/MC, respectively. No weight loss was observed for the HXPE cups against polished heads, whereas a mean wear rate of 2.4 ± 0.3 mg/MC was found against roughened heads. The tests indicated that e-beam irradiation may effectively reduce the harmful PE wear particles production in total hip arthroplasty.

Akagi et al. [89] investigated the wear of UHMWPE tibial inserts as a function of radiation dose on a six-channel displacement-controlled knee simulator (Shore-Western Manufacturing Inc.) [90, 91]. The UHMWPE inserts used were machined from compression-molded GUR 1050 blocks γ irradiated in air at 0, 25, 35, 50, 75, 100, 150, and 200 kGy followed by annealing at 110 °C for 12 h in nitrogen. The weight changes of the tibial inserts in the simulator tests were linear between 1 and 4 MC. As radiation dose increased from 0 to 200 kGy, the wear rate decreased exponentially with 54, 78, 95, and 98% reductions at 50, 75, 100, and 200 kGy, respectively.

2.6 Clinical Applications of Highly Crosslinked UHMWPE

2.6.1 Hip Implants

Wear-induced periprosthetic osteolysis was a long-term problem associated with total arthroplasty with traditional polyethylene liners [3, 92], causing aseptic loosening and hip instability resulting in revision surgery [93, 94]. The particles released by wear incite an inflammatory cascade that may culminate in bone degradation [95, 96]. Highly crosslinked UHMWPE (HXPE) implants were developed in an effort to reduce wear and osteolysis, thereby increasing survivorship. Traditional polyethylene has been shown to wear down at a faster rate than HXPE implants in joint simulator studies and clinical follow-up reports [7, 8, 22].

Radiographic examination of the patients allows the measurement of femoral head penetration into the acetabular liner. Various methods for radiographic assessment of linear penetration have been proposed and evaluated [97–100]. The majority are computerized measurements of the distance between the femoral head and proximal acetabular cup edges. Penetration of femoral head into total hip acetabular is usually obtained from anterior/posterior (A/P) and lateral projections of the hip. For total knee arthroplasty follow-up, A/P, lateral, and patellofemoral radiographs are usually collected [101]. Femoral head penetration into UHMWPE acetabular components is a result of both removal of UHMWPE from the bearing surface, true wear, and creep of the UHMWPE by plastic and elastic deformation. The penetration for the first 1–2 years is mostly due to creep and wear and beyond, the second year of follow-up penetration is attributed to mostly wear alone. Some studies have ignored the early radiographs and started the analysis using a film after the first postoperative year as the baseline film [85, 102].

In a 10.5-year (range, 8.5–14) follow-up study, 48 primary total hip arthroplasties in 45 patients (mean age, 59.6 years; range, 34–75) using the annealed HXPE acetabular liner (Stryker, Mahwah, New Jersey) combined with a 28-mm cobalt–chrome femoral head (L-fit, Stryker) were reviewed [103]. The liner from UHMWPE with 7.5-Mrad γ radiation and annealing and 3.0-Mrad sterilization (a total dose of 10.5 Mrad) showed a femoral head penetration rate of 1.26 mm and 0.122 mm/y for the first 2–3 years (typical bedding-in period), which plateaued at an annual mean wear rate of 0.05 mm/y (SD, 0.04) for the last 5 years, according to radiograph studies. It is noted that the AP hip and AP pelvis radiographic assessments produced statistically significantly different values. The AP hip total wear and wear rate were higher at 1.38 mm and 0.133 mm/y, respectively, while those were 1.13 mm and 0.109 mm/y, respectively, for the AP pelvis radiographs ($p < 0.05$). In addition, minimal evidence of femoral osteolysis was encountered in 2 of 43 hips (4.6%). No evidence of osteolysis was found around the acetabulum in any of the cases.

Engh Jr. et al. reported a prospective, randomized study of crosslinked and non-crosslinked polyethylene for total hip arthroplasty at a 10-year follow-up [104]. Two hundred thirty THAs in 220 patients are comprised of 116 crosslinked and 114 non-crosslinked liners. Marathon®, which is UHMWPE treated with 5 Mrad of gamma irradiation and melting at 150 °C, was compared with the non-crosslinked UHMWPE, terminally sterilized by gas plasma. The femoral head penetration was evaluated based on two-dimensional radiograph relative to the immediate postoperative (nominal 6-week follow-up) reference view using a validated, computer-assisted technique [97]. Ten-year survivorship with reoperation for any reason is 98.1 ± 2.7% for crosslinked compared to 92.6 ± 5.3% for non-crosslinked. Ten-year survivorship with reoperation for wear-related complications as the endpoint is 100% for crosslinked and 94.7 ± 4.6% for non-crosslinked. The head penetration rate for the non-crosslinked was 0.22 ± 0.11 mm/year compared to 0.06 ± 0.05 mm/year for crosslinked; 91% of non-crosslinked hips and 10% of crosslinked hips had a penetration rate greater than 0.10 mm/year. Based on the slope of a regression line calculated for the unrevised THAs that had minimum 9-year radiographic follow-up and at least three head penetration measurements, the wear rate for the non-crosslinked group was significantly higher than that of the crosslinked group (0.22 ± 0.13 mm/year vs 0.04 ± 0.06 mm/year, $p < 0.001$). The wear rate was greater than 0.10 mm/year for 86% of non-crosslinked hips and 11% of crosslinked hips. Based on the radiographic review, 22% (15/68) of the unrevised THAs with non-crosslinked liners had pelvic and/or femoral osteolytic lesions with a combined area of at least 1.5 cm^2 that we considered clinically important. In contrast, none of the hips with crosslinked liners had clinically important lesions. This difference is statistically significant ($p < 0.001$).

Callary et al. reported a radiostereometric study on 21 patients implanted with total hip acetabular with sequentially irradiated and annealed highly crosslinked polyethylene (X3™) as liners. The patients were between 45 and 80 years old. The median cup size was 54 mm (range, 48–62 mm). All patients had RSA examinations at 1 week, 6 months, and 1, 2, and 5 years. The minimum follow-up was 4.8 years (mean, 5.3 years; range, 4.8–6.2 years). Wear was measured by penetration of the femoral head inside the acetabular component with UmRSA1 software (Version 6.0; RSA Biomedical, Umea, Sweden). The mean amounts of medial, proximal, anterior, 2-D, and 3-D head penetration between 1 week and 5 years were 0.005, 0.018, 0.075, 0.071, and 0.149 mm, respectively. The mean medial, proximal, anterior, 2-D, and 3-D wear rates between 1 and 5 years were −0.002, 0.001, 0.007, −0.002, and −0.007 mm/year, respectively. No patient in this cohort had a medial, proximal, 2-D, or 3-D wear rate more than 0.040 mm/year. The low mean proximal wear rate of 0.001 mm/year (range, −0.062 to 0.027 mm/year) reported in this study is similar to that reported in other radiostereometric analysis studies of first-generation HXPE liners with greater than 5-year follow-up (e.g., 0.001 [105], 0.002 [106], 0.005 [107], and 0.014 mm/year [108]). In a study of 51 hips using the X3 liners [6], the mean 2-D head penetration between 6 weeks and 5 years was 0.072 ± 0.286 mm, which was similar to Callary's result of 0.071 ± 0.052 mm between 1 week and 5 years. However, D'Antonio et al. [6] reported a mean 2-D wear rate of 0.015 mm/year.

Kim et al. [109] conducted a minimum 7-year follow-up study on 113 total hip arthroplasties in 109 Korean patients (mean age 57 years, range, 35–84 years). The acetabular liners were highly crosslinked PE liner (Longevity®; Zimmer). At each follow-up, standardized anteroposterior and lateral radiographs of the pelvis and femur were performed. The linear wear rate and volumetric wear rate were measured in annual radiographs using a computer-assisted method with PolyWare software (Draftware Developers, Vevay, IN) [110]. The mean follow-up period was 7.8 years (range, 7–10.5 years). Clinical and radiographic evaluation was scheduled for 3 months, 6 months, 1 year, and then annually after surgery. At the mean 7.1-year follow-up, the mean amount of PE linear wear rate and volumetric wear rate was 0.031 ± 0.012 mm/year and 38.6 ± 18.31 mm^3/year. For the first year after surgery, a steep increase in both liners and volumetric wear rate of about 0.085 mm and 65.7 mm^3 was seen, followed by leveling with a tendency of gradual increase over the course of the study. The mean PE linear wear rate and volumetric wear rate in the steady phase were 0.006 ± 0.011 mm/year and 3.49 ± 4.62 mm^3/year. The measured wear rates were lower than the osteolysis threshold [2, 111] of 0.1 mm/year. None of 113 hips had been revised for mechanical failure of HXPE liner, and there were no hips that showed component loosening at the latest follow-up. Osteolysis around the acetabular prostheses was observed in 12 hips (10.6%). One case of osteolytic lesion appeared after 5 years from index operation and localized around screw fixation site, which was round morphology and 3 by 3 mm in size, but after 4 more years of follow-up, the increment of osteolysis did not exceed 1 mm. Three cases of osteolysis were found during the 7-year follow-up, and the morphology of lesions was all focal and linear appearance and less than 3 mm in width. Another eight cases of lesion were found after the 9-year follow-up. The longest follow-up that had osteolysis was 10 years, which started at 9-year follow-up, and no definitive radiographic change was discovered on 1-year follow-up.

Clohisy et al. [112] compared the incidence of acetabular osteolysis in young patients with conventional versus highly crosslinked polyethylene. Five-year follow-up studies (average, 7.2 years; range, 5.1–10.9 years) on a cohort of 48 patients with

HXPE (mean, 46.5 years) and 50 with CPE (mean, 43.2 years) showed the head penetration based on CT and plain radiographs. Osteolysis was identified on CT in 12 of 50 (24%) hips with CPE liners and only one of 48 (2%) hips with HXPE, representing a 92% reduction in the incidence of osteolysis. HXPE liners are highly effective in reducing the incidence of osteolysis during 5–9-year follow-up. The average head penetration rate in patients demonstrating osteolysis (0.17 ± 0.12 mm/year) on CT scan was greater than those with no evidence (0.07 ± 0.08 mm/year), indicating that the dose of wear debris is critical in the osteolysis process.

2.6.2 Knee Implants

Despite the success of highly crosslinked polyethylene in total hip arthroplasty, it has however not been widely accepted for knee implants due to concerns regarding HXPE [113]. The knee differs from the highly congruent ball-and-socket articulation of the hip joint. The biomechanical profile of the TKA articulation, including rolling, sliding, and rotation, leads to more fatigue failure, pitting, and delamination than those seen in highly congruent THA articulation [114]. Wear in knee arthroplasty occurs due to complex geometry of the articulation which involves rolling, sliding, and rotational motion at the bearing surface. This can cause delamination, pitting, and fatigue failure of the PE surface [115]. In contrast, wear in hip arthroplasty occurs mainly due to micro-adhesion and micro-abrasion [116]. Irradiation and thermal treatment reduced strength, fatigue resistance, and fracture toughness of the material, and increased brittleness of the material can adversely affect the integrity of the acetabular locking mechanism [115]. Furthermore, the wear particles in crosslinked polyethylene may be more biologically active because they are smaller than the particles in non-crosslinked polyethylene, which is another unfavorable factor usually considered when deciding whether to use HXPE in THA [117].

The use of highly crosslinked polyethylene in total knee prostheses is still controversial. Clinical studies comparing use of conventional versus highly crosslinked PE are few. Hofmann et al. [114] performed a retrospective review of 100 subjects receiving HXPE and compared them to 100 subjects who received standard polyethylene in the setting of TKA. The standard polyethylene group had a minimum follow-up of 82 months, while the HXPE group had a minimum follow-up of 69 months. The conventional group received compression-molded polyethylene sterilized by γ irradiated in nitrogen. The highly crosslinked group received compression-molded polyethylene irradiated with a dose 9.5 Mrad through e-beam and melted at 150 °C. Radiographic results showed that the conventional group demonstrated 20 TKAs with radiolucencies, with 4 showing a loose tibial component. This group required three revisions related to lose tibial components. The HXPE group had two subjects that demonstrated radiolucencies on radiograph and no subjects with evidence of tibial loosening. There were no reoperations related to osteolysis. The data suggest HXPE in TKA can be used safely at least short- to midterm. One positive provided by them shows no observable wear of HXPE after serving in vivo for 5 years.

Minoda et al. [118] compared the radiographic and clinical results of conventional and highly crosslinked polyethylene in cruciate retaining (CR) total knee prostheses of completely the same design. A total of 113 knees with conventional polyethylene and 89 knees with highly crosslinked polyethylene were evaluated after more than 2 years of operation. Highly crosslinked polyethylene inserts (Prolong®) with 65 kGy of e-beam and remelting were compared with conventional polyethylene inserts sterilized by γ ray at 37 kGy. The differences in the postoperative clinical scores, range of motion, and radiographic results between highly crosslinked and conventional polyethylene with the same design were not significant. No knee exhibited osteolysis or aseptic loosening in both groups. The results are encouraging; however, this study is very early in terms of follow-up.

Meneghini et al. [101] reported radiographic outcomes of 114 consecutive total knee arthroplasties (TKAs) in 83 patients at a mean of 5 years. Among these TKAs, 50 inserts are from conventional polyethylene (N2Vac; Stryker, Mahwah, NJ), which were compression-molded GUR 1020 UHMWPE, packed in nitrogen and gamma-irradiated at 30 kGy; the other 64 inserts are highly crosslinked polyethylene (X3™). There were no revisions for any reason in either group with the exception of lone revision for infection at 3 months in the conventional polyethylene group. There was no radiographic osteolysis or mechanical failures related to the tibial polyethylene in either group. However, the wear information of these inserts has not been reported in this study.

Most recently, there was a multicenter 5-year follow-up study of highly crosslinked versus conventional polyethylene in total knee arthroplasty [119]. Conventional polyethylene (N2Vac; Stryker) tibial inserts γ irradiated at nominal 3 Mrad were used in the first cohort of 168 TKAs. X3™ inserts were used in the second cohort of 168 TKAs. There was no radiographic osteolysis or mechanical failures related to the tibial polyethylene in either group. Zonal analysis indicated no progressive radiolucencies or gaps at final follow-up minimum of 4 years. Although the results of above follow-up reports support comparative safety, longer-term follow-up is warranted to determine if wear resistance and mechanical properties of HXPE are maintained in the knee over greater follow-up intervals.

An opposite voice based on in vivo study also appeared. Hinarejos et al. [120] reported that HXPE does not reduce the wear in total knee arthroplasty. They compared the number, size, and morphology of polyethylene particles in the synovial fluid of patients after primary TKA using a conventional or HXPE inserts. A total of 34 patients were analyzed with 17 conventional and 17 HXPE inserts. Conventional polyethylene inserts sterilized to 30 kGy were compared with HXPE insert machined from X3. Knee arthrocentesis was performed 1 year after surgery in sterile conditions. The average concentration of polyethylene particles/mL for CPE and HXPE was $1.53 \times 10^6 \pm 1.02 \times 10^6$ and $1.26 \times 10^6 \pm 0.90 \times 10^6$, respectively. The total number of polyethylene particles was $16.47 \times 10^6 \pm 15.97 \times 10^6$ and $17.50 \times 10^6 \pm 18.05 \times 10^6$ for CPE and HXPE group, respectively. The size of the isolated polyethylene particles was highly variable, from 0.1 μm up to 30 μm. The rate of particles greater than 1 μm in diameter was 2.6% in the CPE group and 3.7% in the HXPE group; 78.8% of particles were round shaped. The ratio of non-spherical particles was 22.6% in the CPE group and 19.2% in the HXPE group. These results indicate that there are no significant differences in the concentration, size, or morphology of polyethylene particles between groups. Moreover, they have found a great variability in the number of particles between individuals, suggesting that in vivo polyethylene wear depends on many factors.

2.7 Conclusions

The highly crosslinked UHMWPE (HXPE) has been in clinical use around the globe for the past two decades. The experience with HXPE shows a progressive development in understanding the relationship between UHMWPE structure (crosslink density and crystallinity) and properties (strength, ductility, and wear).

Gamma (γ) or electron beam (e-beam) irradiation generated free radicals to form crosslinks in UHMWPE. Some of the free radicals, such as allyl macroradicals, can survive in irradiated samples for a few years at room temperature. Free radicals trapped in crystalline phase react with diffused oxygen, leading to chain scission and the deterioration of mechanical properties. To eliminate irradiation-generated free radicals, remelting UHMWPE at temperatures above the melting point or annealing them below melting point is a common method.

Crosslinking of UHMWPE results in elevated crystallinity, but subsequent remelting results in lower crystallinity than that before crosslinking. On the other hand, crosslinking and annealing favor a maintenance of high crystallinity as measured by DSC. Crosslinking decreased the fatigue crack propagation (FCP) resistance in compact tension tests, the ultimate tensile strength and elongation at break values in tensile tests, and impact toughness in Izod impact tests.

The crosslinking of UHMWPE results in a restriction of chain mobility in the amorphous region and therefore reduces the overall plasticity of the polymer. The annealed UHMWPE had an overall increased FCP resistance than the remelted UHMWPE. This may related to the loss of crystalline content after postirradiation melting resulted in loss of plastic deformation ability of UHMWPE.

Crosslinked UHMWPE has demonstrated superior wear properties during pin-on-disk wear tests and in vitro joint simulator tests. Crosslinking of UHMWPE molecules decreased the chain mobility needed for the large-scale plastic deformation and slow down the surface fibril formation, hence improving the wear resistance.

From the follow-up study using radiographic examination, HXPE total hip implants in joint simulator have shown a slower wear rate than traditional polyethylene. From the follow-up study using CT scan, HXPE hip liners are highly effective in reducing the incidence of osteolysis. Clinical study of HXPE total knee implants is very early in terms of follow-up. Long-term follow-up is needed to determine if wear resistance is maintained in the knee over greater follow-up intervals.

References

1. Charnley J (1970) Total hip replacement by low-friction arthroplasty. Clin Orthop Relat Res 72:7–21
2. Dumbleton JH, Manley MT, Edidin AA (2002) A literature review of the association between wear rate and osteolysis in total hip arthroplasty. J Arthroplast 17:649–661
3. Harris WH (1995) The problem is osteolysis. Clin Orthop Relat Res 311:46–53
4. Harris WH (2001) Wear and periprosthetic osteolysis – the problem. Clin Orthop Relat Res 393:66–70
5. Cooper RA, McAllister CM, Borden LS, Bauer TW (1992) Polyethylene debris-induced osteolysis and loosening in uncemented total hip arthroplasty. J Arthroplast 7:285–290
6. Kovacik MW, Gradisar IA Jr, Haprian JJ, Alexander TS (2000) Osteolytic indicators found in total knee arthroplasty synovial fluid aspirates. Clin Orthop Relat Res 379:186–194
7. McKellop H, F-w S, Lu B, Campbell P, Salovey R (1999) Development of an extremely wear-resistant ultra high molecular weight polythylene for total hip replacements. J Orthop Res 17:157–167
8. Muratoglu OK, Bragdon CR, O'Connor DO, Jasty M, Harris WH (2001) A novel method of cross-linking ultra-high-molecular-weight polyethylene to improve wear, reduce oxidation, and retain mechanical properties – recipient of the 1999 HAP Paul Award. J Arthroplast 16:149–160
9. Griffith M, Seidenstein M, Williams D, Charnley J (1978) Socket wear in Charnley low friction arthroplasty of the hip. Clin Orthop Relat Res (137):37–47
10. Charnley J, Halley DK (1975) Rate of wear in total hip replacement. Clin Orthop Relat Res 112:170–179
11. McKellop HA, Shen FW, Campbell P, Ota T (1999) Effect of molecular weight, calcium stearate, and sterilization methods on the wear of ultra high molecular weight polyethylene acetabular cups in a hip joint simulator. J Orthop Res 17:329–339
12. Hopper RH, Young AM, Orishimo KF, Engh CA (2003) Effect of terminal sterilization with gas plasma or gamma radiation on wear of polyethylene liners. J Bone Joint Surg Am 85:464–468
13. Kurtz SM, Muratoglu OK, Evans M, Edidin AA (1999) Advances in the processing, sterilization, and crosslinking of ultra-high molecular weight polyethylene for total joint arthroplasty. Biomaterials 20:1659–1688
14. Besong A, Tipper J, Ingham E, Stone M, Wroblewski B, Fisher J (1998) Quantitative comparison of wear debris from UHMWPE that has and has not been sterilised by gamma irradiation. J Bone Joint Surg (Br) 80:340–344
15. Premnath V, Harris W, Jasty M, Merrill E (1996) Gamma sterilization of UHMWPE articular implants: an analysis of the oxidation problem. Biomaterials 17:1741–1753
16. Currier BH, Currier JH, Mayor MB, Lyford KA, Van Citters DW, Collier JP (2007) In vivo oxidation of γ-barrier–sterilized ultra–high-molecular-weight polyethylene bearings. J Arthroplast 22:721–731
17. Collier JP, Sperling DK, Currier JH, Sutula LC, Saum KA, Mayor MB (1996) Impact of gamma sterilization on clinical performance of polyethylene in the knee. J Arthroplast 11:377–389
18. Kurtz SM, Hozack WJ, Purtill JJ, Marcolongo M, Kraay MJ, Goldberg VM, Sharkey PF, Parvizi J, Rimnac CM, Edidin AA (2006) 2006 Otto Aufranc Award Paper: significance of in vivo degradation for polyethylene in total hip arthroplasty. Clin Orthop Relat Res 453:47–57
19. Kurtz SM, Rimnac CM, Hozack WJ, Turner J, Marcolongo M, Goldberg VM, Kraay MJ, Edidin AA (2005) In vivo degradation of polyethylene liners after gamma sterilization in air. J Bone Joint Surg Am 87:815–823

20. Gencur SJ, Rimnac CM, Kurtz SM (2003) Failure micromechanisms during uniaxial tensile fracture of conventional and highly crosslinked ultra-high molecular weight polyethylenes used in total joint replacements. Biomaterials 24:3947–3954
21. Medel FJ, Kurtz SM, Parvizi J, Klein GR, Kraay MJ, Rimnac CM (2011) In vivo oxidation contributes to delamination but not pitting in polyethylene components for total knee arthroplasty. J Arthroplast 26:802–810
22. Kurtz SM, Gawel HA, Patel JD (2011) History and systematic review of wear and osteolysis outcomes for first-generation highly crosslinked polyethylene. Clin Orthop Relat Res 469:2262–2277
23. Kurtz SM, Manley M, Wang A, Taylor S, Dumbleton J (2002) Comparison of the properties of annealed crosslinked (crossfire) and conventional polyethylene as hip bearing materials. Bull Hosp Jt Dis 61:17–26
24. Horii F, Zhu Q, Kitamaru R, Yamaoka H (1990) Carbon-13 NMR study of radiation-induced crosslinking of linear polyethylene. Macromolecules 23:977–981
25. Muratoglu OK (2009) Highly crosslinked and melted UHMWPE. In: Kurtz SM (ed) UHMWPE biomaterials handbook, 2nd edn. Academic, Boston, pp 197–204
26. Muratoglu OK, Bragdon CR, O'Connor DO, Jasty M, Harris WH, Gul R, McGarry F (1999) Unified wear model for highly crosslinked ultra-high molecular weight polyethylenes (UHMWPE). Biomaterials 20:1463–1470
27. Costa L, Carpentieri I, Bracco P (2008) Post electron-beam irradiation oxidation of orthopaedic UHMWPE. Polym Degrad Stab 93:1695–1703
28. Bracco P, Brunella V, Luda M, Zanetti M, Costa L (2005) Radiation-induced crosslinking of UHMWPE in the presence of co-agents: chemical and mechanical characterisation. Polymer 46:10648–10657
29. DeVries K, Smith R, Fanconi B (1980) Free radicals and new end groups resulting from chain scission: 1. y-irradiation of polyethylene. Polymer 21:949–956
30. Igarashi M (1983) Free-radical identification by ESR in polyethylene and nylon. J Polym Sci Polym Chem Ed 21:2405–2425
31. Lacoste J, Carlsson D (1992) Gamma-, photo-, and thermally-initiated oxidation of linear low density polyethylene: a quantitative comparison of oxidation products. J Polym Sci A Polym Chem 30:493–500
32. Perez E, Vanderhart D (1988) A 13C CP-MAS NMR study of irradiated polyethylene. J Polym Sci B Polym Phys 26:1979–1993
33. Randall JC (1990) Carbon 13 NMR of gamma-irradiated polyethylenes. In: Güven O (ed) Crosslinking and scission in polymers. Springer Netherlands, Dordrecht, pp 57–82
34. Dole M, Milner D, Williams TF (1958) Irradiation of polyethylene. II. Kinetics of unsaturation effects. J Am Chem Soc 80:1580–1588
35. Brunella V, Bracco P, Carpentieri I, Paganini M, Zanetti M, Costa L (2007) Lifetime of alkyl macroradicals in irradiated ultra-high molecular weight polyethylene. Polym Degrad Stab 92:1498–1503
36. Bhateja SK, Duerst RW, Aus EB, Andrews EH (1995) Free radicals trapped in polyethylene crystals. J Macromol Sci Part B Phys 34:263–272
37. Costa L, Luda MP, Trossarelli L, Brach del Prever EM, Crova M, Gallinaro P (1998) Oxidation in orthopaedic UHMWPE sterilized by gamma-radiation and ethylene oxide. Biomaterials 19:659–668
38. Costa L, Bracco P (2016) Mechanisms of cross-linking, oxidative degradation, and stabilization of UHMWPE. In: Kurtz SM (ed) UHMWPE biomaterials handbook, 3rd edn. William Andrew Publishing, Oxford, pp 467–487
39. Jahan MS (2016) ESR insights into macroradicals in UHMWPE. In: Kurtz SM (ed) UHMWPE biomaterials handbook, 3rd edn. William Andrew Publishing, Oxford, pp 668–692
40. Brandrup J, Immergut EH, Grulke EA, Abe A, Bloch DR (1989) Polymer handbook. Wiley, New York

41. Llorente M, Mark J (1980) Model networks of end-linked poly (dimethylsiloxane) chains. 8. Networks having cross-links of very high functionality. Macromolecules 13:681–685
42. Dijkstra DJ, Hoogsteen W, Pennings AJ (1989) Cross-linking of ultra-high molecular weight polyethylene in the melt by means of electron beam irradiation. Polymer 30:866–873
43. Gent AN, Vickroy VV (1967) Elastic behavior, birefringence, and swelling of amorphous polyethylene networks. J Polym Sci Part A2 5:47–61
44. Flory PJ, Rehner J (1943) Statistical mechanics of cross-linked polymer networks I. Rubberlike elasticity. J Chem Phys 11:512–520
45. Oral E, Malhi AS, Muratoglu OK (2006) Mechanisms of decrease in fatigue crack propagation resistance in irradiated and melted UHMWPE. Biomaterials 27:917–925
46. Luo Y, Wang G, Lu Y, Chen N, Jiang B (1985) Location of radiation-induced crosslinks and damage in melt-crystallized polyethylene. Radiat Phys Chem 25:359–365
47. Yeh GSY, Chen CJ, Boose DC (1985) CPS K 843 radiation-induced crosslinking: effect on structure of polyethylene. Colloid Polym Sci 263:109–115
48. Jahan MS, King MC, Haggard WO, Sevo KL, Parr JE (2001) A study of long-lived free radicals in gamma-irradiated medical grade polyethylene. Radiat Phys Chem 62:141–144
49. Costa L, Luda MP, Trossarelli L, Brach del Prever EM, Crova M, Gallinaro P (1998) In vivo UHMWPE biodegradation of retrieved prosthesis. Biomaterials 19:1371–1385
50. Sutula LC, Collier JP, Saum KA, Currier BH, Currier JH, Sanford WM, Mayor MB, Wooding RE, Sperling DK, Williams IR (1995) The Otto Aufranc Award: impact of gamma sterilization on clinical performance of polyethylene in the hip. Clin Orthop Relat Res (319):28–40
51. Wang A, Yau S-S, Essner A, Herrera L, Manley M, Dumbleton J (2008) A highly crosslinked UHMWPE for CR and PS total knee arthroplasties. J Arthroplast 23:559–566
52. Kester MA, Herrera L, Wang A, Essner A (2007) Knee bearing technology: where is technology taking us? J Arthroplast 22:16–20
53. Costa L, Jacobson K, Bracco P, Brach del Prever EM (2002) Oxidation of orthopaedic UHMWPE. Biomaterials 23:1613–1624
54. Rueda DR, Hidalgo A, Calleja FJB (1978) An i.r. study of the "amorphous" phase in melt crystallized polyethylene. Spectrochim Acta Part A 34:475–480
55. Medel FJ, Pena P, Cegoñino J, Gómez-Barrena E, Puertolas J (2007) Comparative fatigue behavior and toughness of remelted and annealed highly crosslinked polyethylenes. J Biomed Mater Res Part B 83:380–390
56. Fu J, Ghali BW, Lozynsky AJ, Oral E, Muratoglu OK (2010) Ultra high molecular weight polyethylene with improved plasticity and toughness by high temperature melting. Polymer 51:2721–2731
57. Gomoll A, Wanich T, Bellare A (2002) J-integral fracture toughness and tearing modulus measurement of radiation cross-linked UHMWPE. J Orthop Res 20:1152–1156
58. Zhao Y, Luo Y, Jiang B (1993) Effect of irradiation on crystallinity and mechanical properties of ultrahigh molecular weight polyethylene. J Appl Polym Sci 50:1797–1801
59. Baker DA, Bellare A, Pruitt L (2003) The effects of degree of crosslinking on the fatigue crack initiation and propagation resistance of orthopedic-grade polyethylene. J Biomed Mater Res Part A 66A:146–154
60. Gencur SJ, Rimnac CM, Kurtz SM (2006) Fatigue crack propagation resistance of virgin and highly crosslinked, thermally treated ultra-high molecular weight polyethylene. Biomaterials 27:1550–1557
61. Kurtz SM, Villarraga ML, Herr MP, Bergström JS, Rimnac CM, Edidin AA (2002) Thermomechanical behavior of virgin and highly crosslinked ultra-high molecular weight polyethylene used in total joint replacements. Biomaterials 23:3681–3697
62. Yasuniwa M, Tsubakihara S, Yamaguchi M (1997) Lamellar thickening of polyethylene under high pressure. J Polym Sci B Polym Phys 35:535–543
63. Oral E, Godleski-Beckos C, Ghali BW, Lozynsky AJ, Muratoglu OK (2009) Effect of cross-link density on the high pressure crystallization of UHMWPE. J Biomed Mater Res Part B 90B:720–729

64. Simis KS, Bistolfi A, Bellare A, Pruitt LA (2006) The combined effects of crosslinking and high crystallinity on the microstructural and mechanical properties of ultra high molecular weight polyethylene. Biomaterials 27:1688–1694
65. Butler MF, Donald AM, Ryan AJ (1998) Time resolved simultaneous small- and wide-angle X-ray scattering during polyethylene deformation—II. Cold drawing of linear polyethylene. Polymer 39:39–52
66. Lin L, Argon A (1994) Structure and plastic deformation of polyethylene. J Mater Sci 29:294–323
67. Puértolas JA, Medel FJ, Cegoñino J, Gomez-Barrena E, Ríos R (2006) Influence of the remelting process on the fatigue behavior of electron beam irradiated UHMWPE. J Biomed Mater Res Part B 76B:346–353
68. Pruitt LA (2005) Deformation, yielding, fracture and fatigue behavior of conventional and highly cross-linked ultra high molecular weight polyethylene. Biomaterials 26:905–915
69. Bistolfi A, Turell MB, Lee YL, Bellare A (2009) Tensile and tribological properties of high-crystallinity radiation crosslinked UHMWPE. J Biomed Mater Res Part B 90:137–144
70. Urriés I, Medel FJ, Ríos R, Gómez-Barrena E, Puértolas JA (2004) Comparative cyclic stress–strain and fatigue resistance behavior of electron–beam- and gamma-irradiated ultrahigh molecular weight polyethylene. J Biomed Mater Res Part B 70B:152–160
71. Galeski A (2003) Strength and toughness of crystalline polymer systems. Prog Polym Sci 28:1643–1699
72. Brooks NWJ, Mukhtar M (2000) Temperature and stem length dependence of the yield stress of polyethylene. Polymer 41:1475–1480
73. Laurent MP, Johnson TS, Crowninshield RD, Blanchard CR, Bhambri SK, Yao JQ (2008) Characterization of a highly cross-linked ultrahigh molecular-weight polyethylene in clinical use in total hip arthroplasty. J Arthroplast 23:751–761
74. Jasty M, Goetz DD, Bragdon CR, Lee KR, Hanson AE, Elder JR, Harris WH (1997) Wear of polyethylene acetabular components in total hip arthroplasty. An analysis of one hundred and twenty-eight components retrieved at autopsy or revision operations. J Bone Joint Surg Am 79:349–358
75. Bragdon C, O'Connor D, Lowenstein J, Jasty M, Syniuta W (1996) The importance of multidirectional motion on the wear of polyethylene. Proc Inst Mech Eng Part H 210:157–165
76. Wang A, Stark C, Dumbleton J (1996) Mechanistic and morphological origins of ultra-high molecular weight polyethylene wear debris in total joint replacement prostheses. Proc Inst Mech Eng Part H 210:141–155
77. Wang A (2001) A unified theory of wear for ultra-high molecular weight polyethylene in multi-directional sliding. Wear 248:38–47
78. O'Connor D, Bragdon C, Burke D, Jasty M, Lowenstein J, Harris W (1995) A 12 station upright hip simulator wear testing machine employing oscillating motion replicating the human gait cycle. In: 21st annual meeting of The Society for Biomaterials, San Francisco, 18–22 March 1995
79. Essner A, Sutton K, Wang A (2005) Hip simulator wear comparison of metal-on-metal, ceramic-on-ceramic and crosslinked UHMWPE bearings. Wear 259:992–995
80. Muratoglu OK, Rubash HE, Bragdon CR, Burroughs BR, Huang A, Harris WH (2007) Simulated normal gait wear testing of a highly cross-linked polyethylene tibial insert. J Arthroplast 22:435–444
81. Bragdon CR, Jasty M, Muratoglu OK, O'Connor DO, Harris WH (2003) Third-body wear of highly cross-linked polyethylene in a hip simulator1. J Arthroplast 18:553–561
82. Bragdon CR, Jasty M, Muratoglu OK, Harris WH (2005) Third-body wear testing of a highly cross-linked acetabular liner: the effect of large femoral head size in the presence of particulate poly(methyl-methacrylate) debris. J Arthroplast 20:379–385
83. Burroughs BR, Rubash HE, Harris WH (2002) Femoral head sizes larger than 32 mm against highly cross-linked polyethylene. Clin Orthop Relat Res 405:150–157

84. Sorbie C (2003) Arthroplasty in the treatment of subcapital hip fracture. Orthopedics 26:337–341
85. Saikko V, Calonius O, Keränen J (2002) Wear of conventional and cross-linked ultra-high-molecular-weight polyethylene acetabular cups against polished and roughened CoCr femoral heads in a biaxial hip simulator. J Biomed Mater Res 63:848–853
86. Hall R, Unsworth A, Siney P, Wroblewski B (1996) Wear in retrieved Charnley acetabular sockets. Proc Inst Mech Eng, Part H 210:197–207
87. Hall RM, Siney P, Unsworth A, Wroblewski BM (1997) The effect of surface topography of retrieved femoral heads on the wear of UHMWPE sockets. Med Eng Phys 19:711–719
88. Sychterz CJ, Engh CA Jr, Swope SW, DE MN, Engh CA (1999) Analysis of prosthetic femoral heads retrieved at autopsy. Clin Orthop Relat Res 358:223–234
89. Akagi M, Asano T, Clarke IC, Niiyama N, Kyomoto M, Nakamura T, Hamanishi C (2006) Wear and toughness of crosslinked polyethylene for total knee replacements: a study using a simulator and small-punch testing. J Orthop Res 24:2021–2027
90. Kawanabe K, Clarke IC, Tamura J, Akagi M, Good VD, Williams PA, Yamamoto K (2001) Effects of A–P translation and rotation on the wear of UHMWPE in a total knee joint simulator. J Biomed Mater Res 54:400–406
91. Tamura J, Clarke IC, Kawanabe K, Akagi M, Good VD, Williams PA, Masaoka T, Schroeder D, Oonishi H (2002) Micro-wear patterns on UHMWPE tibial inserts in total knee joint simulation. J Biomed Mater Res 61:218–225
92. Harris WH (2009) The first 50 years of total hip arthroplasty: lessons learned. Clin Orthop Relat Res 467:28–31
93. Parvizi J, Wade FA, Rapuri V, Springer BD, Berry DJ, Hozack WJ (2006) Revision hip arthroplasty for late instability secondary to polyethylene wear. Clin Orthop Relat Res 447:66–69
94. Geerdink CH, Grimm B, Vencken W, Heyligers IC, Tonino AJ (2009) Cross-linked compared with historical polyethylene in THA: an 8-year clinical study. Clin Orthop Relat Res 467:979–984
95. Green TR, Fisher J, Matthews JB, Stone MH, Ingham E (2000) Effect of size and dose on bone resorption activity of macrophages by in vitro clinically relevant ultra high molecular weight polyethylene particles. J Biomed Mater Res 53:490–497
96. Fang H-W, Yang C-B, Chang C-H, Huang C-H, Liu H-L, Fang S-B (2006) The potential role of phagocytic capacity in the osteolytic process induced by polyethylene wear particles. J Int Med Res 34:655–664
97. Martell JM, Berdia S (1997) Determination of polyethylene wear in total hip replacements with use of digital radiographs*. J Bone Joint Surg 79:1635–1641
98. Crockarell JR Jr, Snearly CM (2012) Accuracy and precision of two computer-assisted methods of radiographic wear measurement in total hip arthroplasty. J Arthroplast 27:37–40
99. Rubash H, Sinha R, Paprosky W, Engh C, Maloney W (1999) A new classification system for the management of acetabular osteolysis after total hip arthroplasty. Instr Course Lect 48:37–42
100. Geerdink CH, Grimm B, Vencken W, Heyligers IC, Tonino AJ (2008) The determination of linear and angular penetration of the femoral head into the acetabular component as an assessment of wear in total hip replacement: a comparison of four computer-assisted methods. J Bone Joint Surg (Br) 90-B:839–846
101. Meneghini RM, Lovro LR, Smits SA, Ireland PH (2015) Highly cross-linked versus conventional polyethylene in posterior-stabilized total knee arthroplasty at a mean 5-year follow-up. J Arthroplast 30:1736–1739
102. Geller JA, Malchau H, Bragdon C, Greene M, Harris WH, Freiberg AA (2006) Large diameter femoral heads on highly cross-linked polyethylene: minimum 3-year results. Clin Orthop Relat Res 447:53–59

103. Snir N, Kaye ID, Klifto CS, Hamula MJ, Wolfson TS, Schwarzkopf R, Jaffe FF (2014) 10-year follow-up wear analysis of first-generation highly crosslinked polyethylene in primary total hip arthroplasty. J Arthroplast 29:630–633
104. Engh CA Jr, Hopper RH Jr, Huynh C, Ho H, Sritulanondha S, Engh CA Sr (2012) A prospective, randomized study of cross-linked and non-cross-linked polyethylene for total hip arthroplasty at 10-year follow-up. J Arthroplast 27:2–7.e1
105. Digas G, Kärrholm J, Thanner J, Herberts P (2007) 5-year experience of highly cross-linked polyethylene in cemented and uncemented sockets: two randomized studies using radiostereometric analysis. Acta Orthop 78:746–754
106. Röhrl SM, Nivbrant B, Nilsson KG (2012) No adverse effects of submelt-annealed highly crosslinked polyethylene in cemented cups. Acta Orthop 83:148–152
107. Kraay MJ, Moore RD, Martell JM, Rimnac CM (2010) Reassessment of computerized wear measurement for total hip arthroplasty with correction for projectional image distortion: a brief follow-up report. J Bone Joint Surg 92:1858–1867
108. Callary SA, Campbell DG, Mercer G, Nilsson KG, Field JR (2013) Wear of a 5 megarad cross-linked polyethylene liner: a 6-year RSA study. Clin Orthop Relat Res 471:2238–2244
109. Lee J-H, Lee BW, Lee B-J, Kim S-Y (2011) Midterm results of primary total hip arthroplasty using highly cross-linked polyethylene: minimum 7-year follow-up study. J Arthroplast 26:1014–1019
110. D'Antonio JA, Capello WN, Ramakrishnan R (2012) Second-generation annealed highly cross-linked polyethylene exhibits low wear. Clin Orthop Relat Res 470:1696–1704
111. Dowd JE, Sychterz CJ, Young AM, Engh CA (2000) Characterization of long-term femoral-head-penetration rates : association with and prediction of osteolysis*. JBJS 82:1102–1102
112. Mall NA, Nunley RM, Zhu JJ, Maloney WJ, Barrack RL, Clohisy JC (2011) The incidence of acetabular osteolysis in young patients with conventional versus highly crosslinked polyethylene. Clin Orthop Relat Res 469:372–381
113. Rodriguez JA (2008) Cross-linked polyethylene in total knee arthroplasty. Oppos J Arthroplast 23:31–34
114. Hodrick JT, Severson EP, McAlister DS, Dahl B, Hofmann AA (2008) Highly crosslinked polyethylene is safe for use in total knee arthroplasty. Clin Orthop Relat Res 466:2806–2812
115. Collier JP, Mayor MB, McNamara JL, Surprenant VA, Jensen RE (1991) Analysis of the failure of 122 polyethylene inserts from uncemented tibial knee components. Clin Orthop Relat Res (273):232–242
116. McKellop HA, Campbell P, Park S-H, Schmalzried TP, Grigoris P, Amstutz HC, Sarmiento A (1995) The origin of submicron polyethylene wear debris in total hip arthroplasty. Clin Orthop Relat Res 311:3–20
117. Wright TM (2005) Polyethylene in knee arthroplasty: what is the future? Clin Orthop Relat Res 440:141–148
118. Minoda Y, Aihara M, Sakawa A, Fukuoka S, Hayakawa K, Tomita M, Umeda N, Ohzono K (2009) Comparison between highly cross-linked and conventional polyethylene in total knee arthroplasty. Knee 16:348–351
119. Meneghini RM, Ireland PH, Bhowmik-Stoker M (2016) Multicenter study of highly cross-linked vs conventional polyethylene in total knee arthroplasty. J Arthroplast 31:809–814
120. Hinarejos P, Piñol I, Torres A, Prats E, Gil-Gómez G, Puig-Verdie L (2013) Highly crosslinked polyethylene does not reduce the wear in total knee arthroplasty: in vivo study of particles in synovial fluid. J Arthroplast 28:1333–1337

Chapter 3
Oxidation in Prosthetic UHMWPE

L. Costa

Abstract In this chapter, we will consider the oxidation process involved in high-energy radiation sterilization of UHMWPE, as generally used in the field of orthopedics. The macroradicals produced during irradiation, their stability in various morphological phases, and their reactivity with chain imperfections, as well as the effect of oxygen and additives in the cyclic radical process, will be discussed. Particular attention will be paid to the oxidation process.

Keywords UHMWPE · Oxidation · Gamma sterilization · Mobility macroalkyl radicals · Cyclic radical reactions

3.1 Introduction

Ultrahigh molecular weight polyethylene (UHMWPE) is a biomaterial that has been used in prosthetic implants for over 50 years. Sir John Charnley, the father of prosthetic surgery, was the first to use polyethylene (PE) in orthopedic field in 1962. William Waugh states that this new biomaterial was chosen for "low friction arthroplasty" and that this occurred thanks to Craven, a laboratory technician and colleague of Charnley who was helping him to investigate and prepare prosthetic components. Charnley started using Teflon (polytetrafluoroethylene or PTFE) acetabular components at the end of the 1950s, after having tested their biocompatibility on himself by surgically inserting a piece of PTFE into his thigh and observing the reaction. As a matter of fact, Charnley always tried materials out on himself before implanting them in his patients. However, PTFE proved to be a failure from the clinical point of view in 1962 despite giving good results in Charnley's soft tissue. The material rapidly became worn in patients and led to serious granuloma causing

L. Costa (✉)
Dipartimento di Biotecnologie Molecolari e Scienze per la Salute, University of Turin, Turin, Italy
e-mail: luigi.costa@unito.it

J. Fu et al. (eds.), *UHMWPE Biomaterials for Joint Implants*, Springer Series in Biomaterials Science and Engineering 13,
https://doi.org/10.1007/978-981-13-6924-7_3

osteolysis, which forced the surgeon to remove prostheses. Charnley openly declared his failures and warned his fellow surgeons against using polymers as a combining material. It was sometime after this that a "plastics" salesman knocked on the door of Charnley's hospital to show him one of his materials, high-density polyethylene with high molecular weight. Charnley was still suffering from the disappointment of PTFE's failure and did not like the new polymer at first. Craven, on the other hand, set to work testing its resistance to friction in the "pendulum," and, 3 weeks later, Charnley observed in amazement how the new material behaved so much better than the PTFE. From that moment on, PE and prosthetics have walked hand in hand [1].

Despite UHMWPE meeting all the requirements for its ASTM F648 designation as well as having all the mechanical and friction resistance properties necessary for use in the orthopedic field, an important clinical problem has come to light after some years of positive results; scientific reviews have, more and more frequently, reported the occurrence of implant failure caused by the rapid and serious deterioration of polyethylene components only a few months after implantation and in relatively inactive people.

Earlier, prosthetic components were sterilized just as surgical instruments were, but, as of the end of the 1960s, the industrialization of prosthetic component production led to radiation sterilization being used despite the negative effect that radiation had on PEs being described in the literature [2].

In 1979, Nusbam and Rose found high oxidation levels, an increase in the synovial fluid absorption, and a decrease in mechanical properties when analyzing films of virgin UHMWPE and those sterilized with gamma ray at 25 kGy in air [3].

A comparison of UHMWPE prosthetic components, explanted after some years and that had been sterilized with either gamma radiation in air or ethylene oxide (EtO), highlighted the diffusion of the non-polar components of synovial fluid in both types. The absence of oxidation in prostheses sterilized with EtO and the very variable levels of oxidation in components sterilized with gamma ray indicate that our bodies are not able to oxidize virgin UHMWPE [4, 5].

This chapter will consider the UHMWPE oxidation as induced by high-energy radiation during the sterilization process at temperatures below 40 °C and make comparisons with literature data on the oxidation processes of PEs.

3.2 What Is Ultrahigh Molecular Weight Polyethylene?

The commercial UHMWPE that is currently used in the orthopedic field is produced by Ticona, the engineering polymers business of Celanese Corporation [6]. Its physicochemical and mechanical properties are described in depth on the website of the UHMWPE organization [7]. The direct molding of its powders into prosthetic components is not efficient because of its ultra-high molecular weight and very high viscosity. Instead, UHMWPE powders are either transformed into a sheet in a compression molding process and then cut into bars or directly formed into cylindrical

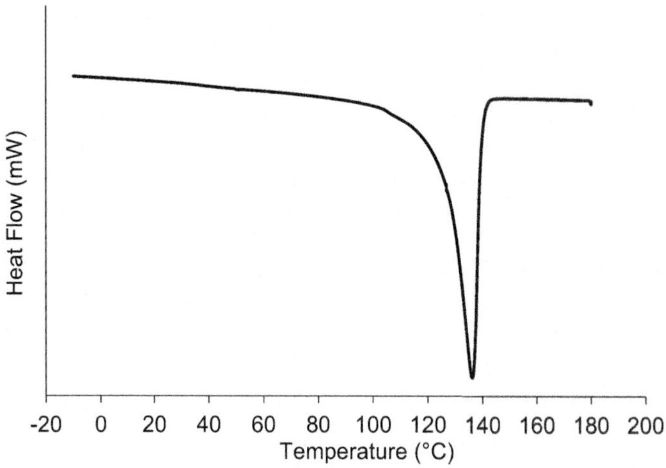

Fig. 3.1 Representative DSC heating curve of a consolidated UHMWPE

bars in a ram extrusion process. Consolidation occurs at temperatures above 200 °C and the material is then slowly cooled down to room temperature. There are effectively only two companies that perform this transformation: Orthoplastics [8] and Meditech [9].

The transformation processing conditions are the intellectual property of the two companies. The material then undergoes annealing treatment by the same companies. Lastly, the material is mechanically machined into its final forms by the producer of the prosthetic component.

The characteristics of UHMWPE bars do not differ greatly from one to the other, while the transformation process, from powder to bars, leads to a variation in the physicochemical properties [7]. The recrystallization of the molten material leads to a substance that has physicochemical properties that are quite different from those of the original UHMWPE powder.

A representative DSC (Differential Scanning Calorimetry) curve of a consolidated UHMWPE is shown in Fig. 3.1. The DSC scan is conducted at 10 °C/min in an inert nitrogen environment. In the diagram, heat flow or the rate of melting as a function of temperature is reported. First, smaller crystallites start to melt and then the larger ones follow. The melting interval is usually between 90 and 145 °C, while its maximum melting rate occurs at 135 °C. The UHMWPE crystallinity is over 70% in the powder form, while because of the difficulty to crystallize during the transformation process, the size of the crystallites and the crystallinity decrease to values of 45–50%.

Crystal lamella thickness is higher in UHMWPE than in other types of PE [10]. The crystalline phase is reduced in relative quantity during the recrystallization process, and a third phase, called the interphase, is formed which displays characteristics that are a mix of the crystalline and amorphous phases [11, 12].

Vinyl chain terminal groups are present in the amorphous phase. Oxygen and nitrogen in the air can slowly diffuse into the amorphous phase and possibly also the interphase of UHMWPE.

The solubility of O_2 in polyethylene has been reported to be around 0.4–1 mmol/L, as derived from permeability values [13–15]. The concentrations of ROOH inside the orthopedic prostheses irradiated with different doses of γ-ray in O_2 saturated conditions have recently been determined constant through spectroscopic measurements after reaction with NO [16, 17]. It can be supposed that all the oxygen, present in the UHMWPE amorphous phase, reacted with the secondary alkyl macroradicals (R"°) and produce all oxidized products (bound and free ROOH, ketones, and acid, etc..). The concentration of the oxidized products is at least 5 mmol/kg [16, 17]; therefore, the concentration of O_2 is at least 5 mmol/kg.

The elimination of oxygen from a prosthetic component is a slow, also under high vacuum, and difficult process. All prosthetic components always contain oxygen prior to sterilization.

3.3 UHMWPE Sterilization

As with all implantable biomaterials, prosthetic components must be sterilized before implantation. The biomaterial may pick up pathological agents (bacteria, spores, or viruses) during its production, but this is only possible on the surface for prosthetic components made from UHMWPE. These components are produced at temperatures above 200 °C and diffusion into the polymer mass is not possible because of the dimensions of bacteria, meaning that only surface sterilization is necessary.

Sterilization can be carried out using a number of different processes [7]. The material can be treated with ethylene oxide or with plasma sterilization where radical species, obtained from the decomposition of peroxides by plasma, destroy the structure of any pathogenic agent present on the UHMWPE surface.

3.3.1 Interaction of High-Energy Radiation with UHMWPE and Macroradical Production

UHMWPE may be treated with radiation at each or both stages of its production: during cross-linking where the material is typically heat-treated above the peak melting point in order to eliminate the macroradicals present in the crystalline phase and/or interphase and, secondly, during prosthesis sterilization at the end of its manufacture.

The first interaction between the polymer and radiation is similar regardless of whether photons or electrons are used. In both cases, interactions will produce new electrons with similar energies, for which subsequent interactions with UHMWPE are equal and will form the same products [18–27]. They penetrate into the amorphous, interphase and crystal phases, cleave covalent bonds, and thus produce free radicals.

The homolytic cleavage of C-C and C-H bonds is endothermic and the initiation rate is a function of the amount of energy supplied to the chemical bonds, instead of the physical state of the polymer. High-energy photons, i.e., γ-ray or X-ray, and electrons (e-beam) possess energy of more than three orders of magnitude higher than the chemical bonds.

γ-Rays are photons which can penetrate the polymer mass by up to several meters, depending on the material characteristics. γ-Ray dose rates are normally in the order of kGy/h. Doses of around 25 kGy, which is the dose frequently used for the sterilization of polymeric materials, are needed to achieve significant changes in a polymer [28, 29], meaning that irradiation times of several hours are required. The rate of formation of radicals is generally constant during γ-irradiation [30–33]. It is possible to study the overall process of initiation, propagation, and termination of radical cycles by working at temperatures slightly above RT. Carlsson asserts that "γ-irradiation is a valuable method for triggering oxidation because it is very controlled as the rate of radical formation is constant during irradiation and halts immediately after irradiation stops" [31].

In contrast, electrons can only penetrate a few mm to a few cm into UHMWPE, as a function of their energy. They are typically produced at very high dose rates, in the order of 5–10 kGy/s, meaning that the irradiation times for a dose of 25 kGy are therefore only a few seconds.

E-beam technology allows the operator to achieve the required irradiation doses in just a few seconds and stop the production of macroradicals by halting the irradiation. Consequently, working at RT, where thermally formed products are stable and reactions in the amorphous phase are terminated [34], means that it is possible to study the transfer of R"° from the crystalline to the amorphous phases as well as the termination of the oxidation cycle.

The interactions of γ-photons and electrons with PE, which occur via complex energy transfer processes [18, 19, 26, 27], lead to the cleavage of C-C and C-H bonds (cleavages a and b in Scheme 3.1), in all three phases of UHMWPE, and either the formation of two primary macroradicals (R'°) (Scheme 3.1, Reaction 3.1) or a hydrogen atom (H°) and a secondary macroradical (R"°) (Scheme 3.1, Reaction 3.3). This second case can lead to the formation of trans-vinylene double bonds and

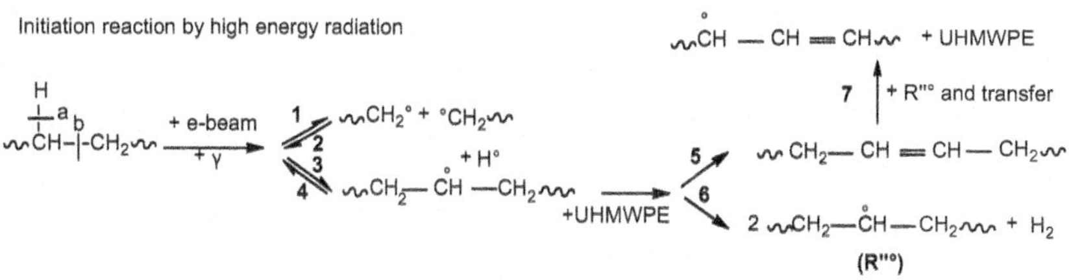

Scheme 3.1 Interaction of UHMWPE with high energy radiation

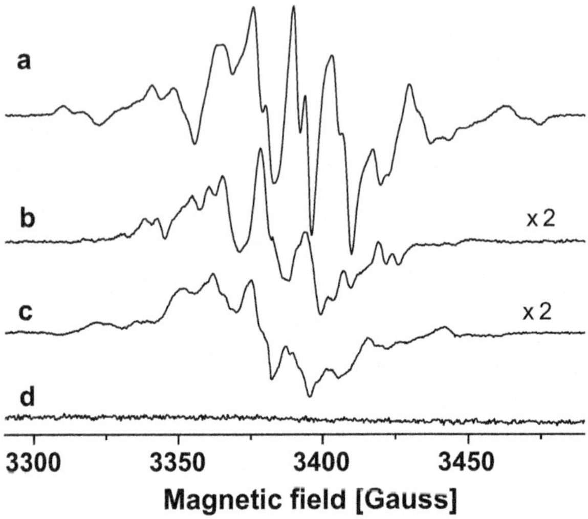

Fig. 3.2 EPR spectra of UHMWPE, HDPE, LDPE, and VLLDPE irradiated with e-beam in an inert nitrogen environment at 60 kGy. In order to compare the intensities, a multiplying factor was applied. (**a**) The radical R''° of UHMWPE and HDPE, 30 min after irradiation. (**b**), The Rall° present in the interphase or in the crystalline phase of the UHMWPE and HDPE in 24 h after the irradiation. (**c**) Rall°in the interphase or the crystalline phase of LDPE after 30 min from the end of irradiation. (**d**) Registered after 30 min from the term irradiation on the sample VLLDPE (Engage)

H_2 [35–38], that is, in a termination reaction (Scheme 3.1, Reaction5). Otherwise, the remaining macroradicals move within UHMWPE (Scheme 3.1, Reaction 6).

3.3.2 Evolution and Decay of Alkyl Macroradicals

Alkyl macroradicals, whether formed by irradiation or H° transfer, may be primary, secondary, tertiary or allyl macroradicals. The rate of their formation is a function of the concentration of the (CH_3, CH_2, and CH) groups present in the polymer chain, the energy of the bond, and the steric hindrance. In the case of UHMWPE, they are principally of the R''° (Fig. 3.2a) type because of the large excess of CH_2 groups, as compared to the CH_3 and CH.

R'° groups formed from the cleavage of C-C bonds in UHMWPE have only been detected at low temperatures [23, 39]. No increase in the presence of methyl groups is detected after the irradiation of UHMWPE, nor is the disproportionation of two primary radical species observed, which is probably due to the steric hindrance and consequently the poor mobility of chain ends. The only possible reaction is thus termination via the coupling of two macroradicals, which reforms a C-C bond (Scheme 3.1, Reaction 2) [40].

Tertiary alkyl macroradicals have recently been detected in UHMWPE [41, 42]. They are kinetically stable at room temperature because the subsequent abstraction of H is sterically hindered. Their formation may be due to the transfer of H from an R''° group to a tertiary carbon already present in the amorphous or interphase of the original polymer due to branches (short or long) on the chain. Otherwise, they may be formed via the consumption of vinyl double bonds, in the so-called "Y-type" cross-linking reaction (Scheme 3.4, Reactions 18, 19).

Allyl macroradicals ($R_{all}°$) (Figs. 3.2b, c) are formed via an intra- or intermolecular H-transfer from R''° to a vinylene double bond that is produced during irradiation (Scheme 3.1, Reaction 7), forming a $R_{all}°$ [35–38]. All spectra are detected after the irradiation of the PE amorphous phase [10] (Fig. 3.2d). It must therefore be assumed that they are formed in the crystalline or interphase regions. This leads us to assume that Reactions 5 and 7 in Scheme 3.1 occur in the crystal phases of UHMWPE but that $R_{all}°$ are stabilized by the steric hindrance in the crystalline and/or interphase regions. $R_{all}°$ are so stable that highly crystalline samples of UHMWPE must be heated to 90° before UHMWPE starts to melt and the allyl decay rate becomes significant [37, 38].

It must be noted that alkyl macroradicals are detectable after irradiation in both the amorphous phase and the crystalline phase, but only at −196 °C – the temperature of liquid nitrogen – whereas they are only detectable in the crystalline phase after relatively short irradiation times at RT.

In conclusion, the alkyl macroradicals formed during UHMWPE irradiation are principally of the R''° type. They are detectible in the crystalline phase after irradiation at RT, while they are found in the amorphous phase only if produced and detected at low temperatures. They are very reactive in the amorphous phase and have very short lifetimes: less than 10^{-3} s at RT. R'''° and $R_{all}°$, absent after irradiation in the amorphous phase, are most likely formed mainly in the interphase during the rapid migration of R''° from the crystalline to the amorphous phase [10].

3.3.3 Thermodynamic and Kinetic Stability of Macroradicals

Radicals (also called free radicals) are neutral atoms or molecules that possess an unpaired valence electron. The unpaired electron makes radicals chemically reactive, with few exceptions. Their freedom of movement in all directions means that free radicals have very short lifetimes in the gas phase or solution, typically less than 10^{-5} s, which indicates high thermodynamic and kinetic reactivity [23–27]. Kinetic stability, or persistence, is often more important than thermodynamic stability in determining radical lifetimes. We must also distinguish radicals obtained in gas, liquids, and solid matrices (whether crystalline, glassy, or composite). In solution, the so-called "stable" radicals are stabilized by their chemical structure, i.e., by resonance stabilization, with half-lives typically in the order of several seconds. Solid phase radicals, and in particular polymeric macroradicals, formed by irradiation and trapped in the crystalline phase, are much more persistent. Their persistence mainly depends on the steric protection of the unpaired electron, the degree of crystallinity, crystal dimensions and temperature [10, 43–49].

3.3.4 Transfer Mobility and Macroradical Hopping During the Irradiation Process

The small H° radicals formed during γ-ray and e-beam treatments have dimensions in the order of angstroms, meaning that they are also highly mobile in the crystalline phase. In fact, H° can move in all the phases where the polymer chains are organized and does so via a rapid transfer reaction and transformation into H_2, which proceeds through the abstraction of an H-atom from a polymer chain (less than 10^{-5} s) and the formation of R"° (Scheme 3.1, Reaction 5).

°OH radicals, produced during the oxidation process, can also move easily in the amorphous polymer phase, while their high reactivity and the high concentration of CH_2 bonds enable very quick reactions with the polymer, via the abstraction of an H atom, to form R"° and H_2O.

H° and OH° are therefore the species with the highest transfer speeds and play the key role in the degradation and oxidation phenomena of UHMWPE.

Chemical reactions take place in a solid polymer phase when the two reactants meet; that is they are at a distance of a few angstroms at which they can interact [50, 51]. The collision chance is related to the mobility, reactant dimensions, and the resulting steric hindrance. The mobility of saturated species in polyolefins in general and in our case, in UHMWPE, depends on the diffusion of molecules throughout the polymer and therefore the time it takes for two species to meet is a function of the diffusion rate.

After the irradiation of HDPE by γ-rays, by treating the generated H_2 with D_2, Dole has reported that HD is formed and that "the exchange probably involves the hydrogen atoms closest to the free radicals or on the nearest neighbor sites, and provides a new mechanism for free radical migration" [52].

Sohma [46] has reported on the irradiation process, "There is experimental evidence to indicate that free radicals are produced uniformly in a polymer in the first stage of γ-irradiated decay and preferentially in the amorphous regions ... These facts require that free radicals in polymer matrices migrate from the original site (crystalline phase) to the amorphous phase ... A molecular mechanism for radical migration was demonstrated via the analysis of the changes in the ESR spectrum of PE. An unpaired electron migrates in a polymer matrix just like the elastic collision of molecules in a gas phase until it is stabilized ...".

Although the possibility is often neglected, R"° in the crystalline phase may migrate very slowly [10], along a polymer chain, or between chains, via an inter- or intra-molecular H-transfer mechanism to form R_{all}°. It has been reported in the literature that only macroradical R"° may migrate within UHMWPE via the transfer ("hopping") of hydrogen [43, 44, 46].

Dole suggests that R"° transfer between two neighboring molecules is more likely than migration along the chain [18]. This hypothesis has been confirmed by calculation [49]. R"° groups take around 10 h to travel across the 25 nm thick UHMWPE crystallites [53]. The transfer of radicals has also been described using the Tune effect (hopping) [54].

The amorphous UHMWPE phase contains chain imperfections (end-groups, branches, other groups that have formed during the polymerization reaction or during induced degradation, for example, via interaction with photons or electrons) related to the polymer chain. Such groups have the possibility of rotational movement, but little translational freedom, so that they are fixed in position.

Oxygen and any additives added during processing are able to migrate within the amorphous phase, but are usually insoluble in the crystal phase and are excluded there. Migration is controlled by the diffusion coefficient (*D*), which is different in the three solid UHMWPE phases.

3.3.5 *Formation, Evolution, and Decay of Oxygenated Macroradicals (ROO° and RO°)*

Oxygen is able to migrate within the amorphous phase, but is usually insoluble in the crystal phase, meaning that it is excluded from the crystalline phase [55]. However, no information on the presence of oxygen in the interphase is currently available.

Peroxy macroradicals in UHMWPE are mainly formed through reaction, in the amorphous phase and the interphase, between oxygen and R"° (Scheme 3.5, Reaction 22), although this reaction is not thermodynamically favorable [56]. In fact, hydrogen abstraction has high activation energy (60 kJ mol^{-1}) [56, 57]. ROO° is generally stable when produced and stored at −78 °C. It reacts slowly, under heating, with the polymer chain via the abstraction of an H-atom to form ROOH + R"° during the propagation cycle (Scheme 3.5, reaction 22).

Some ROO• radicals are stable in HDPE and UHMWPE in spite of the high concentration of available hydrogen atoms. Their stability may be attributed to steric hindrance, instead of thermodynamic stability.

ROOH is one of the most important products in the oxidation of PEs and UHMWPE [30–34, 58]. The ROOH production reaction is possible, with abstraction, by means of an intramolecular cycle, as described for PE by Chien [63]. The decomposition of ROOH consists of two clearly resolvable processes [34, 59], but is not induced by radicals.

Peroxide radicals are not easy to identify despite of their half-life of 10 min at RT [2, 57]. Their EPR spectra overlap with those of polydienes and can be easily confused [31]. Reactions with ^{17}O have demonstrated that the species present during the irradiation process are generally ROO°, instead of polyenyl radicals [60]. However, EPR, UV, and FTIR analyses enable the identification of the species [61].

R"O° macroradicals can be formed during the interaction between a peroxide macroradical R"OO° and an R"° radical, leading to the formation of 2 R"O° (Scheme 3.2, Reaction 14) [54].

Literature studies demonstrate [62, 63] that the alkoxy radical can undergo beta-scission and thus form a primary alkyl radical and CO (Scheme 3.2, Reactions 9–10), which is found during the irradiation of UHMWPE when oxygen is present in the polymer. This reaction is in competition with H-atom abstraction to form alcohols. (Scheme 3.2, Reaction 8).

Reaction 10 in Scheme 3.2 causes cleavage of the polymer chain. The activation energy of the reaction is only 24–50 kJ mol^{-1}, which is surmountable at RT and much less than the energy of the C-C bonds (347 kJ mol^{-1}) in the polymer chain.

There is no possibility for the peroxy and alkoxy macroradicals linked to the polymer chain to transfer since their mobility is locally restricted across the polymer chain segments.

3.4 Cyclic Radical Processes During Polymer Interaction with High Energy Radiation

The degradation of polymer materials, in particular for polyolefins such as UHMWPE, involves cyclic radical chain reactions, which are classified as:

- Initiation produces hydrogen radicals (H°), small radicals (r°), alkyl radicals (R°) or alkoxide macroradicals (RO°) and °OH during the oxidation process, which reacts with UHMWPE to give R"°. The formation of radicals takes place via the homolytic breakage of a C-H or C-C bond. This is induced by the exposure of the polymer to high-energy photons, X-rays, and electrons. The process may be defined as radiation-induced degradation. In the presence of oxygen, the processes are defined as high-energy radiation-induced oxidation (see Sects. 3.3.1 and 3.3.2).
- Propagation, in which R"° reacts with polymer chains, chain imperfections, additives, and oxygen that meet as they move within the polymers, or along the polymer chain, forming a neutral macromolecule and/or a new R°, in a single or over two steps, which continues the propagation reaction.
- Termination, which involves the elimination of two radical species and the formation of a neutral product, which is possibly stable at RT.

If the degradation processes, especially oxidation processes, take place at temperatures below 40 °C, branching reactions may be skipped altogether due to the very slow thermal decomposition of hydroperoxides and peroxides [34, 64].

3.4.1 Termination in the Oxidation Process

When the initiation reaction, i.e., the production of R”°, ceases in the presence of O_2 as occurs during and after e-beam or γ-ray irradiation, following the post-oxidation process induced by macroradicals that come from the crystalline phase, the termination reaction will cause the oxidation propagation cycle for nonvolatile products to cease after some hours at RT [34]. The γ-ray irradiation-induced UHMWPE oxidation propagation process was found to have a constant rate; as the initial rate is constant, it follows that the termination rate is constant as well.

Radical termination requires the interaction of two macroradicals and produces non-radical products.

The main macroradicals present in the oxidation process are R”°, ROO° and R’CO°. The concentration of R’CO° is significantly lower than others because of its rapid decomposition in the presence of R’° (Scheme 3.2, Reaction 8–10).

R’CO° and R”OO° are fixed on the polymer chain and can also abstract H° from a polymer chain, forming alcohol or hydroperoxides, with the formation of a new R”° which continues the propagation reaction (Scheme 3.5, Reaction 21, 22).

The possible termination reactions are reported in Scheme 3.3.

The reaction (Scheme 3.3, Reaction 13) brings about the elimination of two alkyl radicals either via their combination, with the formation of a saturated product, or via disproportionation to give two non-radical products, one of which is saturated and the other unsaturated, as may occur in the termination of a polymerization reaction in solution. There are a number of results to show that R”° are stable for more than 15 h in HDPE and UHMWPE after irradiation with γ- rays [2] and e-beam in inert atmospheres, whereas they decay rapidly in the presence of O_2 [65]. Solid state ^{13}C NMR analyses of HDPE show that cross-linking is only of the Y-type, i.e.,

β scission alkoxy macroradicals

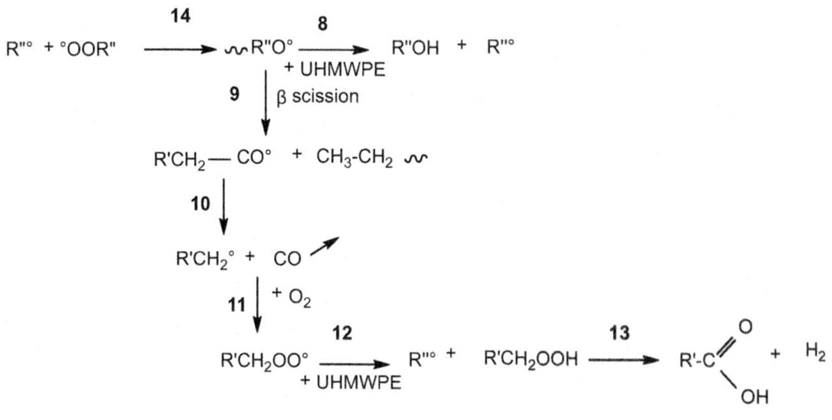

Scheme 3.2 Evolution of the alkoxy radicals

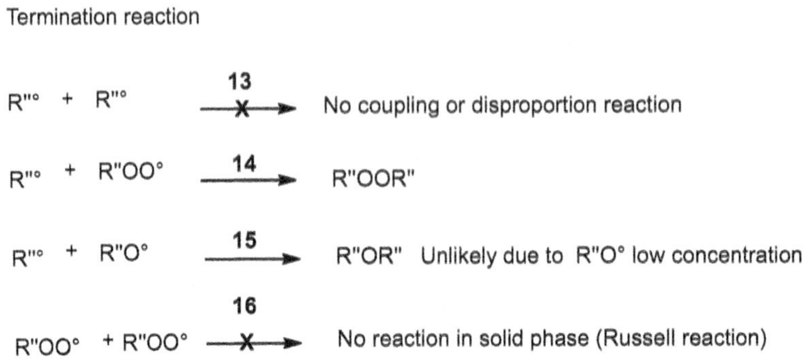

Scheme 3.3 Termination reactions

reactions with vinylic and vinylidene double bonds in the amorphous phase, but not of the X-type, i.e., via the coupling of two R''° macroradicals [40, 66–69].

This implies that the R''° radicals formed are not able to react with each other during the irradiation process at RT. The coupling reaction between R''°, which leads to their elimination, is therefore not feasible, at least for a radical population that is formed in this way.

The termination reaction for thermal oxidation (Scheme 3.3, Reaction 16) is generally accepted as a Russell reaction between two small peroxy radicals with a high diffusion rate in solution, which proceeds via unstable tetraperoxide [70, 74] with the production of one molecule of ketone and one alcohol.

In a more recent article [71], a new citation is reported [72], which highlights the fact that the three termination reactions in solution (Scheme 3.3, Reactions 13, 14, 16) become possible in function of the O_2 pressure exerted on the alkyl radicals produced by 2-6 dimethyl hepta 2-5 diene. Reaction 6 in Scheme 3.3 is possible in solution but not in the solid phase at low O_2 pressure, up to 5 mm Hg. The reaction is easier at high pressures above 100 mm Hg (Scheme 3.3, Reaction 14), and the Russell reaction (Scheme 3.3, Reaction 16) is more likely in the presence of oxygen alone [73].

A number of works have reported that the relative immobility and stability of the peroxy radicals makes the bimolecular Russell termination strongly unfavorable in the solid state at RT [30, 56, 60, 72].

The macroradicals of HDPE in vacuum after irradiation decay slowly, while the macroradicals alkyl decay fast in the presence of oxygen with the same trend of macroradicals peroxidic with possible formation of peroxy groups [74]. It is more likely that the termination occurs between the peroxy radical, fixed on the polymeric chain, and the alkyl macroradical which migrates rapidly through the polymeric chain, with the formation of dialkylperoxides. This reaction is not kinetically inhibited and is thermodynamic feasible. Nevertheless, it is important to note that Carlsson wrote "as yet there is no truly reliable analytical method to detect the low level of dialkyl peroxides expected from polyolefin oxidation" [31]. PEs that have been irradiated with γ -rays in air display a significant concentration of peroxides [75].

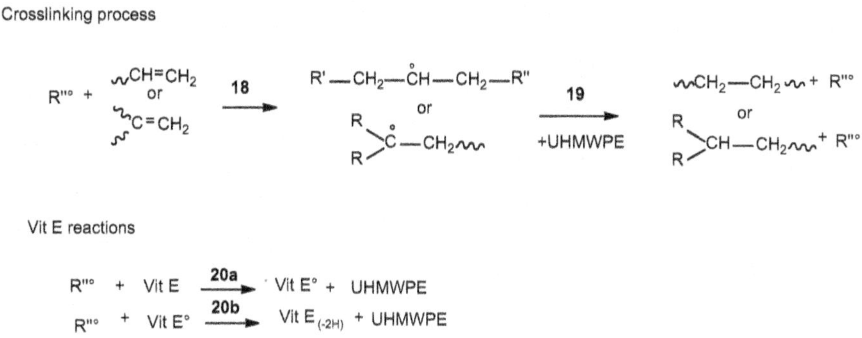

Scheme 3.4 Cross-linking and additive reactions with R"°

The processes behind the thermal decomposition of peroxides are complex and do not always produce radical species [76]. It has been reported that the decomposition of peroxides can lead to the formation of esters via a mechanism that does not involve radical species. This reaction will reduce the rate of initiation under conditions where hydroperoxide decomposition to radicals is concurrent since esters are thermally stable products [77].

3.4.1.1 Termination Reaction of the Oxidation Process in the Presence of a Stabilizer

When a phenolic stabilizing additive as Vitamin E (VitE) is used, R"° reacts with it [40, 78–80] to form a stable or permanent radical of that additive (VitE°). (Scheme 3.4, Reaction 20a) The R"° radicals may then react with the radical VitE°. The coupling reaction is possible because there is no steric hindrance, otherwise the abstraction of a second H atom from the additive molecule may occur with the formation of a double bond (Scheme 3.4, Reaction 20 b) [40]. In this case, there is a decrease in the amount of radical species R"° with the effect of permanent stabilization. In some cases, VitE (-2H) can react with R"° and bond permanently to the polymer in a termination reaction [81, 82].

3.4.2 Propagation Reactions

The propagation reaction is the key reaction in the cyclic radical process; R"° reacts with chain imperfections, with oxygen and with the additives present in the polymer, forming a saturated molecule and another macroradical species in one step, or ROOH in two.

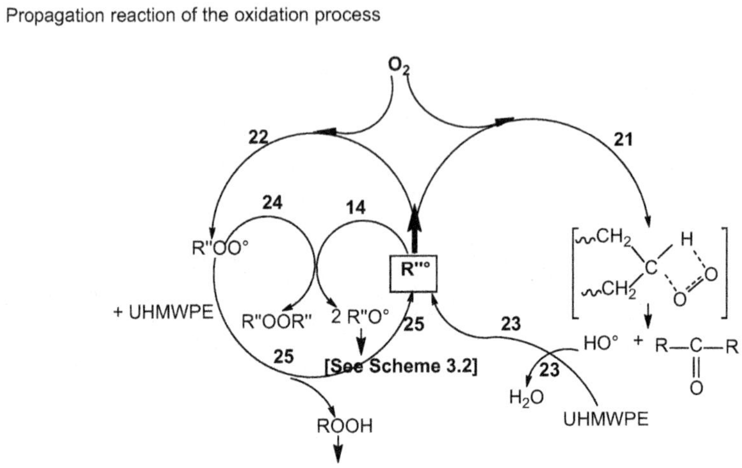

Scheme 3.5 The propagation reactions between oxygen and secondary macroalkyl radicals

The possible, competing propagation reactions are as follows:

R"° + (double bonds in the chain) → cross-link + new radical (Scheme 3.4, Reaction 18)

R"° + O_2 → oxidation products + free radical (Scheme 3.5, Reactions 21 and 22)

R"° + ADH → R"H + AD° and R"° + AD° → R"H + $ADH_{(-2H)}$ (Scheme 3.4, Reactions 20a and 20b)

The rates of these steps can be described as follows:

$V_1 = k_{18}$ [R"°] [double bonds]
$V_2 = k_{21}$ (or k_{22}) [R"°] [O_2]
$V_3 = k_{20a \text{ and } 20b}$ [R"°] [ADH]

Reactions 18, 21, and 22 are exothermic and thermodynamically feasible.

Reactions 20a and 20b are an H-atom abstraction. It is a process which occurs rapidly at RT and is thermodynamically feasible [40, 81–83].

3.4.2.1 Propagation Reaction with Chain Imperfections, Vinyl, Vinylidene, and Vinylene Double Bonds

When UHMWPE is irradiated during the sterilization or cross-linking processes, the vinyl double bonds that are fixed on the polymer chain in the amorphous phase, monosubstituted and non-sterically hindered, immediately decrease in number via an addition reaction with the formation of cross-linkages, according to the number of double bonds and irradiation dose (Scheme 3.4, Reactions 18–19). The reaction, which is similar to polymerization via an addition to a double bond, is exothermic. The same reaction can take place, although less efficiently, even with 1,1-substituted vinylidene double bonds, despite them being partially sterically hindered However, this is prevented for the 1,2- di-substituted vinylene double bonds, which are formed during irradiation, for steric reasons [84].

During irradiation, it is possible to detect the occurrence of decreasing numbers of double bonds and the corresponding increase in cross-linkages up until all the vinyl double bonds are consumed [40].

3.4.2.2 Propagation Reaction in the Presence of Stabilizing Additives

Molecules are generally chosen to be stabilizing additives because they contain easily abstractable H-atoms and produce AD° radicals which are stable toward reinitiation. Klemchuk [81] has reported on the transformation of a number of additives present in PP after irradiation with gamma-rays in the presence of air.

Typical phenolic chain-breaking antioxidants of the ADH type, such as vitamin E, can react with the R"° produced by γ-ray irradiation and e-beams, as well as with natural polyphenols, both in the presence and absence of oxygen, as reported in the literature (Scheme 3.4, Reaction 20) [81, 83]. The disappearance of vitamin E is slightly more effective in an inert environment than in the air. This decrease can be driven by the termination reaction caused by the presence of oxygen (Scheme 3.4, Reaction 20a and 20b).

However, it is only a propagation reaction if the AD° radical is able to reinitiate R"° formation; otherwise, if AD° reacts again with R"°, it is a termination (or retardation) reaction, as in Scheme 3.4, Reaction 20b.

3.4.3 Reaction with Oxygen: Oxidation

Bolland [85] studied the oxidation of hydrocarbons in solution at the end of the 1940s and this study has been applied to the polymer oxidation processing then. Essentially, it is believed that hydroperoxides are the primary products. These then decompose, leading to the formation of all other oxidized products, to the branching of the kinetic chain and the consequent auto-acceleration of oxidation. As described earlier, this assumes that oxidation takes place under conditions where hydroperoxides are sufficiently instable toward decomposition at finite rates. This may not be entirely valid for oxidation in the condensed phase of UHMWPE at low temperatures and particularly for oxidation when induced by γ-ray or e-beam irradiation.

The oxidation behavior of products formed during the γ-irradiation of LLDPE in air [30, 31, 86], after the γ-irradiation of HDPE [31–33], and after the e-beam irradiation of several PEs [10, 34], has all been studied. All the processes were conducted at temperatures below 40 °C, at which temperature hydroperoxides are thermally stable [10, 31–34].

In those papers [10, 31–34], the formation rate of all species during irradiation was found to be proportional to the imparted dose of γ-rays and thus to the amount of macroradical products. It follows that, as expected, all oxidized species are produced in proportion to [R"°], instead of [ROOH].

The rate of product formation after e-beam irradiation was always proportional to R"° [10, 34]. However, since there was no on-going initiation, only the transfer of R"° from the crystalline phase, the formation rate reduced over time as the termination reaction was on-going and the same relationship between species was maintained.

Oxygen has measurable mobility, via simple diffusion, within the polymeric mass. Oxygen and the polymeric macroradical (R°) meet in the amorphous phase, and perhaps in the interphase, thanks to the high transfer rate of R"° and oxygen slow mobility, as described by its diffusion coefficient (Scheme 3.5, Reactions 21 and 22). All reactions are exothermic and there are no steric hindrance problems as oxygen is a small molecule, which allows for reactions with R"°.

In Scheme 3.5, Reactions 21 and 22 show the possible reactions between R"° and oxygen to form ketones, ROO°, and a new R"°.

When LLDPE and HDPE are irradiated with γ-rays in air, in an oxygen excess [10, 31–34, 86], the dose rate is constant as is the oxidized product formation rate during irradiation, as long as the temperature is kept low enough that the kinetic chain branching through hydroperoxide decomposition is negligible. Consequently, the rate of production of R"° is constant, and v_{in} (the initiation rate) is the same as long as irradiation is maintained.

The reactions are therefore a function, not of the distribution of species (for example ROOH), but of the high transfer rate of the polymer macroradical R"°.

The rate of a chemical process, regardless of the phase where it takes place, is proportional to the concentration of the species which reacts and can be described by the Arrhenius relationship.

Arrhenius extrapolations assume that a chemical reaction is controlled by the reaction rate constant *k*, given by

$$k = Aexp^{(-Ea/RT)}$$

where E_a is the Arrhenius activation energy, *R* the gas constant, *T* the absolute temperature and *A* the pre-exponential factor which measures the likelihood and effectiveness of an encounter between reacting species. The E_a of radical processes is minimal, [56] while the pre-exponential factor is a function of the movement rate of the two species and their collision cross-section, which is influenced by steric hindrance, and therefore the possibility of an encounter.

Two important parameters to consider here are the pre-exponential factor, *A*, which is the possibility of an effective collision occurring between the two reacting species, R"° and the other species, and E_a that gives the thermodynamic ease of reaction.

Once consumed locally, oxygen penetrates into the polymer from the air outside according to diffusion laws. The O_2 concentration will vary, according to distance from the external surface, if oxygen is consumed faster than it can migrate. For this reason, oxidation in the prosthesis is non-uniform on many occasions [16, 17, 22].

The polyethylene oxidation products reported in the two conditions, γ-ray and e-beam irradiation, are the same with only some slight variations in concentration: unbound and bound hydroperoxides, ketones, and, to a lesser extent, acids, alcohols, esters, and volatile products (H_2O, H_2, CO, and CO_2) [32, 86, 87].

In contrast to the work by Bolland [85], all oxidation products are achievable without the decomposition of hydroperoxides and are formed solely by reactions between R"° and oxygen and/or via the subsequent product decomposition which even occurs at RT.

Prosthetic components are tested in order to evaluate the oxidative stability of UHMWPE after treatment with gamma rays [88]. The designation gives emphasis to the fact that the method will permit an investigator to compare the oxidative stability of different UHMWPE materials, but it is recognized that this method may not always precisely simulate the degradation mechanisms for an implant during real-time shelf aging and implantation.

3.5 Oxidation Process Critical Products

R"° reacts in the presence of O_2 at RT and forms all oxidation products without activation energy issues. In fact, activation energy is exceeded at RT, but it is likely that different products will form at different rates simply because activation energies are not the same for all reactions, meaning that the ratios between the various species depend on the oxidation reaction temperature.

3.5.1 Ketones

Reaction 21 in Scheme 3.5 shows the formation of ketones from the reaction between R"° and oxygen. OH° is also formed and immediately forms a molecule of water and an R"° upon reaction with UHMWPE. It has been proven that ketones and water form in equimolar amounts [89].

Furthermore, since the process takes place without further chain scissions, it does not induce any variation in molecular mass and only affects minimum changes on the mechanical properties of UHMWPE.

3.5.2 Hydroperoxides

R"° reacts with O_2 to form a peroxy radical. The half-life of hydroperoxide during the thermal degradation of PE at 120 °C is around 30 min in the solid state [34, 59].

R"OOH species are present in the free state and in the bonds with hydroperoxides, ketones, acid, and esters. Secondary oxidized products are consequently formed if more oxygen is available. In contrast, the formation of free hydroperoxide occurs if the initial alkyl macroradical migrates along the polymer chain before reaction with oxygen. The hydrogen-bonded vs. non-hydrogen-bonded hydroperoxide ratio is constant in post-irradiation oxidation, as induced by e-beam irradiation. Therefore, it does not depend on alkyl macroradical concentration or on irradiation dose.

3.5.3 Carboxylic Acids

Acids are produced via mechanisms that have not yet been elucidated [67, 87]. The trends of their formation in the oxidation of UHMWPE appear similar to those reported for ketones and hydroperoxides. Acid formation must involve the cleavage of the polymer chain. Direct splitting of the C-C bonds in the polymer chain at RT is highly unlikely as the bond energy of C-C bonds is at least 320 kJ/mol for the weakest link, the C-C alpha to the double bond. It is thus necessary to propose the breaking of a C-C bond in some species in which the bond energy is low.

Some reactions for the formation of acids are postulated in the literature. In photo-oxidation, acids are the most important products. Photo-oxidation also involves Norrish I photo-reactions, which lead to the formation of an alkoxy radical [67].

The formation of R'° can be induced by the decomposition of an alkoxy macroradical, as seen previously (Scheme 3.2, Reactions 14, 9, and 10). The oxidation of the primary macroradical and subsequent primary hydroperoxide decomposition can then lead to the formation of acids [67, 87].

3.5.4 Sec-Alcohol (R_2CHOH)

Alcohols can be formed via H abstraction, from the polymer chain by an alkoxy radical (Scheme 3.2, Reaction 8), which competes with the β scission reaction. Their presence can be determined by FTIR after their transformation into nitrites by reaction with NO [90, 91]. They are found in oxidation processes, but always at low concentrations.

3.5.5 Esters

Ester and lactone formation occurs during oxidative degradation, although the precise mechanism is not completely clear. These species can be formed during mechanical degradation, e.g., during the microtoming of thin films from UHMWPE rods. Costa et al. have shown that esters can be produced by the decomposition of macroalkyl peroxides that are formed during cutting, in a process which does not produce macroradicals [76, 77].

3.6 Conclusion

Most discussions of oxidation are concerned with the technologically important conditions of the long-term in vivo, where polymer lifetimes may well be in the order of decades. UHMWPE components generally undergo accelerated tests which provide information about the condition of the prosthesis and/or the efficiency of an additive over a short period of time.

When UHMWPE is tested under real conditions and is irradiated with high energy radiations, it produces the same oxidation products as when it undergoes the accelerated tests. The oxidation process is the same, but the test conditions are very different, in particular, the temperature is at least 30 °C higher. The mechanisms of initiation in these conditions involve the thermal decomposition of hydroperoxide traces and subsequent branching reactions, which are virtually impossible in the in vivo oxidation process.

In the radiation-induced oxidation of UHMWPE, R"° that is formed in the crystalline phase during irradiation can move relatively quickly between polymer chains and also into the interphase and/or amorphous phase. During this process, they can react with vinylene double bonds produced during irradiation and form allyl macroradicals in the crystalline and interphase which are stable for a long periods of time due to resonance stability and steric hindrance.

Alkyl macroradicals in the amorphous phase, whether migrating from the crystalline regions or directly formed there by radiation, may encounter reactive imperfection groups in the chains, particularly vinyl double bonds, which take part in propagation reactions leading to cross-linking and the formation of new R"°.

R"° may also react with oxygen in conventional propagation reactions. Oxygen is consumed, while ketones, hydroperoxides and oxidized products are produced along with new R"°. The consumed oxygen can be replaced by diffusion from the outside. This process is linked to sample thickness and oxygen diffusion in the polymer.

Typical stabilizing additives, of the H-acceptor type, react principally with alkyl and peroxy radicals. The balance here depends on the relative concentrations and the mobility of the R"OO° and R° species. In an effective stabilizer, the initial AD° product for interaction with R" is too stable to reinitiate the oxidation cycle, although it usually reacts further with a radical in a termination step and forms new products.

Antioxidant additives that are present in our body, such as vitamin E, are ideal for stabilizing prosthetic UHMWPE. The mechanism for antioxidant stabilization is generally the same in the body's normal mechanisms and in vivo, because degradation products are the same and are also biocompatible.

In conclusion, the best sterilization systems for prosthetic UHMWPE are ethylene oxide and plasma gas, and these are even better if stabilized against oxidation with vitamin E.

References

1. Li S (1998) In: Callaghan JJ, Rosenber A, Rubash HE (eds) Polyethylene. The Adult Hip Lippincott-Raven Publishers, Philadelphia, pp 105–122
2. Ohnishi S, Sugimoto S, Nitta I (1963) Electron spin resonance study of radiation oxidation of polymers. IIIA. Results for polyethylene and some general remarks. J Polym Sci Part A Polym Chem 1:605–623
3. Nusbam HJ, Rose RM (1979) The effects of radiation sterilisation on the properties of UHMWPE. J Biomed Mater Res 13:557–576
4. Costa L, Luda MP, Trossarelli L, Brach del Prever EM, Crova M, Gallinaro P (1998) Oxidation in orthopaedic UHMWPE sterilized by gamma-radiation and ethylene oxide. Biomaterials 19:659–668
5. Costa L, Bracco P, Brach del Prever E, Luda MP, Trossarelli L (2001) Analysis of products diffused in UHMWPE prosthesis components *in vivo*. Biomaterials 22:307–315
6. https://www.celanese.com/engineered-materials/products/GUR-UHMW-PE/What-is-UHMW-PE.aspx
7. www.UHMWPE.org
8. http://www.orthoplastics.com/
9. http://www.quadrantplastics.com/eu-en/products/meditech-R-life-science-grades/implantable-polymers.html
10. Carpentieri I et al (2011) Post-irradiation oxidation of different polyethylenes. Polym Degrad Stab 96:624–629
11. Peacock AJ (2000) Chapter 4: Morphology and crystallization of polyethylene. In: Handbook of polyethylene. Structures, properties and applications, 67th edn. Marcel Dekker, New York
12. Barron D, Birkinshaw C (2008) Ultra-high molecular weight polyethylene – evidence for a three-phase morphology. Polymer 49(13–14):3111–3115
13. Michaels AS, Bixler HJ (1961) Solubility of gases in polyethylene. J Polym Sci 50:393–412
14. Pauly S (1999) Permeability and diffusion data. In: Brandrup J, Immergut EH, Grulke EA (eds) Polymer handbook IV ed. Wiley, New York, p IV/545
15. Billingham NC (1990) Physical phenomena in the oxidation and stabilisation of polymers. In: Pospisil J, Klemchuck PP (eds) Oxidation inhibition in organic materials. CRC Press, Boca Raton, pp 249–298
16. Bracco P, Brach del Prever EM, Cannas M, Luda MP, Costa L (2006) Oxidation behaviour in prosthetic UHMWPE components sterilised with high energy radiation in a low oxygen environment. Polym Degr Stab 91:2030–2036
17. Bracco P, Brunella V, Luda MP, Brach del Prever EM, Zanetti L, Costa L (2006) Oxidation behaviour in prosthetic UHMWPE components sterilised with high energy radiation in the presence of oxygen. Polym Degr Stab 91:3057–3064
18. Dole M (ed) (1972) The irradiation chemistry of macromolecules, Vol I. Academic, New York
19. Dole M (ed) (1973) The irradiation chemistry of macromolecules, Vol II. Academic, New York
20. Makhlis FA (ed) (1975) Radiation physics and chemistry of polymers. Willey, New York
21. Pospisil J, Klemchuck PP (eds) (1990) Oxidation inhibition in organic materials, vol II. CRC Press, Boca Raton
22. Clegg D, Collyer A (eds) (1991) Irradiation effects on polymers. Elsevier Applied Science, New York
23. Ranby B, Rabek JF (eds) (1977) ESR spectroscopy in polymer research. Springer, Berlin
24. Ivanov VS (ed) (1992) Radiation chemistry of polymers. VSP Utrech, Utrech
25. Woods RJ, Pikaev AK (eds) (1994) Applied radiation chemistry. Radiation processing. Wiley, New York
26. Carlsson DJ, Chmela S (1990) Polymer and high-energy irradiation: degradation and stabilisation. In: Scott G (ed) Mechanism of polymer. Degradation and stabilisation. Elsevier Applied Science, London, pp 109–134

27. Carlsson DJ (1993) Degradation and stabilisation of polymers subjected to high energy radiation. In: Scott G (ed) Atmospheric oxidation and antioxidants. Elsevier, Amsterdam, pp 495–528
28. Harrison N (1991) Radiation sterilisation and food packaging. In: Clegg D, Collyer A (eds) Irradiation effects on polymers. Elsevier Applied Science, New York, pp 319–344
29. Halls NA (1991) Gamma-irradiation processing. In: Clegg D, Collyer A (eds) Irradiation effects on polymers. Elsevier Applied Science, New York, pp 253–296
30. Decker C, Mayo FR, Richardson H (1973) Aging and degradation of polyolefins. III. Polyethylene and ethylene–propylene copolymers. J Polym Sci, Polym Chem Ed 11:2879–2898
31. Mallego LJ, Carlsson DJ, Deschenes L (2001) Post-γ-irradiation reactions in vitamin E stabilised and unstabilised HDPE. Nuclear Instr Method Phys Res B 185:283–297
32. Lacoste J, Carlsson DJ (1992) Gamma-, photo-, and thermally-initiated oxidation of linear low density polyethylene: a quantitative comparison of oxidation products. J Polym Sci A Polym Chem 30(3):493–500
33. Mallégol J, Carlsson DJ, Deschenes L (2001) Antioxidant effectiveness of vitamin E in UHPDPE and tetradecane at 32°C. Polym Degrad Stab 73:269–280
34. Costa L, Carpentieri I, Bracco P (2008) Post electron-beam irradiation oxidation of orthopaedic UHMWPE. Polym Degrad Stab 93:1695–1703
35. Wen WY, Johnson DR, Dole M (1974) Radiation chemistry of polyethylene XII. Alkyl radical decay kinetics in single crystalline and extended-chain samples of linear polyethylene. Macromolecules 7:199–204
36. Williams TF, Dole M (1958) Irradiation of polyethylene. III. Influence of temperature and phase physical and inorganic chemistry. J Am Chem Soc 81:2919–2926
37. Waterman DC, Dole M (1970) The radiation chemistry of polyethylene. X kinetics of the conversion of alkyl to ally free radicals. J Phys Chem 74:1913–1922
38. Patel VM et al (1978) Radiation chemistry of polyethylene. XIV Allyl radical decay kinetic in different type of polyethylene. J Poly Sci Polym Phys Ed 16:467–458
39. Igarashi M (1983) Free-radical identification by ESR in polyethylene and nylon. J Polym Sci Part A-1, Polym Chem 21:2405–2425
40. Costa L, Bracco P (2016) Mechanism of crosslinking and oxidative degradation and antioxidant stabilization of UHMWPE. In: Kurtz S (ed) UHMWPE biomaterials handbook, 3rd edn. Elsevier, Amsterdam, pp 467–487
41. Brunella V, Paganini MC (2011) Electron beam radiation effects on UHMWPE: an EPR study. Magn Reson Chem 49:562–569
42. Paganini MC, Brunella V, Chiesa M (2012) HYSCORE and Davies ENDOR study of irradiated ultra high molecular weight polyethylene. Magn Reson Chem 50:615–619
43. Nara S, Shimada S, Kashiwabara H, Sohma J (1968) Relation of the decay of free radicals in irradiated polyethylene to the molecular motion of the polymer and the configurations of the free radicals. J Polym Sci Part A-2 6:1435–1449
44. Butiagin PJ (1972) The decay of free radicals in polymer media. Pure Appl Chem 30:57–65
45. Grillee D, Ingold KU (1976) Persistent carbon-center radicals. Acc Chem Res 9:13–19
46. Sohma J (1983) Radical migration as an elementary process in degradation. Pure Appl Chem 55:1595–1961
47. Hicks RG (2007) What's new in stable radical chemistry? Org Biomol Chem 5:1321–1338
48. Brustolon M, Giamello E (2009) Electron paramagnetic resonance. A practitioner's toolkit. Wiley, Hoboken
49. Jones RA (1998) A note on H-atom transfer and alkyl radical migration in polyethylene crystallites: MNDO saddle-point energies in a model n-heptane crystal. Radiat Phys Chem 53:19–23
50. Waite TR (1958) General theory of bimolecular reaction rates in solids and liquids. J Chem Phys 28:103–109

51. Williams F (1972) Early process in radiation chemistry and the reaction of intermediates. In: Dole M (ed) The irradiation chemistry of macromolecules. Vol I. Academic, New York, pp 21–29
52. Dole M, Böhm GGA, Waterman DC (1969) Mechanism and kinetics of radical reactions in polymers. Eur Polym J 5(Supp):93–104
53. Gray HB, Winkler JR (2003) Electron tunneling through proteins. Q Rev Biophys 36:341–372
54. Lazar M, Rychly J, Klimo V, Pelikan P, Valko L (1989) Free radicals in chemistry and biology, Chapter 3. CRC Press, Boca Raton, p 51
55. Winslow FH, Hellman MU, Matreyek W, Stills SM (1966) Autoxidation of semicrystalline polyethylene. Polym Eng Sci 6:273–278
56. Gryn'ova G, Hodgson JL, Coote ML (2011) Revising the mechanism of polymer autooxidation. Org Biomol Chem 9:480–490
57. Ingold KU (1969) Peroxy radicals. Acc Chem Res 2:1–9
58. Carlsson DJ, Bazan G, Chmela S, Wiles DM, Russell KE (1987) Oxidation of solid polyethylene films: effects of backbone branching. Polym Degrad Stabil 19:195–292
59. Chien J (1968) Polymer reactions. II. Thermal decomposition of polyethylene Hydroperoxides. J Poly Sci Part A 6:375–379
60. Alam T:M, Celina M, Assink RA, Clough RL, Gillen KT (2001) ^{17}O NMR investigation of oxidative degradation in polymers under γ-irradiation. Radiat Phys Chem 60:121–127
61. Kasser MJ, Silverman J, Al-Sheikhly M (2010) EPR simulation of Polyenyl radicals in ultrahigh molecular weight polyethylene. Macromolecules 43:8862–8867
62. Wypych G (2015) Handbook of UV degradation and stabilization, 2nd edn. Chem Tec Publishing, Toronto
63. Hartley GH, Guillet JE (1968) Photochemistry of ketone polymers. I. Studies of ethylene-carbon Monoxide copolymers. Macromolecules 1:165
64. Lacoste J, Carlsson DJ, Falicki S, Wiles DM (1991) Polyethylene hydroperoxide decomposition products. Polym Degrad Stab 34:309–325
65. Seguci T, Tamura N (1973) Mechanism of decay of alkyl radicals in irradiated polyethylene on exposure to air as studies by electron spin resonance. J Phys Chem 77:40–44
66. Randall JC (1988) Carbon13 NMR gamma-irradiated polyethylene. In: Guven O (ed) Crosslinking and scission in polymers. Kluwer Academic Publishers, Dordrecht, pp 57–76
67. Perez E, Vanderhart DL (1983) A ^{13}C CP-MAS NMR study of irradiated polyethylene. J Polym Sci B Polym Phys 26:1979–1993
68. Smedberg A, Hjertberg T, Gustafsson B (1997) Crosslinking reactions in an unsaturated low density polyethylene. Polymer 38:4127–4138
69. Bracco P, Brunella V, Luda MP, Zanetti M, Costa L (2005) Radiation-induced crosslinking of UHMWPE in the presence of co-agents: chemical and mechanical characterization. Polymer 46:10648–10657
70. Russel JA (1957) Deuterium-isotope effects in the autoxidation of alkyl hydrocarbon. Mechanism of the interaction of peroxy radicals. J Am Chem Soc 79:3871–3877
71. Russell JA (1959) Fundamental process of autoxidation. J Chem Edu 36:111–118
72. Bateman L, Morris AL (1953) The autoxidation of 2,6 dimethylhepta 2,5 diene. Trans Faraday Soc 49:1026–1032
73. Alam TM, Celina M, Collier JP, Currier BH et al (2004) γ-irradiation of ultrahigh-molecular-weight polyethylene: electron paramagnetic resonance and nuclear magnetic resonance spectroscopy and imaging studies of the mechanism of subsurface oxidation. J Polym Sci A Polym Chem 42:5929–5941
74. Onishi S, Sugimoto S, Nitta I (1963) Electron spin resonance study of radiation oxidation of polymer. IIIA. Results for polyethylene and some general remarks. J Polym Sci Part A 1:605–623
75. Niki E, Decker C, Mayo FR (1973) Aging and degradation of polyolefins. I. Peroxide-initiated Oxidations of Atactic Polypropylene. J Polym Sci Polym Chem Ed 11:2813–2845

76. Sanchez J, Myers TN (2000) Peroxides and peroxide compounds, organic peroxides. Kirk-Othmer encyclopedia of chemical technology. Wiley, New York, pp 1230–1235
77. Costa L, Luda MP, Trossarelli L (1997) Ultra high molecular weight polyethylene: I. Mechano-oxidative degradation. Polym Degrad Stab 55:329–338
78. Klemchuk PP (1993) Protecting polymers against damage from gamma radiation. Radiat Phys Chem 41(1–2):165–172
79. Breese KD, Lamèthe JF, DeArmitt C (2000) Improving synthetic hindered phenol antioxidants: learning from vitamin E. Polym Degrad Stab 70(1):89–96
80. Bracco P, Brunella V, Zanetti M, Luda MP, Costa L (2007) Stabilisation of ultra-high molecular weight polyethylene with vitamin E. Polym Degrad Stab 92(12):2155–2162
81. Costa L, Carpentieri I, Bracco P (2009) Post electron-beam irradiation oxidation of orthopaedic Ultra-High Molecular Weight Polyethylene (UHMWPE) stabilized with vitamin E. Polym Degrad Stab 94(9):1542–1547
82. Badertscher RP, Lerf R, Delfosse D, Adlhart C (2012) Grafting of a-tocopherol upon γ-irradiation in UHMWPE probed by model hydrocarbons. Polym Degrad Stab 97(11):2255–2261
83. Oral E, Neils AL, Rowell SL, Lozynsky AJ, Muratoglu OK (2013) Increasing irradiation temperature maximizes vitamin E grafting and wear resistance of ultrahigh molecular weight polyethylene. J Biomed Mater Res Part B 101:436–440
84. Shinde A, Salovey R (1985) Irradiation of ultrahigh-molecular-weight polyethylene. J Polym Sci A, Polym Phys 23(8):1681–1689
85. Bolland JL (1949) Kinetics of olefin oxidation. Q Rev Chem Soc 3(1):1–21
86. Matsuo H, Dole M (1959) Irradiation of polyethylene IV. Oxidation effects. J Phys Chem 63:837–842
87. Costa L, Luda MP, Trossarelli L (1997) Ultra high molecular weight polyethylene – II. Thermal- and photo-oxidation. Polym Degrad Stab 58(1–2):41–54
88. Designation: F 2003 – 02. Standard practice for accelerated aging of ultra-high molecular weight polyethylene after gamma irradiation in air
89. Petruj J, Marchal J (1980) Mechanism of ketone formation in the thermooxidation and radiolytic oxidation of low density polyethylene. Radiat Phys Chem 16:27–36
90. Carlsson DJ, Brousseau R, Zhang C, Wiles DM (1987) Polyolefin oxidation: quantification of alcohol and hydroperoxide products by nitric oxide reactions. Polym Degrad Stab 17(4):303–318
91. Carlsson DJ, Brousseau R, Can Z, Wiles DM (1988) Identification of products from polyolefin oxidation by derivatization reactions. Chemical reactions on polymers. American Chemical Society, Washington, DC, pp 376–389

Chapter 4
Natural Polyphenol-Stabilized Highly Cross-Linked UHMWPE for Joint Implants

Jun Fu

Abstract Radiation cross-linked ultra-high molecular weight polyethylene (UHMWPE) with high mechanical strength, wear resistance, and oxidative stability is critical for the long-term performance and life span of joint implants. The use of antioxidants and/or radical scavengers has proven efficient in stabilizing cross-linked UHMWPE against oxidation, whereas challenges remain to explore new methods to offer not only oxidative stability but also superior wear resistance and mechanical properties. This chapter introduces the use of natural polyphenols with three hydroxy groups to stabilize irradiated UHMWPE. Dodecyl gallate (DG) and gallic acid (GA) are blended with UHMWPE and consolidated by compression moulding prior to e-beam irradiation. The polyphenols show a slight phenol loss and offer improved oxidation stability. Such a strong antioxidation potency even shows a significant protection against the oxidative challenges. The antioxidation mechanisms have been investigated. The tensile and impact properties of these polyphenol-stabilized highly cross-linked UHMWPE after accelerated ageing in accordance with ASTM (American Society of Testing Materials) F2003 are superior to those of the irradiated and remelted UHMWPE. Pin-on-disc (POD) wear tests of these materials demonstrate low wear comparable to highly cross-linked and remelted UHMWPE. The oxidation and antioxidation mechanisms with the presence of polyphenols are investigated by analysing the oxidation products and oxidation kinetics. The antioxidants are potent to protect the polymer from oxidation with the presence of unsaturated lipids, which is encouraging for the prevention of the in vivo oxidation of UHMWPE implants related to the synovial lipids.

Keywords Polyphenol · Oxidation · Antioxidation · Stabilization · Squalene · Mechanical properties · Wear

J. Fu (✉)
Ningbo Institute of Materials Technology and Engineering, Chinese Academy of Sciences, Ningbo, China
e-mail: fujun@nimte.ac.cn

J. Fu et al. (eds.), *UHMWPE Biomaterials for Joint Implants*, Springer Series in Biomaterials Science and Engineering 13,
https://doi.org/10.1007/978-981-13-6924-7_4

4.1 Introduction

Highly cross-linked ultra-high molecular weight polyethylenes (UHMWPEs) have showed low wear and been widely used for joint implants [1, 2]. Clinical outcomes have demonstrated significantly reduced wear-related osteolysis and revisions of implants based on highly cross-linked UHMWPEs (XL-PEs) [3]. Usually, highly cross-linked polyethylenes are produced by exposing the polymer to ionizing radiations (e.g. electron beam or γ-ray) to generate free radicals that are prone to recombine in the amorphous phase into three-dimensional networks. However, the free radicals entrapped in the crystalline phase may migrate and react with diffused oxygen in the long term, leading to polymer oxidation and embrittlement. Post-irradiation thermal treatments, i.e. remelting above the melting point or annealing below the melting point of irradiated UHMWPE, are conducted to eliminate or reduce the radicals to low levels in order to improve the oxidation stability of cross-linked UHMWPE [4–6]. However, such thermal treatments usually decrease the crystallinity and ductility of the polymers and thus sacrifice the strength, toughness, and fatigue resistance [7, 8] of the materials and the components [9], which could be detrimental for the joint implants [7, 10]. On the other hand, there are a lot of lipids in the synovial fluid, including squalene, cholesterol, and hexadecanoic and octadecanoic esters of cholesterol, etc. These unsaturated lipids may diffuse into UHMWPE components in vivo [11]. Recently, these absorbed lipids are related to the initiation of in vitro or in vivo oxidation of the cross-linked and remelted polyethylene components [12–14]. Therefore, there is a need to develop alternative methods to stabilize radiation cross-linked UHMWPE for joint implants.

Stabilization of irradiated polyethylene with free radical scavengers or antioxidants (AO) has demonstrated efficient to protect the polymer against oxidative challenge, and thus post-irradiation thermal treatments are avoided [15–19]. Vitamin E (α-tocopherol) is a successful antioxidant applied to stabilize radiation cross-linked UHMWPE. It donates a proton of the phenolic OH to a macroradical and itself becomes a much less reactive phenoxy radical due to the delocalization of unpaired electron over the aromatic ring. The stabilization mechanism of vitamin E and its antioxidation efficiency in irradiated polymers have been studied in the melt [20, 21] and in the solid state [22]. In the melt, the mobility of vitamin E molecules in UHMWPE is favourable for macroradical stabilization beyond that in the solid state, although the proton-donating mechanism remains the same for both cases.

Vitamin E can be either blended with UHMWPE powders before consolidation and irradiation [18, 23–25] or diffuse into irradiated UHMWPE blocks [17, 26]. Irradiation of consolidated VE/UHMWPE blends can consume the phenolic OH group [27]. During irradiation, vitamin E may lose some of the phenolic OH group, particularly as the vitamin E concentration is low (<0.1 wt%) [27, 28]. Such phenol loss is attributed to the reaction with macroradicals upon irradiation and may be associated with the loss of its antioxidation activity. At low vitamin E concentrations, the minimal protective vitamin E concentration is dependent on the irradiation dose [29]. On the other hand, the termination of free radicals by vitamin E, especially with high vitamin E content (>0.3 wt%), can hinder cross-linking and may even increase chain scission [24, 30]. Such side effects are not wanted for the wear-resistant materials. Alternatively, diffusion of vitamin E into irradiated UHMWPE does not hinder cross-linking or change the chemistry of vitamin E, yielding high cross-link density (and thus low wear [26]) and antioxidation activity, as well as high strength and fatigue toughness of the irradiated UHMWPE [17, 26]. This process, however, usually involves homogenization (100–120 °C) below the melting temperature of UHMWPE for a long time [31].

Many natural polyphenols are widely used as antioxidants in food and pharmaceutics due to their free radical scavenging capability [32, 33]. The multiple phenolic groups are able to stabilize or terminate the free radicals. In this chapter, the use of representative polyphenols to stabilize UHMWPE for high-dose irradiation is presented. The preparation of polyphenol-stabilized UHMWPE, the stability of polyphenols against high-dose irradiation, and the potency to protect the radiation cross-linked polymer from adverse oxidative challenging will be presented and discussed. The polyphenol-stabilized polyethylene shows outstanding mechanical strength and toughness, high cross-link density, and low wear rates. Besides, the antioxidation mechanism has been investigated. It is found efficient to depress the in vitro oxidation with the presence of unsaturated lipids.

4.2 Preparation of Polyphenol-Stabilized Highly Cross-Linked UHMWPE

Two representative natural polyphenols with three phenolic OH groups each, i.e. gallic acid (GA) and dodecyl gallate (DG), are used. The chemical structures of these polyphenols are showed in Fig. 4.1. The three phenolic OH groups, in comparison to the single one for vitamin E, are supposed to provide more protection of UHMWPE from oxidative challenging. Similar to vitamin E, DG has an aliphatic tail, which is compatible with the polyethylene chains. In contrast, the GA molecules may be less compatible with the UHMWPE matrix. The polyphenols are respectively blended with UHMWPE resins and then press consolidated at elevated temperatures and high pressure. The polyphenols are homogeneously distributed

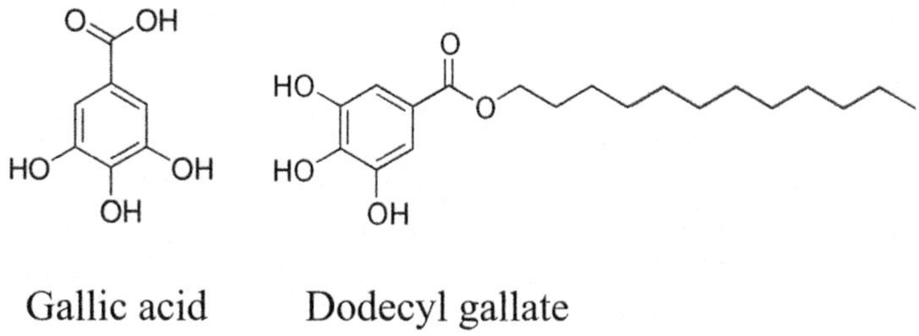

Fig. 4.1 The chemical structures of gallic acid and dodecyl gallate

in the consolidated preform with a brown colour, with the colour depending on the polyphenol content. The presence of polyphenols with contents less than 1 wt% does not significantly influence the crystallinity or melting point of the preforms. The ultimate tensile strength (UTS), elongation at break (EAB), and double-notched Izod impact strength (I_z) values are not significantly different from those of virgin UHMWPE preforms prepared at the same consolidation conditions.

The polyphenols are stable against the high-temperature consolidation, as confirmed by using Fourier transform infrared (FTIR) spectroscopy. Figure 4.2 compares the FTIR spectra of the antioxidants and the antioxidant/UHMWPE consolidates. After consolidation, the phenolic group band of vitamin E at 1206 cm^{-1} remains unchanged (Fig. 4.2a). In contrast, the phenolic group shoulder at 1205 cm^{-1} for GA increases and becomes equal to that at 1210 cm^{-1} (Fig. 4.2b), suggesting a change in the phenolic ring structure. Meanwhile, the phenolic band of DG becomes broad from 1180 to 1280 cm^{-1} (Fig. 4.2c), indicating an isomerization of the phenol group during consolidation.

Subsequently, electron beam at 10 MeV is used to irradiate the UHMWPE blends to total doses of 50, 100, and 150 kGy at a dose rate of 25 kGy/pass. During irradiation, the polymer temperature is raised to no higher than 70 °C, which is much lower than the melting point of the polymer crystals. As a result, the polyphenol/polyethylene blends are cross-linked.

After irradiation, the phenolic group band of vitamin E 1206 cm^{-1} vanishes (Fig. 4.2a), indicating the phenol loss upon irradiation [22, 27, 28]. In contrast, the phenolic bands at 1209 cm^{-1} for GA remain unchanged (Fig. 4.2b), while that at 1246 cm^{-1} for DG shifts to 1205 cm^{-1} (Fig. 4.2c). Different from the phenol loss of vitamin E, no phenol loss is observed for GA, while structural changes of the phenolic groups take place for DG upon irradiation. These results suggest a strong anti-irradiation potency of the polyphenols, with most of the phenolic OH groups surviving the irradiation and hopefully preserving the antioxidation activity.

The shift in FTIR bands during consolidation and irradiation suggests that the polyphenols are able to form resonant structures after losing one or more protons. As the protons are subtracted from the phenolic groups, the phenolic rings are converted into quinones and unsaturated ketones. In contrast to vitamin E, the three phenolic OH groups offer better tolerance to irradiation, and the formed resonant structures may provide more potency to protect UHMWPE from oxidation.

4.3 Oxidation Stability

Despite the structural changes of the polyphenols, the prepared high-dose irradiated polyphenol/UHMWPE consolidates show outstanding resistance against aggressive oxidative challenges. Accelerated ageing at 70 °C and 5 atm oxygen according to ASTM F2003 does not result in detectable oxidation by FTIR for all the irradiated AO/UHMWPE blends with doses up to 150 kGy, while severe oxidation occurs for virgin polyethylene receiving the same doses.

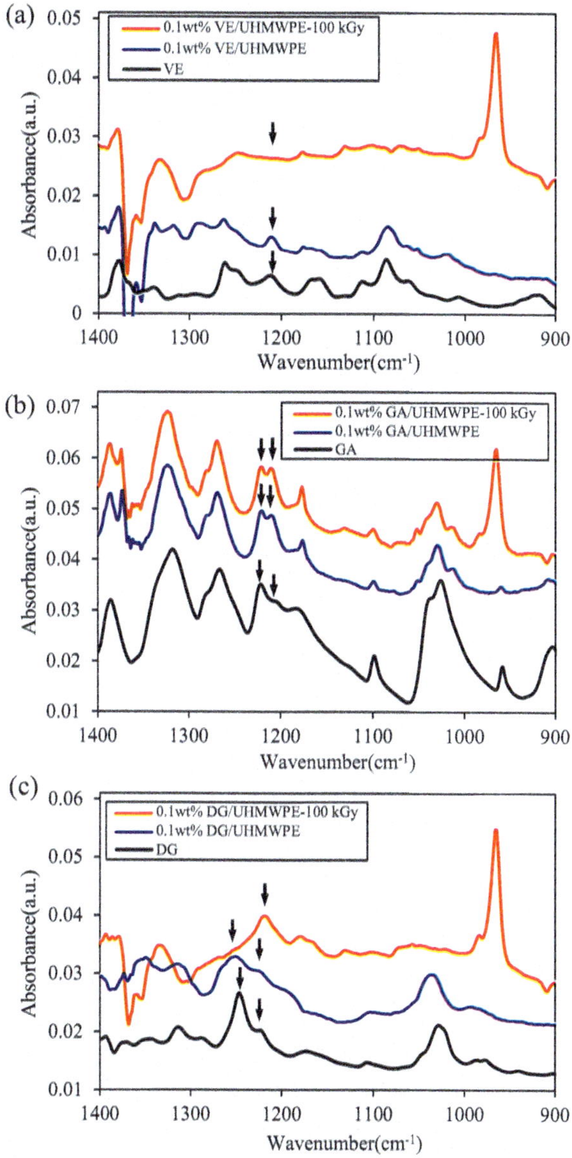

Fig. 4.2 Representative FTIR spectra of (a) vitamin E (VE)/UHMWPE, (b) gallic acid (GA)/UHMWPE, and dodecyl gallate (DG)/UHMWPE blends after consolidation and subsequent e-beam irradiation [34]. (Reproduced by permission of the Royal Society of Chemistry)

To compare the antioxidation potency of the polyphenols to protect UHMWPE with vitamin E, the irradiated AO/UHMWPE thin films are aggressively challenged at accelerated oxidation at 120 °C in air for a long time [35], until severe oxidation occurs. The oxidation of UHMWPE generates carbonyl groups, which show a representative absorbance at about 1718 cm^{-1} (Fig. 4.3). This band is used to characterize the oxidation degree of the polymer. The band area is normalized to that of the internal standard at 1368 cm^{-1} to quantify the oxidation level (Fig. 4.3) or

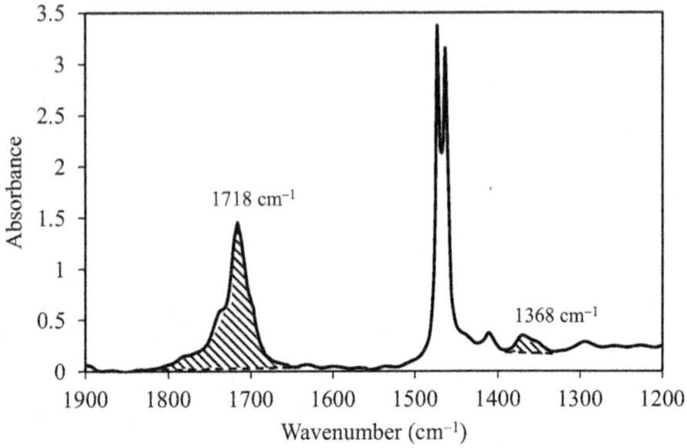

Fig. 4.3 The oxidation index is defined as the area ratio of the band at 1718 cm^{-1} (1680–1820 cm^{-1}, carbonyl stretching) to that at 1368 cm^{-1} (1335–1390 cm^{-1}, polyethylene end methyl stretching)

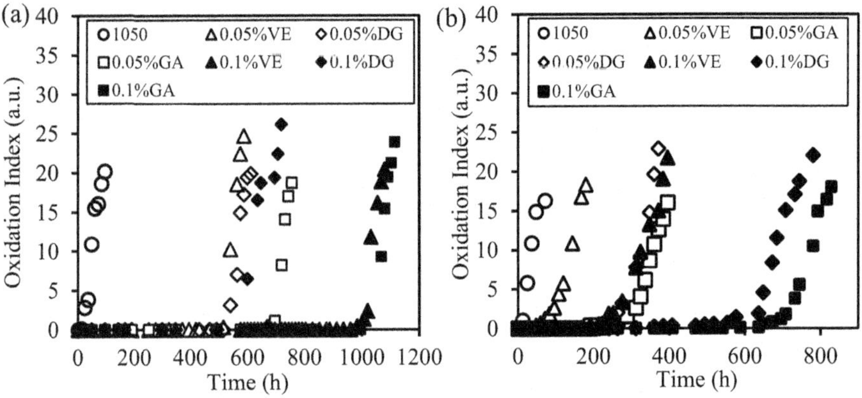

Fig. 4.4 The evolution of oxidative index during accelerated ageing of e-beam irradiated antioxidant/UHMWPE blends at 120 °C over time: (**a**) 0 kGy, (**b**) 100 kGy [34]. (Reproduced by permission of the Royal Society of Chemistry)

the oxidation index. The oxidation indices of the samples are tracked over time. The oxidation initiation is defined as the oxidation index increases to 0.5.

The virgin UHMWPE is rapidly oxidized at high temperature in air, with an initiation time of 4 h. The presence of 0.05 wt% vitamin E significantly postpones the oxidation initiation to 500 h and even 1000 h with 0.1 wt% vitamin E. The oxidation initiation is further delayed with the presence of 0.05 wt% gallic acid (700 h) and dodecyl gallate (500 h) (Fig. 4.4a).

Irradiation at 100 kGy, for example, significantly shortens the $t_{ini.}$ values (Fig. 4.4b). The $t_{ini.}$ of the 0.1 wt% VE/UHMWPE is largely shortened to about 250 h, while that of the 0.1 wt% DG/UHMWPE is decreased to about 640 h after 100 kGy irradiation. With an even lower antioxidant concentration (0.05 wt%), the $t_{ini.}$ values of the irradiated GA/UHMWPE and DG/UHMWPE blends are about 300 h and 250 h, respectively, much larger than that of the irradiated VE/UHMWPE (about 50 h). After initiation, the oxidation rates (slopes of the profiles [36]) are approximately the same until the samples approached an oxidation plateau. These results suggest the possibility to achieve excellent antioxidation activity by using a low antioxidant concentration (e.g. 0.05 wt%) but high radiation dose (e.g. 100 kGy). This advantage is very important for application of such highly cross-linked UHMWPE as joint implants.

Such adverse oxidative challenging is much harsher than that regulated by ASTM F2003. Thus, the above results suggest a great potency of the polyphenols to protect highly cross-linked UHMWPE against oxidation, which is greater than that of vitamin E. Therefore, it is reasonable to believe that the polyphenol-stabilized highly cross-linked UHMWPE may be more stable on-shelf and in vivo. The stability of the polyphenols against high-dose irradiation may endure terminal radiation sterilization, which has almost been abandoned for the highly cross-linked UHMWPE and even vitamin E UHMWPE components to avoid any additional generation of free radicals and consumption of vitamin E activity.

4.4 Mechanical Properties

Radiation cross-linking also reduces the strength and toughness of polyethylene. Post-irradiation thermal treatments used for the first-generation highly cross-linked UHMWPE stabilize the polymers but decrease the crystallinity and crystal size and thus further decrease the strength and toughness of the materials. The loss in mechanical strength, toughness, and fatigue resistance may cause rim fracture of acetabular liners at short time after implantation [37].

The tensile properties of the irradiated UHMWPE with 0, 0.05, and 0.1 wt% antioxidants are systematically investigated. Figure 4.5 compares the UTS values of the irradiated UHMWPE and AO/UHMWPE materials before and after ageing. The vitamin E/UHMWPE shows higher UTS than virgin UHMWPE, while DG and GA do not significantly influence the tensile properties in comparison to virgin

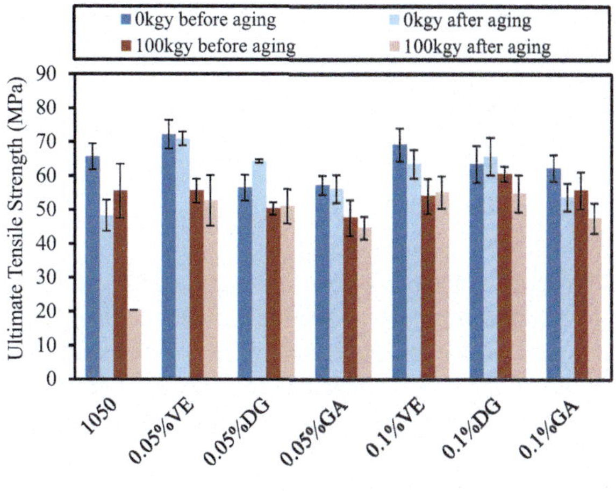

Fig. 4.5 Ultimate tensile strength of UHMWPE and antioxidant/UHMWPE blends before and after 100 kGy irradiation and those before and after oxidation

UHMWPE. For example, the UTS is 72.1 ± 4.2 MPa for 0.05 wt% vitamin E/UHMWPE and 69.2 ± 4.8 MPa for 0.1 wt% vitamin E/UHMWPE, higher than that for virgin UHMWPE 65.6 ± 3.8 MPa. Meanwhile, the UTS is 65.8 ± 2.2 MPa for 0.05 wt% DG/UHMWPE and 63.5 ± 5.3 MPa for 0.1 wt% DG/UHMWPE, which are close to that for virgin UHMWPE. The GA/UHMWPE blends show a slightly lower UTS, probably due to the less compatibility of GA with UHMWPE.

E-beam irradiation significantly decreases the UTS and the elongation at break (EAB) of consolidated UHMWPE and AO/UHMWPE blends due to the formation of cross-links. After 100 kGy irradiation, the UTS decreases to 55.5 ± 7.9 MPa for virgin UHMWPE, 55.6 ± 3.5 MPa for 0.05 wt% VE/UHMWPE, 51.2 ± 0.7 MPa for 0.05 wt% DG/UHMWPE, 60.6 ± 2.2 MPa for 0.1 wt% DG/UHMWPE, and 53.7 ± 3.0 MPa for 0.1 wt% GA/UHMWPE.

Accelerated ageing at 70 °C and 5 atm oxygen for 14 days decreases the UTS and EAB of the virgin and 100 kGy irradiated UHMWPE. The aged 100-kGy irradiated UHMWPE became too brittle to tensile. In contrast, slight or no significant decreases in the EAB ($p = 0.04$) or UTS ($p = 0.6$) values were induced by ageing the 0.05 wt % VE/UHMWPE for 2 weeks. Ageing of the 100 kGy irradiated 0.05 wt% VE/UHMWPE decreases the UTS to 50.9 ± 7.2 MPa (100 kGy, $p = 0.36$). The phenol loss in the 0.05 wt% VE/UHMWPE after irradiation may offer less protection of polyethylene against oxidation [36]. With higher vitamin E concentration (e.g. 0.1 wt%), no significant changes occurred in the tensile properties after ageing for the irradiated materials.

In contrast, the 0.05 wt% gallic acid/UHMWPE shows no significant decreases in UTS ($p = 0.35$) or EAB ($p = 0.12$) after ageing. Similar stability against oxidation is also observed for irradiated 0.05 wt% GA/UHMWPE. No significant reduction in the mechanical properties is observed for the irradiated DG/UHMWPE after ageing.

The double-notched Izod impact strength (Z_I) of the consolidated UHMWPE was 130 ± 12.3 kJ/m^2, while those of the AO/UHMWPEs were in the range of 110–135 kJ/m^2. E-beam irradiation decreased the Z_I values of virgin UHMWPE to 81 ± 1.9 kJ/m^2 (50 kGy) and 56 ± 1.5 kJ/m^2 (100 kGy), which are lower than those of the irradiated AO/UHMWPEs. Less decrease in the Z_I values was induced by radiation cross-linking for AO/UHMWPE than that for virgin UHMWPE. After accelerated ageing, the Z_I values decreased for 100-kGy irradiated UHMWPE, whereas no significant changes were seen for 100-kGy irradiated AO/UHMWPE.

4.5 Cross-Link Density

The extent of cross-linking is measured by using a term cross-link density. It is determined by using a swelling method. Most polymers are able to dissolve in solvents. However, the cross-linked polymers do not dissolve in solvents since the polymer chains are connected to form a 3D network. Instead, the network uptakes the solvent at given conditions. The maximum solvent uptake is related to the properties of the polymer chains and the network density. Upon solvent uptake, the mass or volume of the polymer network increases. The gravimetric or volumetric swelling ratio is used to calculate the cross-link density, according to the Flory theory. The cross-link density is calculated using the following equations:

$$d_x = \frac{\ln\left(1 - q_{\text{eq}}^{-1}\right) + q_{\text{eq}}^{-1} + Xq_{\text{eq}}^{-2}}{V_1\left(q_{\text{eq}}^{-1/3} - q_{\text{eq}}^{-2}\right)} \tag{4.1}$$

$$X = 0.33 + \frac{0.55}{q_{\text{eq}}} \tag{4.2}$$

where the specific volume of xylene, V_1, was 136 cm^3/mol and the equilibrium swelling degree q_{eq} can be measured.

To determine the cross-link density of irradiated UHMWPE, small cubes are immersed in boiling xylene until equilibrium. The xylene uptake is measured for the calculation of the cross-link density. For virgin UHMWPE, the cross-link density increases from 0.009 mol/dm^3 to 0.177, 0.286, and 0.337 mol/dm^3 with 0, 50, 100, and 150 kGy.

It is known that the presence of vitamin E inhibits the cross-linking because of its reaction with free radicals [24, 30]. With 100 kGy irradiation, the cross-link density decreases from 0.286 for virgin UHMWPE to 0.242, 0.219, and 0.161 mol/dm^3 with vitamin E concentration increasing from 0.05 to 0.1 and 0.3 wt% (Fig. 4.6).

In contrast, the presence of polyphenols does not cause severe decreases in cross-link density even at high concentrations (Fig. 4.6). The cross-link densities of irradiated DG/UHMWPE are very close to that of the virgin UHMWPE receiving the same radiation doses, although the cross-link density slightly decreases at DG concentrations above 0.3 wt% (Fig. 4.6a). Importantly, for each antioxidant concentration and radiation dose, the cross-link density of irradiated DG/UHMWPE is

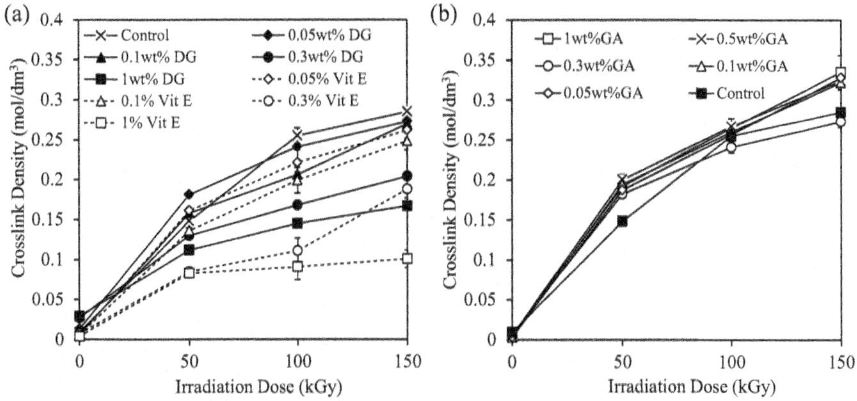

Fig. 4.6 Cross-link density vs irradiation dose for the e-beam irradiated UHMWPE [34]. (Reproduced by permission of the Royal Society of Chemistry)

higher than that of irradiated VE/UHMWPE. For example, with 100 kGy irradiation, the cross-link density is 0.168 ± 0.004 mol/dm^3 for 0.3 wt% DG/UHMWPE, 0.111 ± 0.031 mol/dm^3 for 0.3 wt% VE/UHMWPE, and 0.255 ± 0.010 mol/dm^3 for virgin UHMWPE. With 150 kGy, the corresponding cross-link densities are increased to 0.204, 0.188, and 0.285 mol/dm^3.

Interestingly, the cross-link densities of irradiated GA/UHMWPE are independent on the GA content from 0.05 wt% to 1.0 wt% and close to or even a little higher than those of the irradiated UHMWPE (Fig. 4.6b). These results suggest that the GA molecules do not actively react with the free radicals to influence the cross-link density, which may be related with the less affinity of GA with the UHMWPE matrix.

Practically, a terminal irradiation dose close to 100 kGy is preferable to achieve wear resistance similar to 100 kGy irradiated and remelted UHMWPE in clinical use. To this end, the vitamin E and dodecyl gallate concentrations of 0.10 wt% or below are recommended to achieve adequate cross-link density and wear resistance that are needed for novel joint implants with low wear and antioxidation stability.

4.6 Wear Rates

In general, the knee or hip experience multi-axial or multi-directional wear. To evaluate the wear behaviour of UHMWPE, hip and knee simulators have been designed to mimic the gaiting conditions of human beings. Before this sophisticated evaluations, however, a simplified method, or pin-on-disc (POD) test has been widely employed to pre-evaluate the wear of UHMWPE materials. Here, a home-built pin-on-disc (POD) wear tester (Fig. 4.7) is designed to drive bidirectional movement of the polymer pins on the CoCrMo discs at a constant load with bovine serum as lubricant. The pin samples of UHMWPE with different formulations are tested up to 5 million cycles (MC) to examine the effect of radiation cross-linking and antioxidant type and concentration on the wear rate.

Fig. 4.7 A custom-built pin-on-disc test machine

Table 4.1 The pin-on-disc wear rates of the irradiated and remelted UHMWPE, irradiated vitamin E/UHMWPE, irradiated gallic acid/UHMWPE, and irradiated dodecyl gallate/UHMWPE [34]

Materials	Irradiation dose (kGy)	Thermal treatment	Pin-on-disc wear rate (mg/MC)
UHMWPE	0	N.A.	12.38 ± 2.28
	50	Remelt	6.21 ± 0.81
	75	Remelt	2.23 ± 0.31
	100	Remelt	1.68 ± 0.25
	150	Remelt	0.91 ± 0.58
0.05 wt% vitamin E	100	N.A.	2.05 ± 0.22
0.1 wt% vitamin E	100	N.A.	2.56 ± 0.45
0.05 wt% gallic acid	100	N.A.	1.65 ± 0.32
0.1 wt% gallic acid	100	N.A.	2.69 ± 0.57
0.05 wt% dodecyl gallate	100	N.A.	2.29 ± 0.31
0.1 wt% dodecyl gallate	100	N.A.	4.41 ± 0.61

The virgin UHMWPE shows an average wear rate of 12.38 ± 2.28 mg/MC. After 50 kGy irradiation and remelting, the wear rate significantly decreases to 6.21 ± 0.81 mg/MC. As the dose is further increased to 75, 100, and 150 kGy, the wear rate decreases to 2.23 ± 0.31, 1.68 ± 0.25, and 0.91 ± 0.58 mg/MC (Table 4.1).

The radiation cross-linked UHMWPE with antioxidants also shows highly reduced wear rate. For example, the wear rate of 100 kGy irradiated 0.05 wt% GA/UHMWPE (1.65 mg/MC) is comparable to that of 100 kGy irradiated and remelted polyethylene (1.68 mg/MC), close to that of the 100 kGy irradiated and remelted by Muratoglu et al. [38]. Most other highly cross-linked AO/UHMWPEs show POD wear rates about 2 mg/MC. It is noted that no post-irradiation thermal treatments are needed to stabilize the irradiated AO/UHMWPE since the antioxidants are potent to protect the irradiated polymer from oxidation.

There has been a consensus that radiation cross-linking reduces the polyethylene wear by creating rigid three-dimensional network and decreasing the deformation and plastic flow of the semi-crystalline polymer [38, 39]. This mechanism governs a unified wear model for the first-generation highly cross-linked UHMWPE [38]. That is, the wear rate decreases with increasing cross-link density. This mechanism applies to the antioxidant-stabilized cross-linked UHMWPEs. In the wear rate-cross-link density plot, all the wear rate data of irradiated AO/UHMWPEs fall along with the fitting line of the irradiated and remelted UHMWPEs (Fig. 4.8).

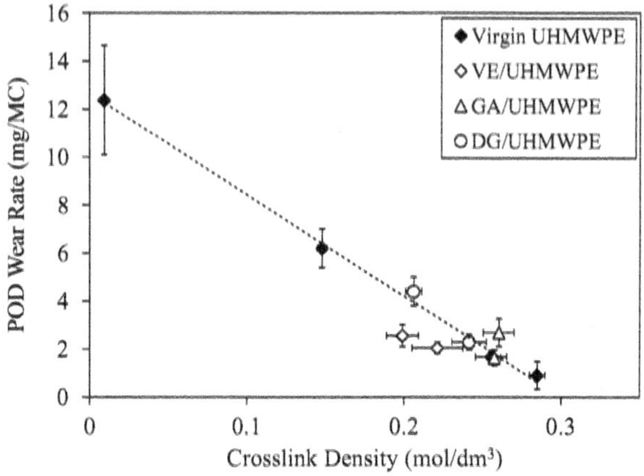

Fig. 4.8 The dependence of POD wear rate on the cross-link density of irradiated and remelted polyethylene, together with the irradiated polyethylene with vitamin E, gallic acid, and dodecyl gallate [34]. (Reproduced by permission of the Royal Society of Chemistry)

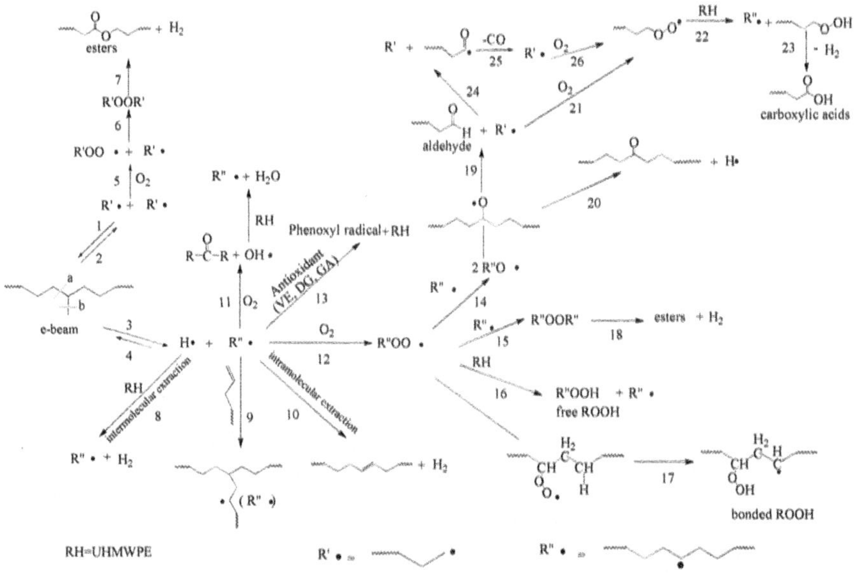

Scheme 4.1 Schematic illustrations of the oxidation processes of irradiated UHMWPE with the presence of antioxidants [41]. (Copyright Elsevier 2014)

4.7 Oxidation and Antioxidation Mechanisms

The oxidation of UHMWPE has been discussed in Chap. 3. Here, the mechanism of antioxidant to interrupt the oxidation procedure is analysed and discussed. It is known that the oxidation of irradiated UHMWPE starts from the primary and secondary macro-alkyl radicals with the presence of oxygen and free radicals to generate peroxyl radicals. The resulted radicals further react with polyethylene chains, alkyl macroradicals or other macroradicals (Scheme 4.1), which is usually known as the oxidation cascade. As a result, the polymer chains are scissioned and the material becomes brittle. During oxidation, many oxidation products are formed, including ketones, hydroperoxides, lactones, esters, carboxylic acids (isolated and associated), ketoacids, etc. [32]. These oxides are formed through different routes, as illustrated in Scheme 4.1. Therefore, the analysis of the oxidation products is performed to investigate the oxidation and particularly antioxidation mechanisms with the presence of antioxidants.

Typically, the formed ketones show a characteristic band at 1720 cm^{-1}, the hydroperoxides show broad and weak absorbances at 3420 cm^{-1} (hydrogen bonded) and 3550 cm^{-1} (non-hydrogen bonded), and the lactones, esters, carboxylic acids, and ketoacids show a broad band in the range from 1640 to 1810 cm^{-1}. As the bands of the hydroperoxides are usually overlapped by many other species, which makes it difficult to directly quantify their contents, the hydroperoxides are converted into nitrates through treatments with nitrogen monoxide (NO) [31]. The obtained nitrates show representative absorbance at 1630 cm^{-1}, which is used to determine the hydroperoxide concentration, according to calculations based on its extinction coefficient (1933 L mol^{-1} cm^{-1}) [40].

In order to minimize the interference of sample preparation procedure to the FTIR spectrum intensities, the FTIR spectra are normalized to the intrinsic reference band at 2018 cm^{-1} (CH_2 twisting). In order to further emphasize the absorbance of specific functional groups, particularly of those formed through some special treatment (e.g. ageing), the normalized spectra of samples after the treatments are subtracted by their counterparts before the treatments. Thus, the spectral changes are clearly exposed. For example, the spectra of aged samples are subtracted by those of samples before ageing, which helps indexing the concentration of ketones or other oxidation products in the aged materials. Besides, the spectra of aged samples treated by NO are subtracted by those of samples without NO treatment in order to reveal the concentration of nitrides that are derived from hydroperoxides in the polyethylene.

Despite of the spectral subtraction, the absorbances of some oxidation products overlap each other within a narrow range. For example, the lactones, esters, carboxylic acids (isolated and associated), and ketoacids have IR absorbances overlapping in the range from 1640 to 1810 cm^{-1}, leading to an apparent single band with very subtle shoulders (Fig. 4.9a). In order to figure out the oxidation products, mathematical curve-fitting treatments are conducted to deconvolute the normalized carbonyl bands in the range 1810–1640 cm^{-1}, as introduced by Salvalaggio et al. [42]. The second- and fourth-order differentiation of the band are derived from the spectrum to identify the existence and position of sub-absorbances from possible oxidation products. Figure 4.9 compares a typical spectrum (Fig. 4.9a) and its derivative after the second- and fourth-order differentiation (Fig. 4.9b). The corresponding negative second derivatives and positive fourth derivatives are used to identify the sub-absorbance bands of the oxidation products. Accordingly, the IR band at 1810–1640 cm^{-1} is deconvoluted into 5–7 sub-bands (Fig. 4.9c) centred around 1788 (γ-lactones), 1767 (carboxylic acids, isolated), 1737 (esters), 1714 (carboxylic acids, associated), 1707 (ketoacids, acid), and 1696 cm^{-1} (ketoacids, keto), in addition to that at 1718 cm^{-1} (ketone) [42]. The overlapping sub-absorbances provide a best fitting to the experimental absorbance with a degree of confidence (R^2) larger than 0.999.

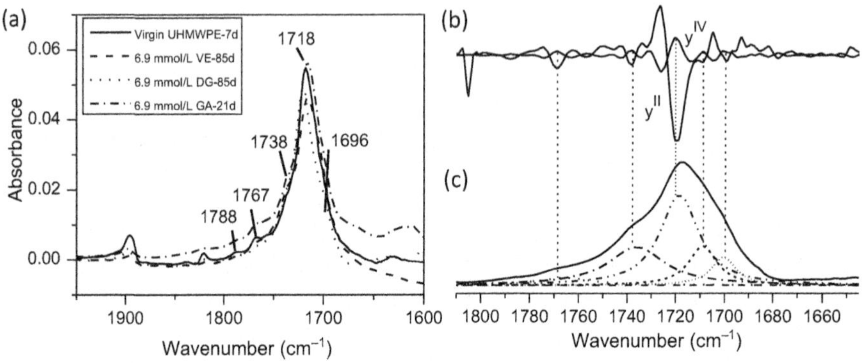

Fig. 4.9 (**a**) Representative IR spectral subtraction of 150 kGy irradiated virgin UHMWPE, antioxidant/UHMWPE blends aged to an identical oxidation level of about 0.05. (**b**) The second (Y^{II}) and fourth (Y^{IV}) differentiation of the band in the range 1640–1810 cm^{-1}. (**c**) A mathematical deconvolution of the band based on the fourth differentiation in (**b**) [41]. (Copyright Elsevier 2014)

The presence of antioxidants significantly influences the oxidation product type and contents. First, the antioxidants highly retard the oxidation of UHMWPE. It took 7, 21, 85, and 85 days for highly cross-linked virgin UHMWPE, GA/UHMWPE, VE/UHMWPE, and DG/UHMWPE with 6.9 mmol/L antioxidants to reach an identical oxidation level of about 0.05 (Fig. 4.9a).

On the other hand, according to the deconvoluted results of the absorbance at 1810–1640 cm^{-1}, the oxidation products of the irradiated virgin, VE/UHMWPE, GA/UHMWPE, and DG/UHMWPE are compared. The ketone levels (at 1720 cm^{-1}) for the DG/UHMWPE are the highest, while those for the other materials are almost the same. Meanwhile, the acid level (the sum of carboxylic acid at 1767 and 1714 cm^{-1} and ketoacid at 1707 cm^{-1}) for the irradiated DG/UHMWPE is significantly lower than those for the irradiated VE/UHMWPE and virgin UHMWPE. The irradiated GA/UHMWPE shows the highest acid level, presumably due to the acid groups in the antioxidant. The formation of carboxylic acids indicates a chain-breaking mechanism, which causes oxidative degradation of UHMWPE [43].

The oxidation kinetics are also influenced by the antioxidants [34]. According to Scheme 4.1, the secondary macro-alkyl radicals, and peroxy radicals are critical intermediates for the formation of ketones, carboxylic acids, and other oxides. Most of these radicals are prone to reactions with the antioxidants (Reaction 13, Scheme 4.1), leading to a decrease in the concentrations of peroxides and ketones.

The presence of antioxidants significantly influences the formation of hydroperoxides. The content of hydroperoxides with an oxidation level of 0.05 is quantitatively investigated. In order to quantify the hydroperoxides, they are converted into nitrates that show strong characteristic absorbances at 1633, 1276, and 859 cm^{-1} for the secondary hydroperoxides and 1649 and 1298 cm^{-1} for the primary hydroperoxides (Fig. 4.10) [40]. The bands at 1633 cm^{-1} for the irradiated VE/UHMWPE and DG/UHMWPE are lower than that for the irradiated UHMWPE, while that for the irradiated GA/UHMWPE was the highest (Fig. 4.10). Post-irradiation oxidation

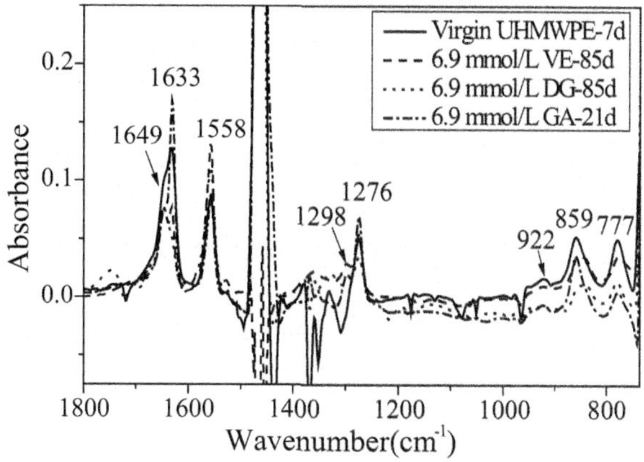

Fig. 4.10 FTIR spectral subtraction of the 150 kGy irradiated virgin and antioxidant-containing UHMWPE after ageing and subsequent NO treatment by those before NO treatment [41]. (Copyright Elsevier 2014)

significantly increases the hydroperoxide concentrations for the 150 kGy irradiated virgin and GA-blended UHMWPE, while those for the 150 kGy irradiated VE/UHMWPE and DG/UHMWPE materials are much lower.

The different behaviour of GA from DG and VE is attributed to the molecular structures. The lyophilic tails of DG and VE make them compatible with the UHMWPE matrix. Thus, during irradiation, vitamin E and dodecyl gallate aggressively react with the macroradicals, and convert themselves into phenoxy radicals with much less reactivity. Thus, the macroradicals on the UHMWPE chains are terminated, and the formation of peroxyl radicals and peroxides is interrupted (Scheme 4.1). Consequently, the concentration of peroxides in the irradiated UHMWPE remains a constant low-level during ageing (4 mmol/L). In contrast, GA is less compatible with the UHMWPE matrix and has less reactivity to the macroradicals, leaving more radicals to react with oxygen to form peroxyl radicals and hydroperoxides (8–10 mmol/L).

The formation of ketones may come from the direct reaction of oxygen with the macroradicals (Reaction 11, Scheme 4.1) and the decomposition of peroxides (Reaction 12, Scheme 4.1). In particular, Reaction 11 is independent of the content of peroxides. Thus, although the peroxides concentration is low, the ketones concentrations are much higher in the aged polymers. In fact, the peroxides are quite stable at room temperature and are unlikely to decompose into ketones (Reaction 12, Scheme 4.1) [43]. The direct formation of ketones may be a major mechanism for the formation of high ketones concentrations [22, 44].

The effect of antioxidants on the oxidation is further investigated by tracing the formation kinetics of ketones in the irradiated UHMWPE with the presence of antioxidants. The ketone concentration in the 150-kGy irradiated VE/UHMWPE after ageing is lower than those for the 150-kGy irradiated DG/UHMWPE, GA/UHMWPE, and virgin UHMWPE. The ketone concentrations are lower for the same AO/UHMWPE blends receiving less radiation dose (60 kGy). The ketone production rate is very high right after irradiation and decays rapidly to about zero in about 250 h. Besides, the ketone formation rate is lower at 60 kGy than that at

150 kGy. The antioxidant concentration has less influence on the ketone concentration and production rate. The formation of ketones takes place mainly during irradiation, where the concentration of macroradicals plays a critical role in the formation of ketones (Reaction 11, Scheme 4.1).

4.8 Stabilization Against Lipid-Related Oxidation

Despite of the in vitro oxidation stability of highly cross-linked and remelted UHMWPE, the in vivo oxidation of polyethylene components after implantation has been reported. Both in vitro and in vivo studies suggest that the unsaturated lipids in synovial fluids diffused into the polyethylene [45] may initiate oxidation of UHMWPE [46]. Such oxidation is basically similar to the oxidation of the polymer chains. It is efficiently prevented by vitamin E. In vitro studies have demonstrated that the highly cross-linked UHMWPE with vitamin E is resistant to oxidative challenging even with the presence of squalene, a model unsaturated lipid used to model the synovial lipids [47].

Encouraged by the outstanding antioxidation potency of the natural polyphenols, it is reasonable to expect a superior protection of highly cross-linked UHMWPE against aggressive oxidative challenging with the presence of lipids. To test this hypothesis, high-dose irradiated polyphenol/UHMWPE blends are doped with squalene at 120 °C for 2 h and challenged with 5 atm oxygen at 70 °C for different durations. First, the squalene uptake for the unirradiated polymers is about 40% higher than for the 90 kGy irradiated materials. Squalene reaches a saturated level on the surface and diffuses into the UHMWPE cubes, forming a gradient profile into the bulk. The penetration depth is about 5 mm for virgin UHMWPE, while it is about 2.8 mm for the 90 kGy irradiated polyethylene. The diffusion of squalene into UHMWPE could be described by Fick's second law of diffusion. A curve-fitting analysis of the diffusion results in a diffusion coefficient of squalene (D_{SQ}) in unirradiated UHMWPE of 3.3×10^{-5} mm^2/s at 120 °C, while those in 90 kGy irradiated UHMWPE and VE/UHMWPE are 1.0×10^{-5} mm^2/s and 2.0×10^{-5} mm^2/s, respectively.

After accelerated ageing, the squalene-doped UHMWPE, FTIR scans across the thin slices show oxidation gradient profiles from the free surface towards the sample bulk (Fig. 4.11), which are exclusively consistent with the squalene profiles. The high carbonyl peak at 1718 cm^{-1} roots from the oxidized products of squalene and UHNMWPE. To exclude the interference from squalene, the samples are extracted by using hot hexane to remove the free species before IR scanning. Thus, the IR spectra of the thin slices after hexane extraction reveal the oxidation levels of the UHMWPE.

The presence of squalene significantly renders very high oxidation levels even with the presence of antioxidants. Figure 4.12 shows the squalene-related oxidation profiles of irradiated AO/UHMWPEs with very low antioxidant concentrations (0.23 mmol/L) after ageing for different times. Without squalene, the antioxidants

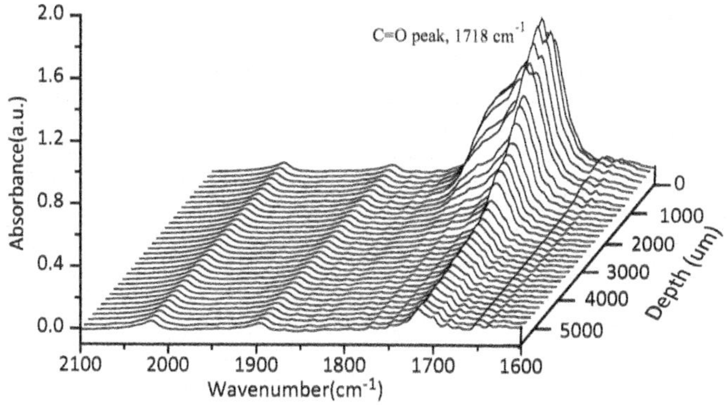

Fig. 4.11 Representative FTIR spectra of thin slices of squalene-doped UHMWPE after oxidation in 5 atm O_2 at 70 °C for 14 days [48]. (Copyright Elsevier 2014)

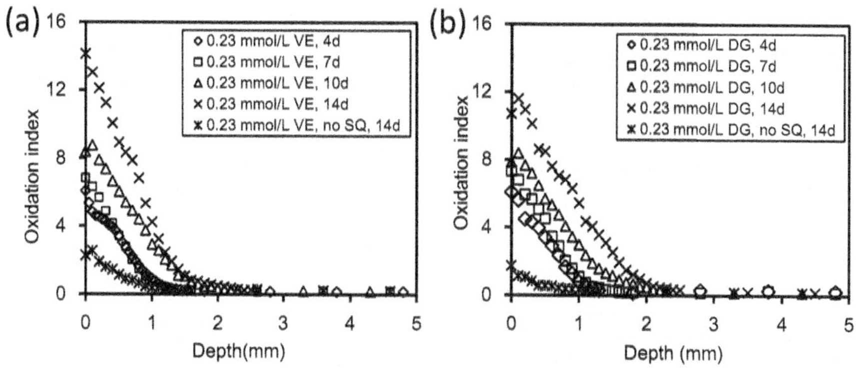

Fig. 4.12 The oxidation profiles of 90 kGy irradiated antioxidant/UHMWPE doped with squalene after accelerated ageing for different durations. (**a**) Vitamin E/UHMWPE and (**b**) DG/UHMWPE [48]. (Copyright Elsevier 2014)

efficiently protect the irradiated UHMWPE from oxidation. Slight surface oxidation (depth ~ 1 mm) occurs after accelerated ageing for 14 days (Fig. 4.12). With the presence of squalene, the oxidation levels on the surface and in bulk increase linearly over time.

The polyphenol antioxidants could significantly depress the squalene-related oxidation, depending on the antioxidant concentration. Figure 4.13 compares the evolution of the average surface oxidation levels over time for the irradiated AO/UHMWPE with different antioxidant concentrations. The squalene uptakes for all these samples are almost the same. The surface oxidation level after accelerated ageing decreases with increasing vitamin E concentration (Fig. 4.13a). With 1.1 mmol/L vitamin E, the oxidation is not initiated until day 10. With 2.3 mmol/L VE, no oxidation takes place until day 14. With further increase in vitamin E concentration to 6.9 mmol/L VE, no oxidation is detectable at day 28 (Fig. 4.13a). Similar prevention of squalene-related oxidation is also observed with the presence

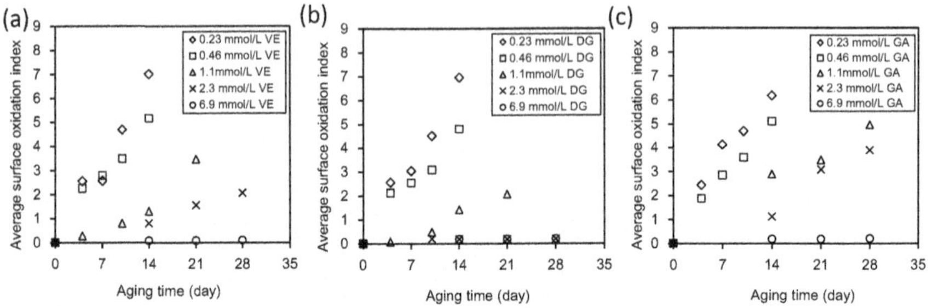

Fig. 4.13 The evolution of average surface oxidation indices over time for the 90 kGy irradiated (**a**) VE/UHMWPE, (**b**) DG/UHMWPE, and (**c**) GA/UHMWPE with different concentrations. The materials are doped with squalene at 120 °C for 2 h [48]. (Copyright Elsevier 2014)

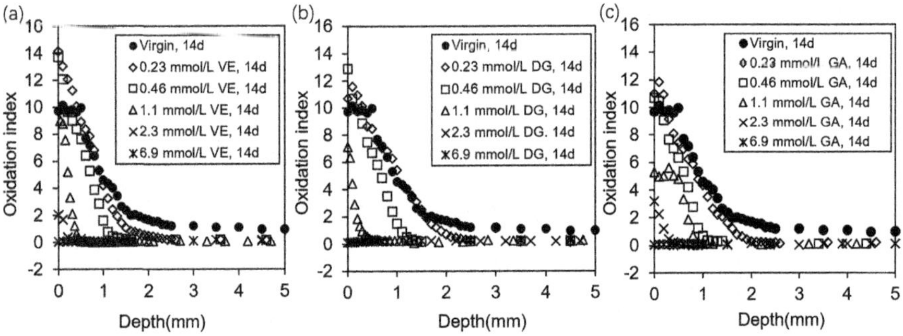

Fig. 4.14 The oxidation profiles of squalene-doped 90 kGy irradiated (**a**) VE/UHMWPE, (**b**) DG/UHMWPE, and (**c**) GA/UHMWPE after accelerated ageing for 14 days with different antioxidant concentrations [48]. (Copyright Elsevier 2014)

of DG (Fig. 4.13b) and GA (Fig. 4.13c). According to ASTM F2003, no oxidation takes place for the 90 kGy irradiated UHMWPE with 2.3 mmol/L antioxidants with the presence of squalene.

In order to figure out the critical antioxidant concentration for the prevention of saturated squalene-related oxidation, the oxidation profiles of the squalene-doped irradiated UHMWPE aged for 14 days are compared in Fig. 4.14. Significant surface oxidation takes place for 90 kGy irradiated UHMWPE with 1.1 mmol/L antioxidants and less, although the oxidation level is lower with the presence of polyphenols (Fig. 4.14b, c). As the antioxidant concentration is 2.3 mmol/L, no detectable oxidation takes place. It is likely that there is a critical antioxidant concentration between 1.1 and 2.3 mmol/L for the prevention of 90 kGy irradiated UHMWPE from oxidation in 14 days. Certainly, this critical value may depend on the irradiation dose. Considering that 65–110 kGy dosages are usually used for the manufacturing of highly cross-linked UHMWPE, the critical antioxidation concentration may be valuable for the material design.

Previous studies have reported the effective stabilization of irradiated high-temperature melted vitamin E/UHMWPE blends doped with squalene although some of vitamin E has changed the chemical structure upon irradiation [47]. The resonant structures of irradiated vitamin E have been reported potent in stabilization of irradiated UHMWPE [27]. The polyphenols are more potent in stabilizing highly cross-linked UHMWPE and even provide prolonged protection of irradiated UHMWPE against aggressive in vitro oxidative challenges at low antioxidant concentrations [49]. Such outstanding antioxidation potency of the natural polyphenols offers superior resistance against oxidation at the presence of squalene [50]. This study is effective in distinguishing the oxidation resistance of these antioxidants at a relatively short experimental time scale. At a given antioxidant concentration, the oxidation initiation time and oxidation penetration depth are two important parameters to assess the potency of the antioxidants by comparing the oxidation levels and oxidation profiles. To this end, the DG/UHMWPE blends showed the lowest oxidation levels and penetration depth while the GA/UHMWPE blends showed the highest oxidation levels and penetration depth among the antioxidant/UHMWPE blends at 1.1 mmol/L (Fig. 4.14). These 90 kGy irradiated materials hardly oxidize with the presence of 2.3 mmol/L DG after accelerated ageing for 14 days (Fig. 4.14). This irradiation dose is close to those for the highly cross-linked UHMWPE materials for clinical use. Thus, the results presented in this study may have clinical relevance in terms of oxidative stability.

4.9 Conclusions

In this chapter, natural polyphenols with three phenolic groups each are used to stabilize radiation cross-linked UHMWPE. The polyphenols are blended with medical grade UHMWPE powders and survive high-temperature consolidation and subsequent high-dose irradiation by electron beam. Despite the structural stability of polyphenols against irradiation, the cross-link densities of the radiation cross-linked polyphenol/UHMWPE blends are higher than those for the irradiated vitamin E/UHMWPE consolidates and close to those for the irradiated virgin UHMWPE. No post-irradiation thermal treatments are needed to achieve oxidative stability. Thus, the irradiated polyphenol/UHMWPE show improved tensile strength and toughness than those of irradiated and remelted UHMWPE. The polyphenols after irradiation show much higher potency against adverse oxidative challenging than vitamin E. Pin-on-disc (POD) tests demonstrate a significant reduction of the wear rates of the irradiated polyphenol/UHMWPE in comparison to the virgin UHMWPE. The wear rate level is comparable to that for the irradiated and remelted UHMWPE. The antioxidation mechanisms are investigated by tracking the oxidation products and their evolution kinetics. Moreover, the presence of polyphenols is very much efficient in squalene-initiated oxidation, which suggests an encouraging performance of the irradiated polyphenol/UHMWPE materials for clinical applications as artificial joint components.

References

1. Kurtz SM (ed) (2009) UHMWPE biomaterials handbook. Ultra-high molecular weight polyethylene in total joint replacement and medical devices, 2nd edn. Elsevier Inc., New York
2. Harris W (1995) The problem is osteolysis. Clin Orthop 311:46–53
3. Kurtz SM, Medel FJ, MacDonald DW, Parvizi J, Kraay MJ, Rimnac CM (2010) Reasons for revision of first-generation highly cross-linked polyethylenes. J Arthroplast 25(6):67–74
4. Muratoglu OK, Bragdon CR, O'Connor DO, Jasty M, Harris WH (2001) A novel method of cross-linking ultra-high-molecular-weight polyethylene to improve wear, reduce oxidation, and retain mechanical properties. Recipient of the 1999 HAP Paul award. J Arthroplasty 16 (2):149–160
5. Kurtz SM, Manley M, Wang A, Taylor S, Dumbleton J (2002) Comparison of the properties of annealed crosslinked (Crossfire) and conventional polyethylene as hip bearing materials. Bulletin (Hospital for Joint Diseases (New York, NY)) 61(1–2):17–26
6. Dumbleton J, D'Antonio J, Manley M, Capello W, Wang A (2006) The basis for a second-generation highly cross-linked UHMWPE. Clin Ortho Relat Res 453:265–271
7. Medel FJ, Pena P, Cegonino J, Gomez-Barrena E, Puertolas JA (2007) Comparative fatigue behavior and toughness of remelted and annealed highly crosslinked polyethylenes. J Biomed Mater Res B Appl Biomater 83(2):380–390
8. Atwood SA, Van Citters DW, Patten EW, Furmanski J, Ries MD, Pruitt LA (2011) Tradeoffs amongst fatigue, wear, and oxidation resistance of cross-linked ultra-high molecular weight polyethylene. J Mech Behav Biomed Mater 4(7):1033–1045
9. Tower SS, Currier JH, Currier BH, Lyford KA, Van Citters DW, Mayor MB (2007) Rim cracking of the cross-linked longevity polyethylene acetabular liner after total hip arthroplasty. J Bone Joint Surg Am 89(10):2212–2217
10. Puertolas JA, Medel FJ, Cegonino J, Gomez-Barrena E, Rios R (2006) Influence of the remelting process on the fatigue behavior of electron beam irradiated UHMWPE. J Biomed Mater Res Part B Appl Biomater 76B:346–353
11. Costa L, Bracco P, Brach del Prever EM, Luda MP, Trossarelli L (2001) Analysis of products diffused into UHMWPE prosthetic components in vivo. Biomaterials 22(4):307–315
12. Currier BH, Van Citters DW, Currier JH, Collier JP (2010) In vivo oxidation in Remelted highly cross-linked retrievals. J Bone Joint Surg 92(14):2409–2418
13. Muratoglu OK, Wannomae KK, Rowell SL, Micheli BR, Malchau H (2010) Ex vivo stability loss of irradiated and melted ultra-high molecular weight polyethylene. J Bone Joint Surg 92 (17):2809–2816
14. Currier BH, Van Citters DW, Currier JH, Carlson EM, Tibbo ME, Collier JP (2013) In vivo oxidation in retrieved highly crosslinked tibial inserts. J Biomed Mater Res B Appl Biomater 101B(3):441–448
15. Tomita N, Kitakura T, Onmori N, Ikada Y, Aoyama E (1999) Prevention of fatigue cracks in ultrahigh molecular weight polyethylene joint components by the addition of vitamin E. J Biomed Mater Res 48(4):474–478
16. Wolf C, Krivec T, Blassing J, Lederer K, Schneider W (2002) Examination of the suitability of a-tocopherol as a stabilizer for ultra-high molecular weight polyethylene used for articulating surfaces in joint endoprostheses. J Mater Sci Mater Med 13:185–189
17. Oral E, Wannomae KK, Hawkins N, Harris WH, Muratoglu OK (2004) Alpha-tocopherol-doped irradiated UHMWPE for high fatigue resistance and low wear. Biomaterials 25 (24):5515–5522
18. Shibata N, Tomita N (2005) The anti-oxidative properties of alpha-tocopherol in gamma-irradiated UHMWPE with respect to fatigue and oxidation resistance. Biomaterials 26 (29):5755–5762
19. Gijsman P, Smelt H, Schumann D (2010) Hindered amine light stabilizers: an alternative for radiation cross-linked UHMwPE implants. Biomaterials 31:6685–6691

20. Almalaika S, Ashley H, Issenhuth S (1994) The antioxidant role of alpha-tocopherol in polymers. 1. The nature of transformation products of alpha-tocopherol formed during melt processing of LDPE. J Polym Sci Part A Polym Chem 32(16):3099–3113
21. Al-Malaika S, Goodwin C, Issenhuth S, Burdick D (1999) The antioxidant role of alpha-tocopherol in polymers II. Melt stabilising effect in polypropylene. Polym Degrad Stab 64 (1):145–156
22. Mallegol J, Carlsson DJ, Deschenes L (2001) Antioxidant effectiveness of vitamin E in HDPE and tetradecane at 32 degrees C. Polym Degrad Stab 73(2):269–280
23. Oral E, Malhi AS, Wannomae KK, Muratoglu OK (2008) Highly cross-linked ultrahigh molecular weight polyethylene with improved fatigue resistance for total joint arthroplasty – recipient of the 2006 Hap Paul Award. J Arthroplasty 23(7):1037–1044
24. Oral E, Godleski Beckos C, Malhi AS, Muratoglu OK (2008) The effects of high dose irradiation on the cross-linking of vitamin E-blended ultrahigh molecular weight polyethylene. Biomaterials 29(26):3557–3560
25. Oral E, Greenbaum ES, Malhi AS, Harris WH, Muratoglu OK (2005) Characterization of irradiated blends of alpha-tocopherol and UHMWPE. Biomaterials 26(33):6657–6663
26. Oral E, Christensen SD, Malhi AS, Wannomae KK, Muratoglu OK (2006) Wear resistance and mechanical properties of highly cross-linked, ultrahigh-molecular weight polyethylene doped with vitamin E. J Arthroplasty 21(4):580–591. https://doi.org/10.1016/j.arth.2005.07.009
27. Bracco P, Brunella V, Zanetti M, Luda MP, Costa L (2007) Stabilisation of ultra-high molecular weight polyethylene with vitamin E. Polym Degrad Stab 92:2155–2162
28. Costa L, Carpentieri I, Bracco P (2009) Post electron-beam irradiation oxidation of orthopedic Ultra-High Molecular Weight Polyethylene (UHMWPE) stabilized with vitamin E. Polym Degrad Stab 94(9):1542–1547
29. Kurtz SM, Dumbleton J, Siskey RS, Wang A, Manley M (2009) Trace concentrations of vitamin E protect radiation crosslinked UHMWPE from oxidative degradation. J Biomed Mater Res A 90A(2):549–563
30. Parth M, Aust N, Lederer K (2002) Studies on the effect of electron beam radiation on the molecular structure of ultra-high molecular weight polyethylene under the influence of alpha-tocopherol with respect to its application in medical implants. J Mater Sci Mater Med 13 (10):917–921
31. Oral E, Wannomae KK, Rowell SL, Muratoglu OK (2007) Diffusion of vitamin E in ultra-high molecular weight polyethylene. Biomaterials 28(35):5225–5237
32. Fiuza SM, Gomes C, Teixeira LJ, Girão da Cruz MT, Cordeiro MNDS, Milhazes N, Borges F, Marques MPM (2004) Phenolic acid derivatives with potential anticancer properties—a structure–activity relationship study. Part 1: Methyl, propyl and octyl esters of caffeic and gallic acids. Bioorg Med Chem 12(13):3581–3589
33. Cirillo G, Kraemer K, Fuessel S, Puoci F, Curcio M, Spizzirri UG, Altimari I, Iemma F (2010) Biological activity of a gallic acid−gelatin conjugate. Biomacromolecules 11(12):3309–3315
34. Fu J, Shen J, Gao G, Hou R, Xu Y, Cong Y, Cheng Y (2013) Natural polyphenol-stabilised highly crosslinked UHMWPE with high mechanical properties and low wear for joint implants. J Mater Chem B 1:4727–4735
35. Wolf C, Macho C, Lederer K (2006) Accelerated ageing experiments with crosslinked and conventional ultra-high molecular weight polyethylene (UHMW-PE) stabilised with α-tocopherol for total joint arthroplasty. J Mater Sci Mater Med 17(12):1333–1340. https://doi.org/10.1007/s10856-006-0608-6
36. Mallégol J, Carlsson DJ, Deschênes L (2001) Antioxidant effectiveness of vitamin E in HDPE and tetradecane at 32°C. Polym Degrad Stab 73(2):269–280
37. Furmanski J, Anderson M, Bal S, Greenwald AS, Halley D, Penenberg B, Ries M, Pruitt L (2009) Clinical fracture of cross-linked UHMWPE acetabular liners. Biomaterials 30 (29):5572–5582

38. Muratoglu OK, Bragdon CR, O'Connor DO, Jasty M, Harris WH, Gul R, McGarry F (1999) Unified wear model for highly crosslinked Ultra-high Molecular Weight Polyethylenes (UHMWPE). Biomaterials 20(16):1463–1470
39. Pruitt LA (2005) Deformation, yielding, fracture and fatigue behavior of conventional and highly cross-linked ultra high molecular weight polyethylene. Biomaterials 26(8):905–915
40. Lacoste J, Carlsson DJ (1992) Gamma-initiated, photo-initiated, and thermally-initiated oxidation of linear low-density polyethylene – a quantitative comparison of oxidation-products. J Polym Sci Part A Polym Chem 30(3):493–500. https://doi.org/10.1002/pola.1992.080300316
41. Shen J, Costa L, Xu Y, Cong Y, Cheng Y, Liu X, Fu J (2014) Stabilization of highly crosslinked ultra high molecular weight polyethylene with natural polyphenols. Polym Degrad Stab 105:197–205
42. Salvalaggio M, Bagatin R, Fornaroli M, Fanutti S, Palmery S, Battistel E (2006) Multicomponent analysis of low-density polyethylene oxidative degradation. Polym Degrad Stab 91(11):2775–2785
43. Costa L, Luda MP, Trossarelli L (1997) Ultra high molecular weight polyethylene-II. Thermal- and photo-oxidation. Polym Degrad Stab 58:41–54
44. Costa L, Carpentieri I, Bracco P (2008) Post electron-beam irradiation oxidation of orthopaedic UHMWPE. Polym Degrad Stab 93(9):1695–1703
45. Costa L, Bracco P, del Prever EB, Luda MP, Trossarelli L (2001) Analysis of products diffused into UHMWPE prosthetic components in vivo. Biomaterials 22(4):307–315
46. Oral E, Ghali BW, Neils A, Muratoglu OK (2012) A new mechanism of oxidation in ultrahigh molecular weight polyethylene caused by squalene absorption. J Biomed Mater Res B Appl Biomater 100B(3):742–751. https://doi.org/10.1002/jbm.b.32507
47. Fu J, Doshi BN, Oral E, Muratoglu OK (2013) High temperature melted, radiation cross-linked, vitamin E stabilized oxidation resistant UHMWPE with low wear and high impact strength. Polymer 54(1):199–209
48. Shen J, Liu X, Fu J (2014) Effect of squalene absorption on oxidative stability of highly crosslinked UHMWPE stabilized with natural polyphenols. Polym Degrad Stab 110:113–120
49. Fu J, Shen J, Gao G, Xu Y, Hou R, Cong Y, Cheng Y (2013) Natural polyphenol-stabilised highly crosslinked UHMWPE with high mechanical properties and low wear for joint implants. J Mater Chem B 1(37):4727–4735
50. Oral E, Ghali BW, Neils A, Muratoglu OK (2012) A new mechanism of oxidation in ultrahigh molecular weight polyethylene caused by squalene absorption. J Biomed Mater Res Part B Appl Biomater 100B(3):742–751

Chapter 5
High-Temperature Melted, Cross-Linked, and Stabilized Ultrahigh Molecular Weight Polyethylene

Jun Fu, Ebru Oral, and Orhun K. Muratoglu

Abstract Highly cross-linked ultrahigh molecular weight polyethylene has been one of the most popular materials for the articular surfaces of joint implants. Major challenges in clinical use include the sacrificed mechanical properties and susceptibility to in vivo oxidation due to the high dose radiation cross-linking and subsequent thermal treatments. It is desired to develop highly cross-linked UHMWPE with improved strength and toughness. This chapter introduces a method to improve the ductility of UHMWPE by melting at very high temperatures to eliminate the structural defects. Subsequently, the high-temperature melted (HTM) UHMWPE was radiation cross-linked to generate low cross-link densities in comparison to virgin UHMWPE receiving the same radiation doses. The obtained radiation cross-linked HTM UHMWPE shows low wear and high strength. This HTM method is also used to vitamin E-stabilized UHMWPE, which shows outstanding strength, oxidation stability, and wear resistance. The clinical relevance is discussed.

Keywords High-temperature melting · UHMWPE · Oxidation · Vitamin E · Structural defect · Wear · Tensile strength · Ductility · Crystal structures · Pin-on-disc

5.1 Introduction

Highly cross-linked ultrahigh molecular weight polyethylene (UHMWPE) with low wear and good mechanical properties has been very successful in artificial joints for clinical use since the first FDA approval in 1998 [1–3]. Midterm follow-up studies

J. Fu (✉)
Ningbo Institute of Materials Technology and Engineering, Chinese Academy of Sciences, Ningbo, China
e-mail: fujun@nimte.ac.cn

E. Oral · O. K. Muratoglu
Harris Orthopaedics Laboratory, Massachusetts General Hospital, Boston, MA, USA

Department of Orthopaedic Surgery, Harvard Medical School, Boston, MA, USA

J. Fu et al. (eds.), *UHMWPE Biomaterials for Joint Implants*, Springer Series in Biomaterials Science and Engineering 13,
https://doi.org/10.1007/978-981-13-6924-7_5

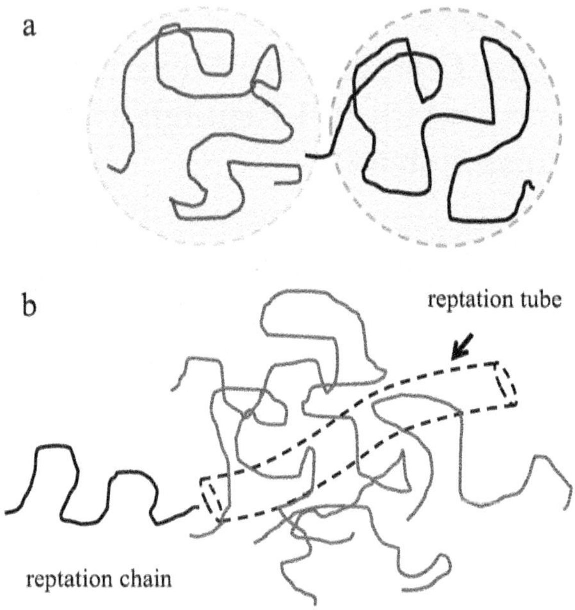

Fig. 5.1 (**a**) Schematic illustration of polymer chain diffusion across the UHMWPE granules and (**b**) polymer chain reptation through the virtual tubes of neighboring chains

on implants based on medium or highly cross-linked UHMWPE have demonstrated a significant reduction in osteolysis and wear-related implant failure [3] and suggest promising long-term performance and longevity of implants based on highly cross-linked UHMWPE [4]. With an increasing population of younger and more active patients, joint implants with improved longevity, reduced wear, and lower risks of early failure are desired [5]. Essentially, the key to these demands relies on the development of new materials with excellent wear resistance, fatigue resistance, strength, and toughness. Current highly cross-linked UHMWPE has low wear, but their strength and toughness are compromised due to the formation of cross-linked networks and post-irradiation thermal treatments [6–8]. It is very important to further improve the strength and toughness of UHMWPE before or after radiation cross-linking.

UHMWPE is known for its very high viscosity in molten state due to its ultrahigh molecular weight ($2{\sim}10 \times 10^6$ g/mol), which makes it difficult to process by using conventional methods such as injection molding. Current processing through compression molding or ram extrusion consolidates UHMWPE powders into blocks at high temperatures (~190–220 °C) [9]. According to reptation theory by de Gennes [10] and the tube model of Doi-Edwards [11], the motion of a single linear chain is described as restricted by the topological constraints ("tube") from the neighboring chains so that this chain can only relax along the tube axis (Fig. 5.1). Based on dynamic thermal analysis, the terminal relaxation time of entangled UHMWPE chains is about 15 h at 180 °C [12], much longer than the conventional processing time (less than 1 h in general). Thus, it is very difficult for polymer chains to completely diffuse across the granular boundaries (Fig. 5.1), inevitably leaving huge amount of fusion defects inside the consolidated batches [13, 14]. Two types of fusion defects are found. Type 1 fusion defects, or incomplete interparticle voids, can be removed by properly controlling the consolidation conditions. However, type 2 fusion defects (incomplete interparticle cohesion by self-diffusion) are difficult to remove. The local strength and toughness near granular boundaries are lower than those of bulk UHMWPE [15]. These fusion defects are potential sites for failure under cyclic loadings. Moreover, upon radiation cross-linking, such structural defects and mechanical weakness could be readily exaggerated, which may further decrease the strength and toughness of the materials. Therefore, one hypothesis is that the elimination of structural defects and better fusion of UHMWPE resins may significantly improve its strength and toughness.

In order to eliminate the structural defects, two methods can be used. The first is to add processing aids to reduce the melt viscosity so that the melted resin particles can fuse efficiently. This has been widely used in UHMWPE industry, but it is forbidden by ASTM F648, which reads "No stabilizers, antioxidants, or processing aids are to be added to the virgin polymer powder during manufacture of a fabricated form (5.1.1)" and "No stabilizers, antioxidants, or processing aids are to be added to the fabricated form during manufacture of the final implant (5.1.2)." Although vitamin E-blended UHMWPE has been clinically used in artificial joints, there are so far no processing aids to be added to facilitate the processing of UHMWPE by lowering the melt viscosity. Therefore, it is not possible to promote granule fusion by adding processing aids.

Another method is to lower the viscosity and accelerate chain diffusion by elevating the processing temperature. The viscosity (η) of linear polymer melt decreases with increasing temperature. The chain diffusion and granule fusion may be accelerated by melting at higher temperatures. For UHMWPE, its ultralong chains may readily degrade at very high temperatures, even before detectible volatiles are formed ($\leq$ 300 °C) [16]. Therefore, optimal melting conditions are critical for effective elimination of fusion defects before UHMWPE has been severely degraded.

In this chapter, we demonstrate the use of high-temperature melting of consolidated UHMWPE to eliminate the fusion defects and manipulate the crystal structures and thus improve the mechanical properties of UHMWPE. The melting temperature window has been determined by thermal gravimetric analysis (TGA) at a series of temperatures over time in order to avoid severe thermal degradation of polymer chains. The as-consolidated UHMWPE blocks are melted at selected temperatures for a certain period in vacuum or nitrogen before spontaneously cooling down to room temperature. The tensile and impact toughness, as well as ductility, have been significantly improved in comparison to conventionally consolidated UHMWPE blocks. These high-temperature melted (HTM) polyethylenes are further cross-linked by electron beam to reduce wear to the levels close to those of highly cross-linked UHMWPEs. Such cross-linked HTM polyethylenes show higher tensile properties and impact strength than those for the first-generation highly cross-linked UHMWPE.

In order to stabilize such high-temperature melted and cross-linked UHMWPE, vitamin E is blended with UHMWPE powders before consolidation. Vitamin E remains its antioxidation potency during high-temperature melting and offers excellent oxidation resistance even after high dose irradiation. Tensile tests, accelerated aging, and wear tests have been conducted, showing encouraging results in comparison to highly cross-linked UHMWPE.

Finally, the clinical relevance of these high-temperature melted, vitamin E-stabilized, and radiation cross-linked UHMWPE is discussed.

5.2 High-Temperature Melting

In order to accelerate self-diffusion of UHMWPE chains across the granular boundaries, compression-molded (CM) UHMWPE is melted at temperatures higher than the consolidation temperature in inert atmosphere to avoid oxidation. At high temperatures, it is well established that UHMWPE chains will start to degrade before volatiles are generated. Such degradation through chain scission usually decreases the molecular weight and viscosity, which benefits the self-diffusion of polymer chains. If the temperature is too high (e.g., > 500 °C), severe degradation may occur, generating volatiles and leading to decomposition of polymer chains [16]. Herein, the melting temperature should be adequate to eliminate the fusion defects, but should not be too high to severely break the chains and deteriorate the mechanical properties and wear resistance.

5.2.1 Determination of the Working Temperatures

In order to determine the working temperature for the melting of consolidated UHMWPE, TGA of CM UHMWPE has been systematically conducted at temperatures from 220 °C to 400 °C with a 20 °C increment in nitrogen atmosphere for up to 24 h with the mass recorded over time. The criteria to determine the working temperature include the following: (1) the weight loss over 12 h should not exceed 5%, (2) the fusion defects should be completely eliminated according to scanning electron microscopy (SEM) images, and (3) the mechanical properties should be significantly enhanced in comparison to the consolidated UHMWPE.

Figure 5.2 shows the TGA traces of compression-molded UHMWPE samples at different high temperatures in nitrogen. The weight loss is less than 3.4% for UHMWPE remained at 320 °C for 24 h. At higher temperatures (e.g., 340 °C), severe degradation occurs. Therefore, temperatures below 320 °C are used to melt consolidated UHMWPE.

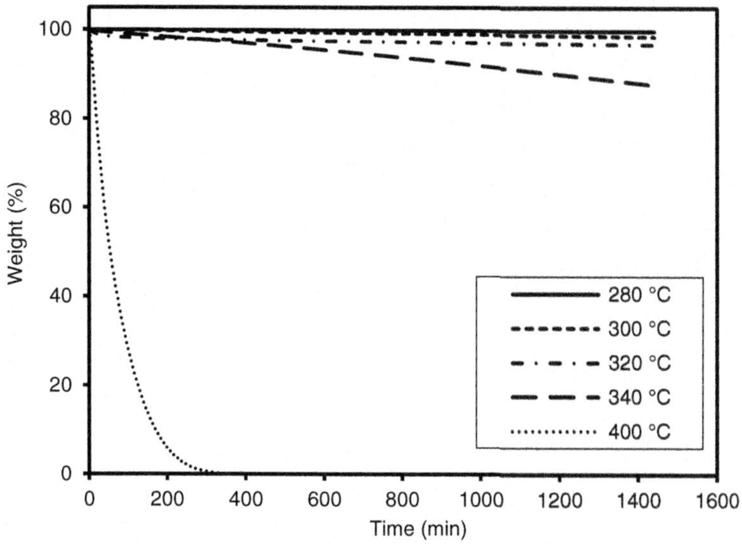

Fig. 5.2 TGA traces of consolidated UHMWPE at different temperatures in nitrogen [17]. (Copyright 2010 Elsevier)

5.2.2 Granule Fusion by High-Temperature Melting

Melting the consolidated UHMWPE at temperatures between 280 °C and 320 °C in inert atmosphere effectively eliminates the fusion defects. After melting at high temperatures, the materials are spontaneously cooled down to room temperature and freeze-fractured in liquid nitrogen to expose the internal structures. The fractured surfaces are investigated by using SEM. Figure 5.3 shows representative SEM images for compression-molded UHMWPE and those melted at 280, 300, and 320 °C for up to 12 h. Numerous fusion defects are found among the granules (Fig. 5.3a). Such fusion defects are eliminated at 280 °C for 2 h, with boundaries discernible (Fig. 5.3b). The granule size is very close to that of the nascent UHMWPE resins. With increasing melting temperature and time, the granule sizes become smaller (Fig. 5.3c, d). These results suggest that the polymer chain diffusion across the granule interfaces is largely accelerated and enhanced at high temperatures over time. When melted at 320 °C for 2 h, for example, the granule boundaries become ambiguous, and no granules are discernible inside these materials (Fig. 5.3e). After further extension of melting time to 5 h, the UHMWPE batch becomes so tough that ductile tearing patterns are observed on the freeze-fractured surface (Fig. 5.3f).

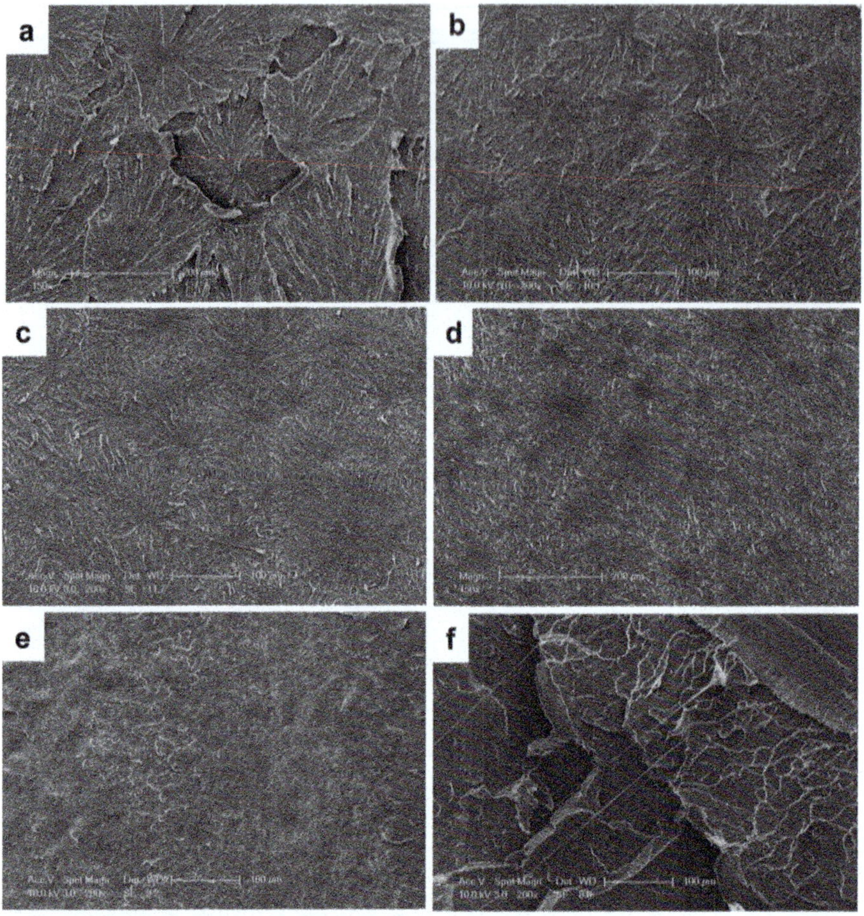

Fig. 5.3 SEM images of freeze-fractured surfaces of (**a**) compression-molded UHMWPE and those melted at (**b**) 280 °C for 2 h, (**c**) 280 °C for 5 h, (**d**) 300 °C for 5 h, (**e**) 320 °C for 2 h, and (**f**) 320 °C for 5 h [17]. (Copyright 2010 Elsevier)

5.2.3 Chain Degradation at High Temperatures

Due to the ultrahigh molecular weight, the polymer chains are prone to degradation at high temperatures. The degradation mechanisms are related to the temperature. At temperatures no higher than 300 °C or so, polymer chains start to break through C-C scissioning, generating free radicals that may terminate or recombine. At much higher temperatures (e.g., above 500 °C), severe and extensive chain scissioning occurs, leading to cascade generation of free radicals and formation of volatiles. Possible thermal degradation mechanisms for UHMWPE in inert atmosphere have been extensively investigated [16]. Herein, the degradation mechanisms of

$$\underset{P}{\sim CH_2CH_2\text{-}CH_2CH_2\sim} \xrightarrow{k_i} \underset{R_p\bullet}{\sim CH_2CH_2\bullet} + \underset{R_p\bullet}{\bullet CH_2CH_2\sim} \tag{5.1}$$

$$\underset{P}{\sim CH_2CH_2\text{-}Y\sim} \xrightarrow{k_{iw}} \underset{R_p\bullet}{\sim CH_2CH_2\bullet} + \bullet Y\sim \tag{5.2}$$

$$\underset{R_p\bullet}{\sim CH_2CH_2CH_2\bullet} \xrightarrow{k_\beta} \underset{R_p\bullet}{\sim CH_2\bullet} + CH_2{=}CH_2 \tag{5.3}$$

$$\underset{R_p\bullet}{\sim CH_2CH_2\bullet} + \underset{P}{\sim CH_2CH_2CH_2CH_2CH_2\sim} \xrightarrow{k_H} \underset{PANE}{\sim CH_2CH_3} + \underset{R_s\bullet}{\sim CH_2CH_2CH_2CH(\bullet)CH_2\sim} \tag{5.4}$$

$$\underset{R_s\bullet}{\sim CH_2CH_2CH_2CH(\bullet)CH_2\sim} \xrightarrow{k_\beta'} \underset{R_p\bullet}{\sim CH_2CH_2\bullet} + \underset{PANE}{CH_2{=}CHCH_2\sim} \tag{5.5}$$

$$\underset{R_P\bullet}{\sim CH_2CH_2CH_2CH_2CH_2(CH_2)_{x\text{-}4}CH_2\bullet} \xrightarrow{k_{1,x}} \underset{R_s'\bullet}{\sim CH_2CH_2CH(\bullet)CH_2CH_2(CH_2)_{x\text{-}4}CH_3} \tag{5.6}$$

Fig. 5.4 Degradation reactions of UHMWPE at high temperatures

UHMWPE at about 300 °C are illustrated in Fig. 5.4, according to studies by Poutsma et al.

Initiation of degradation takes place either by C-C scission in the polyethylene backbone (Eq. 5.1) or by homolysis of "weak links" (Eq. 5.2), generating alkyl free radicals $R_p\bullet$. In UHMWPE, the ultralong chain length and ultralow branch density may favor the C-C homolysis. Thus, the molecular weight and melt viscosity could decrease. The $R_p\bullet$ radicals are prone to propagate through two pathways. (1) β-scission: the free radical generates itself and generates a monomer, without decreasing its degree of polymerization (Eq. 5.3). (2) Abstraction of hydrogen from a polymer chain. In this way, the radical is transferred to another chain, thus generating a PANE and a new secondary free radical $R_s\bullet$ (Eq. 5.4). Such $R_s\bullet$ radicals are prone to propagate through β-scission to generate $R_p\bullet$ and a PENE (Eq. 5.5). Again, such $R_p\bullet$ radicals can undergo intramolecular shift of free radical to form a secondary on-chain radical (Eq. 5.6). Such propagation can proceed through pathways similar to those illustrated in Fig. 5.4. At temperatures below 300 °C, no severe propagation toward generation of volatiles occurs. The free radicals are entrapped by the very high melt viscosity and recombine before degradation into small molecule volatiles. As a result, the molecular weight of UHMWPE slightly decreases, generating PENE or polymer chains with vinyl end groups. For details on the degradation mechanisms, see reference [16] and related work.

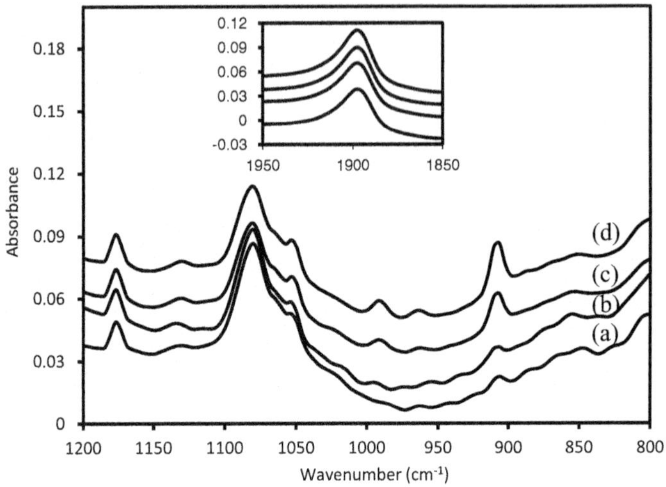

Fig. 5.5 FTIR spectra of CM UHMWPE (**a**) before and after melting at 320 °C for (**b**) 2 h, (**c**) 5 h, and (**d**) 12 h [17]. (Copyright 2010 Elsevier)

According to these mechanisms, unsaturated bonds (e.g., vinyl end groups) are generated as a result of thermal degradation of UHMWPE. Such terminal double bonds show an absorbance at 909 cm^{-1} at Fourier transform infrared (FTIR) spectrum (Fig. 5.5). The band intensity is indicative of chain scission at high temperatures. For CM UHMWPE, the vinyl number is low (Fig. 5.5a). When melted at very high temperatures, the band intensity increases, indicating that the vinyl group content becomes higher than the nascent UHMWPE resins. The vinyl index increases with increasing melting temperature and time. At temperatures below 280 °C, the vinyl content is not significantly different from that before melting, indicating negligible degradation of polymer chains. At higher temperatures, e.g., 300 and 320 °C, the vinyl content becomes much higher. Such an increase in vinyl group content is nearly proportional to the melting temperature and time (Fig. 5.6). Therefore, the vinyl group index calculated according to FTIR spectra is used to quantitatively characterize the degree of degradation. As to be presented below, such degradation significantly enhances the toughness of UHMWPE.

5.2.4 Crystalline Structures of High-Temperature Melted UHMWPE

High-temperature melting changes the crystalline structures of UHMWPE. UHMWPE is a typical semicrystalline polymer comprised of crystalline phase and amorphous phase. In most cases, the polyethylene chains fold into lamellae with thickness of a few to tens of nanometers and lateral dimension from a few to

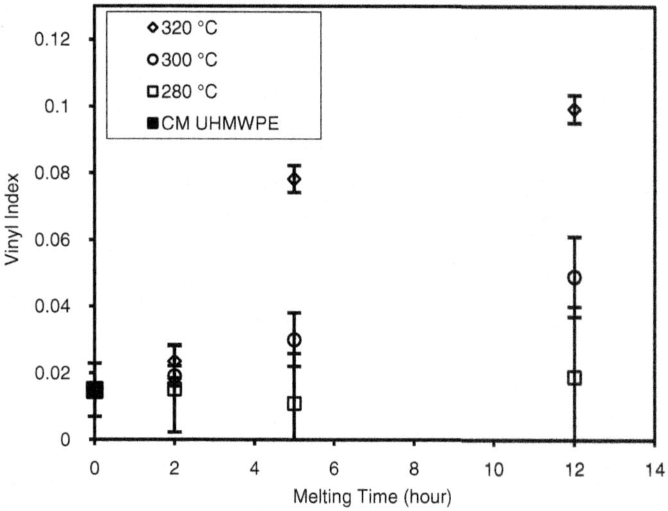

Fig. 5.6 The vinyl index of UHMWPE as a function of melting temperature over time [17]. (Copyright 2010 Elsevier)

hundreds of micrometers. The lamellae pack together to form spherulites with diameters from a few to hundreds of micrometers. Upon loading, the crystalline phase behaves elastically until the rupture of spherulite crystals and the slippage of lamellae and polymer chains therein. At this point, the polymer yields and experiences a plastic deformation. The crystal lamella thickness, spherulite size, and crystallinity are critical to the strength, modulus, and toughness of polymer materials.

The mechanical properties of semicrystalline polymer materials are strongly related to their molecular weight, molecular weight distribution, and crystal structures. In general, processing at temperatures much higher than their peak melting points (T_m) may induce degradation, oxidation, rearrangement, or cross-linking of polymer chains. For ultralong polymer chains, the chain entanglements are enhanced, which has significant influences on the strength and toughness of polymer materials. Moreover, the crystalline lamella size, spherulite dimension, and crystallinity of polymers can be largely altered after high-temperature melting.

UHMWPE has a very strong tendency to fold into crystals upon polymerization [18], generating very high crystallinity for as-synthesized resins. Such resins contain very low entanglements and thus show outstanding ductility [19]. During consolidation at temperatures much higher than its T_m, the polymer chains establish entanglements immediately through a well-known chain explosion process [20]. Such entanglements are unfavorable to the formation of UHMWPE crystals. As a result, the consolidated UHMWPE show crystallinity and melting points lower than those for the nascent UHMWPE powders. For example, the melting point of nascent UHMWPE powders is about 144–147 °C, with a crystallinity of 65% to 75%. After

Table 5.1 Crystallinity (X_c) and peak melting point (T_m) of HTM UHMWPE melted at different temperatures [17]

		T_m (°C)		X_c (%)
Sample code	Description	1st heat	2nd heat	1st heat
CM UHMWPE	No melting	135.1 ± 0.1	133.0 ± 0.1	51.9 ± 2.9
UH 280–2	280 °C, 2 h	133.7 ± 0.2	132.3 ± 0.2	55.2 ± 1.3
UH 280–5	280 °C, 5 h	133.1 ± 0.1	133.0 ± 0.1	54.5 ± 0.7
UH 280–12	280 °C, 12 h	133.5 ± 0.2	132.3 ± 0.3	56.5 ± 0.9
UH 300–2	300 °C, 2 h	133.6 ± 0.1	132.3 ± 0.05	56.9 ± 1.7
UH 300–5	300 °C, 5 h	134.0 ± 0.1	133.0 ± 0.4	60.5 ± 0.4
UH 300–12	300 °C, 12 h	133.7 ± 0.1	132.6 ± 0.2	61.9 ± 1.4
UH 320–2	320 °C, 2 h	133.9 ± 0.3	132.6 ± 0.04	60.6 ± 0.3
UH 320–5	320 °C, 5 h	134.1 ± 0.2	132.9 ± 0.4	61.9 ± 2.1
UH 320–12	320 °C, 12 h	133.7 ± 0.2	132.9 ± 0.7	63.5 ± 2.6

consolidation, the melting point decreases down to 136 °C to 142 °C, with the crystallinity of 50%~55%.

As the melting temperature is increased to 280, 300, and 320 °C, the chain entanglement is accelerated and increased. As a result, the peak melting point (T_m) and crystallinity are further decreased. In comparison to the compression-molded UHMWPE, the T_m of HTM UHMWPE decreases to 133~134 °C, while the crystallinity increases to 55%~62% (Table 5.1). The decreased T_m values are related to thinner lamellae. Despite the subtle difference in T_m, the crystallinity (X_c) is strongly related to the melting temperature and time. As the melting temperature is increased from 280 to 300 and 320 °C, the X_c is increased from about 56% to 60% and 62%. At each temperature, the X_c value increases with longer melting time. The increase in X_c values suggest an enhanced capability of polymer chains to fold into lamellae. At high temperatures, the polymer chains may slightly degrade, as suggested by the linearly increasing vinyl index with temperature and time (Fig. 5.6). The degradation of UHMWPE chains into shorter ones may allow for higher crystallinity.

The changes in chain entanglements and molecular weight significantly affect the crystal structures of UHMWPE. SEM images of UHMWPE before and after high-temperature melting with the amorphous phase etched away show representative crystal spherulites (Fig. 5.7). Different from the granule structures shown in Fig. 5.3, the crystal spherulites are much smaller after high-temperature melting.

Each granule contains several spherulites. There are no obvious boundaries between different spherulites. Within the spherulites of CM UHWMPE, the polyethylene lamellae are randomly distributed (Fig. 5.7b). After high-temperature melting, the spherulites grow into larger sizes. When melted at 300 °C for 12 h, for example, the average spherulite size becomes similar to the granule size (Fig. 5.7c). Some spherulites grow across the granules boundaries. Spherulites across the granule boundaries become prevailing for UHMWPE melted at 320 °C (Fig. 5.7e). At such a high temperature, ring-banded spherulites start to form when

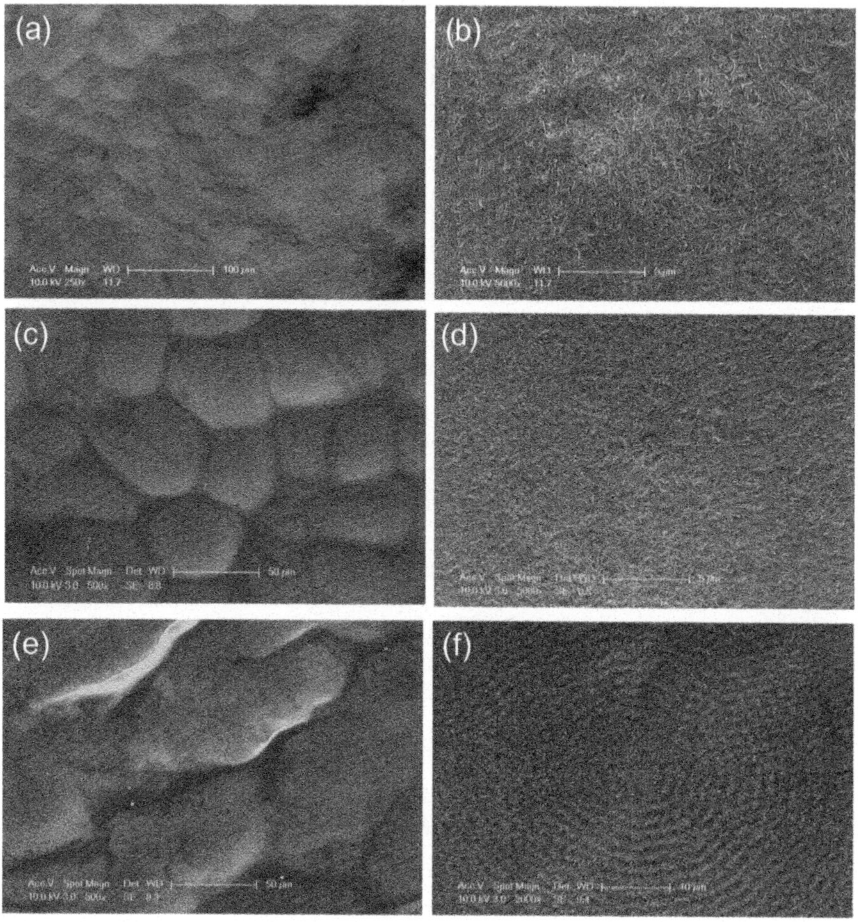

Fig. 5.7 SEM images of (**a, b**) compression-molded UHMWPE and those melted at (**c, d**) 300 °C and (**e, f**) 320 °C for 5 h, with the amorphous phase etched away

the melting time is extended from 2 to 5 and 12 h. Figure 6.7f shows the ring-banded spherulites of UHMWPE melted at 320 °C for 12 h. The granules completely fuse and the spherulites are much larger than the granules. These spherulites are featured with concentric bands from the center (Fig. 5.7f). Such ring-banded spherulites are formed due to the hindered and frustrated growth of UHMWPE lamellae from the nuclei [21].

Figure 5.8 shows representative SEM images of compression-molded UHMWPE before and after melting at 320 °C for 5 h. The polyethylene lamellae in spherulites before melting appear random orientation and distribution (Fig. 5.8a). In contrast, the lamellae within the ring-banded spherulites show periodic twisting from the center of the spherulite (Fig. 5.8b). In average, the lamellae adopt flat-on configuration at the center, which grow outward and gradually twist into edge-on lamellae

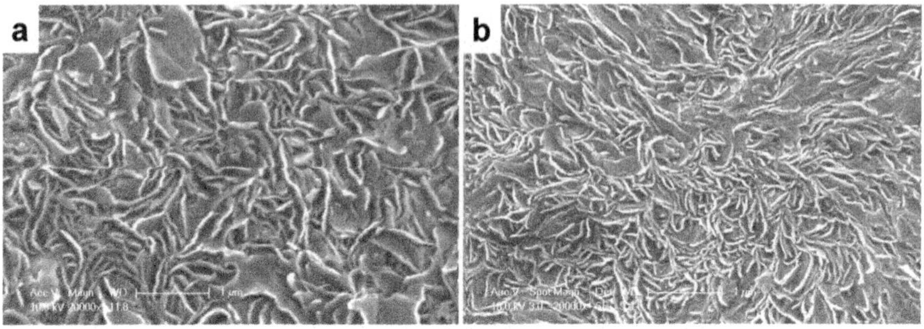

Fig. 5.8 SEM images of crystalline lamellae of (**a**) compression-molded UHMWPE and (**b**) those melted at 320 °C for 5 h, with the amorphous phase etched away [17]. (Copyright 2010 Elsevier)

and so on. It is interesting that different lamellae follow the similar periodicity when growing outward from the center. The twisted growth of lamellae is believed to be due to the frustrated chain folding from the melt. The periodicity indicates an intrinsic tension between the viscous melt and the growth front of the lamellae. However, detailed mechanisms remain open for further investigations. Herein, it is likely that the degraded chains and lowered melt viscosity may account for the formation of ring-banded spherulites. In addition to the ring-banded spherulites, the average lamella thickness after high-temperature melting is less than those in the CM UHMWPE.

5.2.5 Mechanical Properties of High-Temperature Melted UHMWPE

As mentioned before, high-temperature melting promotes the chain entanglements, grows the spherulites, reduces crystal lamella thickness, and improves crystallinity. The structure changes due to high-temperature melting result in significant changes in the strength, toughness, and ductility of UHMWPE. Table 5.2 summarizes the tensile properties of UHMWPE melted at various temperatures, in comparison to that of CM UHMWPE. Melting at 280 °C for 5 h, for example, does not significantly influence the tensile properties. As the temperature is increased to 300 and 320 °C, the elongation at break (EAB) is astonishingly increased from about 400% to 700% and 1100%. During the very high stretching, the translucent gauge gradually becomes opaque, particularly expanding from the gauge center (Fig. 5.9).

This phenomenon suggests the break and reorganization of UHMWPE spherulites under such a high strain. This behavior strongly suggests that the elimination of internal structural defects and promotion of semicrystalline structures of UHMWPE by high-temperature melting significantly improve the toughness, since the probabilities for the formation and propagation of cracks inside UHMWPE are largely reduced. These structural changes largely increase the elongation at break (EAB).

Table 5.2 Tensile and impact properties of UHMWPE melted at different temperatures [17]

Materials	Melting condition	UTS (MPa)	YS (MPa)	EAB (%)	WF (kJ/m^2)	Impact strength (kJ/m^2)
UHMWPE	N.A.	59 ± 4.5	21 ± 0.8	401 ± 15	3242 ± 444	127 ± 6.7
UH 280–2	280 °C, 2 h	56 ± 1.2	20	444 ± 1	3314 ± 167	N.A.
UH 280–5	280 °C, 5 h	52 ± 6.3	20 ± 0.5	473 ± 36	3120 ± 637	N.A.
UH 280–12	280 °C, 12 h	55 ± 4.7	20 ± 0.6	513 ± 16	3673 ± 550	175 ± 9.5
UH 300–2	300 °C, 2 h	53 ± 3.3	20	457 ± 20	3266 ± 379	123 ± 5.8
UH 300–5	300 °C, 5 h	60 ± 3.3	21 ± 0.6	546 ± 37	4691 ± 653	140 ± 3.9
UH 300–12	300 °C, 12 h	60 ± 4.5	22 ± 0.6	752 ± 23	6646 ± 991	111 ± 4.5
UH 320–2	320 °C, 2 h	60 ± 2.2	21 ± 0.5	521 ± 24	4638 ± 1133	139 ± 18.8
UH 320–5	320 °C, 5 h	53 ± 1.9	22 ± 0.5	974 ± 46	6651 ± 263	94 ± 3.7
UH 320–12	320 °C, 12 h	43 ± 2.0	23 ± 0.6	1061 ± 71	5726 ± 505	66 ± 5.4

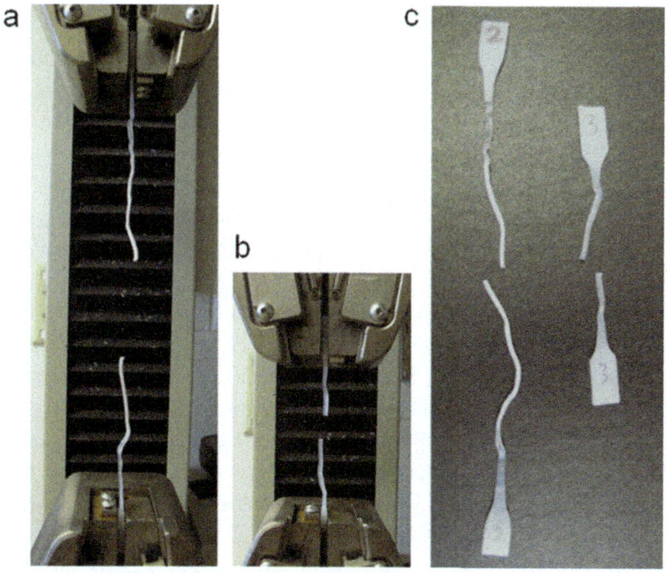

Fig. 5.9 Photographs of tensile-tested CM UHMWPE and that melted at 320 °C for 5 h [17]. (Copyright 2010 Elsevier)

On the other hand, the ultimate tensile strength (UTS) is not significantly affected by high-temperature melting except for melting at 320 °C for 5 and 12 h.

Such ultrahigh elongation indicates an extraordinary ductility of HTM UHMWPE. This is attributed to not only the relatively thinner lamellae thickness but also the elimination of structural defects in the bulk. The tensile toughness is

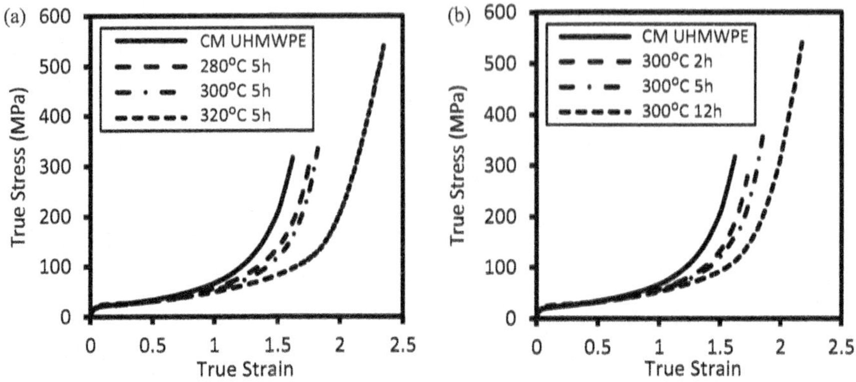

Fig. 5.10 True stress-strain curves of high-temperature melted UHMWPE as a function of (**a**) temperature and (**b**) time [17]. (Copyright 2010 Elsevier)

characterized by work to failure (WF), which is defined by the area underneath the engineering stress-strain curves. With increasing melting temperature and time, the WF values are highly enhanced to twice that for CM UHMWPE (Table 5.2).

The impact strength of UHMWPE is extremely important for joint implants that undergo millions of cyclic loadings in the body. The ASTM F648 standard defines a double-notched impact strength measurement of UHMWPE. After melting at different temperatures, the double-notched impact strength is increased from 127 kJ/m^2 for CM UHMWPE to 175 kJ/m^2 after melting at 280 °C for 12 h. With increasing melting temperature, however, the impact strength starts to decrease to 140, 90 kJ/m^2 and even lower. The elimination of fusion defects significantly improves the impact strength of the polymer. The chain scission and formation of large spherulites may account for the decrease in impact strength.

During tensile deformation, UHMWPE experiences plastic deformation, yielding, plastic flow, strain hardening, and fracture. This procedure is explicitly reflected by the true stress-strain curves. Figure 5.10 shows the true stress-strain curves of HTM UHMWPE. The true ultimate tensile strength monotonically increases with temperature and time from about 300 MPa to over 550 MPa, while the true strain increases from about 1.6 to 1.8, and 2.4. During such a long extension, strain hardening occurs after the plastic deformation and yielding. The strain-hardening modulus is regarded as an intrinsic property of an entangled network [22]. For semicrystalline polymers, the strain hardening is related to both entanglements and sliding and stretching of lamellae [23]. This property is characterized by strain-hardening modulus, which is related with the true stress (σ) and extension ratio (λ) by [22]

$$\sigma = G(\lambda^2 - 1/\lambda) + Y \tag{5.7}$$

where $\lambda = \exp(\varepsilon_t)$ with ε_t as true strain and Y is the flow stress exerted by the crystalline phase including intra- and intermolecular coupling. Thus, the G values are calculated by using the Haward plot (Fig. 5.11a). With increasing melting

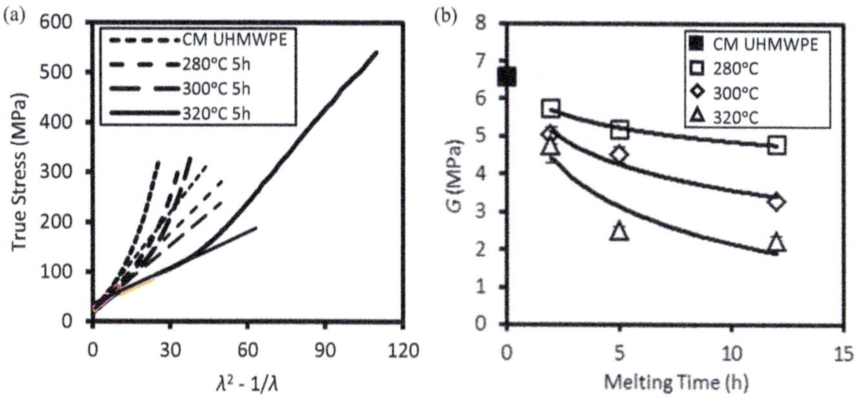

Fig. 5.11 (**a**) Representative Haward plots derived from the true stress-strain curves of CM UHMWPE and those melted at 280, 300, and 320 °C for 5 h. The slopes of the dashed fitted lines represent the strain-hardening modulus, *G*. (**b**) The strain-hardening modulus (*G*) as a function of melting time at 280, 300, and 320 °C [17]. (Copyright 2010 Elsevier)

temperature and time, the *G* value decreases (Fig. 5.11b). The decrease in *G* value indicates an enhancement in the plasticity of the crystalline phase, which is mainly attributed to the chain scissioning and formation of thin lamellae during melting [17].

5.2.6 Wear Properties of High-Temperature Melted UHMWPE

The wear of UHMWPE during articulation against metal or ceramic surfaces is a result of yielding, plastic flow, and fracturing of the polymer [24, 25]. Therefore, the yield strength, strain hardening, ultimate tensile strength, and ductility are critical for the wear of UHMWPE. The yield strength of high-temperature melted UHMWPE is slightly higher than that of compression-molded UHMWPE (Table 5.2). But the strength hardening modulus (*G*) is much lower, while the plastic flow and ductility are higher. These properties may suggest higher wear rates of HTM UHMWPE.

The wear rate of HTM UHMWPE is measured by using a pin-on-disc (POD) wear tester under constant load [26]. Herein, UHMWPE pins (diameter 9 mm, length 13 mm) are machined from HTM UHMWPE, with CM UHMWPE pins as control. Six holders hold the pins and drive them to slide on polished CoCrMo discs in a rectangular pattern with bovine serum as lubricant. The average wear rates of the pins are strongly related to the melting temperature and time (Table 5.3). For UHMWPE melted at 280 °C for 5 and 12 h, the wear rates are 10.7 ± 1.1 and 11.2 ± 0.3 mg/MC (million cycle), close to that of CM UHMWPE (10.2 ± 0.9 mg/MC). After melting at 300 °C for 12 h, the wear rate increases to 12.1 ± 2.9 mg/MC.

Table 5.3 Pin-on-disc (POD) wear rates of high-temperature melted UHMWPE [17]

Materials	Melting condition	POD wear rate (mg/MC)	*p*-value
UHMWPE	N.A.	10.2 ± 0.9	1
UH 280–2	280 °C, 2 h	N.A.	N.A
UH 280–5	280 °C, 5 h	10.7 ± 1.1	0.29
UH 280–12	280 °C, 12 h	11.2 ± 0.3	0.17
UH 300–2	300 °C, 2 h	11.2 ± 2.3	0.55
UH 300–5	300 °C, 5 h	11.3 ± 0.7	0.17
UH 300–12	300 °C, 12 h	12.1 ± 2.9	0.05
UH 320–2	320 °C, 2 h	9.7 ± 2.7	0.78
UH 320–5	320 °C, 5 h	13.0 ± 0.8	0.04
UH 320–12	320 °C, 12 h	15.6 ± 2.5	0.05

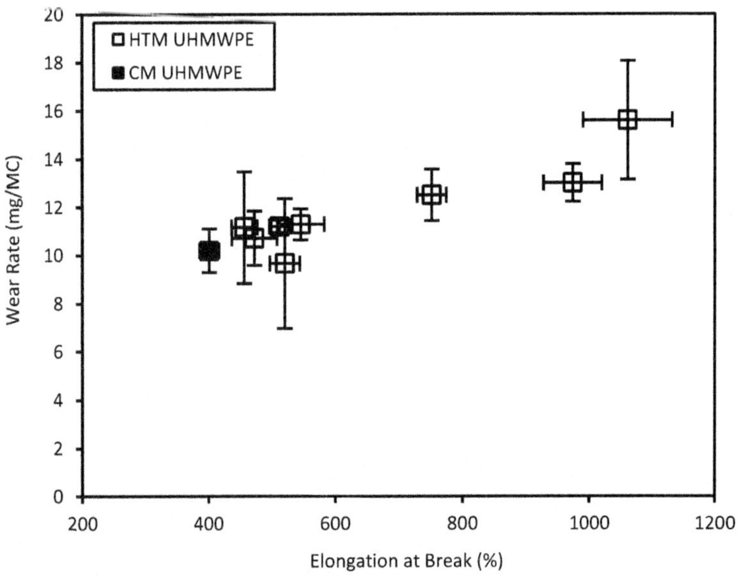

Fig. 5.12 The pin-on-disc (POD) wear rate of HTM UHMWPE as a function of elongation at break [17]. (Copyright 2010 Elsevier)

After melting at 320 °C for 5 and 12 h, the wear rate significantly increases to 13.0 ± 0.8 and 15.6 ± 2.5 mg/MC.

The wear rate is related to the chain scission and the ductility of polyethylene. Figure 5.12 shows the dependence of wear rate on the elongation at break of UHMWPE. The wear rate is approximately linearly dependent on the elongation at break. Besides, it is established that the elongation at break is proportional to the vinyl index [17]. Therefore, the wear rate is linearly related to the vinyl index, or chain scission degree. Essentially, the decrease in strain-hardening modulus and increase in plastic flow of HTM UHMWPE may mainly account for the increase in wear rate.

5.3 High-Temperature Melted, Radiation Cross-Linked UHMWPE

The highly improved ductility and tensile toughness of high-temperature melted (HTM) UHMWPE provide an excellent basis for radiation cross-linking to improve the wear resistance while obtaining improved mechanical strength and toughness. In general, radiation cross-linking is known to generate a three-dimensional network of polymer chains, reducing the plastic flow of polymer chains and thus enhancing the wear resistance [8, 24]. Meanwhile, the severe loss in ductility is related to the loss in fatigue resistance and tensile toughness [8], which are critical for the clinical performance [27] and longevity of joint implants. Herein, the highly enhanced chain entanglements and ductility by high-temperature melting may benefit the terminal properties of UHMWPE after radiation cross-linking. This section intends to investigate the chemical, crystalline structures, cross-linking structures, the mechanical properties, and the wear properties of high-temperature melted and irradiated UHMWPE. Besides, the relation between the structures and properties will be discussed.

5.3.1 *Chemical Structures of HTM-XL PE*

Upon irradiation, the vinyl groups, as well as the CH groups, are converted into free radicals. The free radicals are able to recombine to form new covalent bonds. Meanwhile, trans-vinylene groups form due to irradiation [28]. The trans-vinylene groups show a band at 987 cm^{-1} in FTIR spectrum. Thus, the vinyl end group content is decreased while the trans-vinylene content is increased. FTIR spectra normalized against the internal reference polyethylene skeleton band at 1895 cm^{-1} show significant decreases in the 909 cm^{-1} band and remarkable increases in the 987 cm^{-1} band after exposing HTM UHMWPE to electron beam to 50 and 100 kGy. Figure 5.13 shows representative FTIR spectra of CM UHMWPE and HTM UHMWPE melted at 320 °C for 5 h before and after exposure to 50 and 100 kGy electron beam. First, in comparison to CM UHMWPE, melting at 320 °C for 5 h generates new vinyl end groups, which show a remarkable increase in the 909 cm^{-1} band intensity (Fig. 5.13a, d). After irradiation to 50 and 100 kGy, the vinyl band decreases, while a new band at 987 cm^{-1}corresponding to trans-vinylene groups appears. As the radiation dose increases from 50 to 100 kGy, the vinyl band decreases, while the trans-vinylene band increases.

The cross-link density of HTM UHMWPE shows a strong dependence on the melting process. Figure 5.14 summarizes the cross-link densities of UHMWPE melted at 280, 300, and 320 °C for 2, 5 and 12 h. In general, the cross-link density increases with irradiation dose. On the other hand, the cross-link density of HTM UHMWPE is lower than that for CM UHMWPE at each dose. With increasing melting temperature, the cross-link density is lower for given doses. Similarly, for

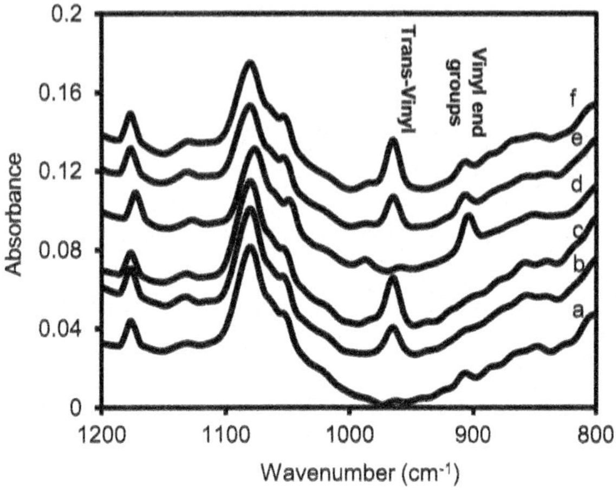

Fig. 5.13 (**a**) FTIR spectra of (a) CM UHMWPE, those irradiated at (**b**) 50, and (**c**) 100 kGy, and UHMWPE melted at 320 °C for 5 h (**d**) before and after irradiation to (**e**) 50, (**f**) 100 kGy [29]. (Copyright 2011 Elsevier)

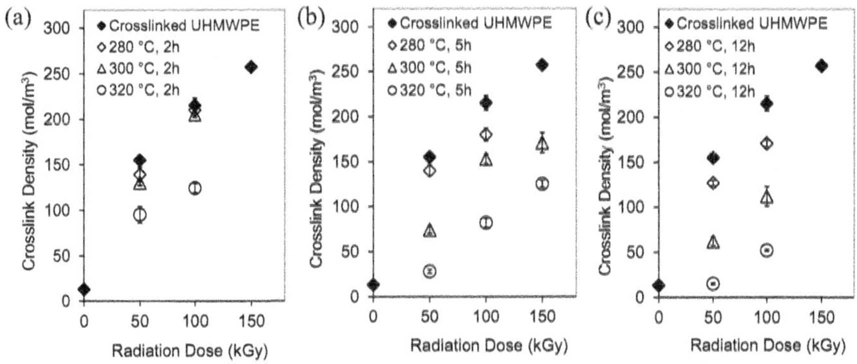

Fig. 5.14 Cross-link densities of high-temperature melted UHMWPE exposed to electron beam to total doses of 50, 100, and 150 kGy. The polyethylene is melted at different temperatures for (**a**) 2, (**b**) 5, and (**c**) 12 h [29]. (Copyright 2011 Elsevier)

specific melting temperatures, the cross-link density is lower with increasing melting time. The cross-link densities of polyethylene melted at 320 °C and irradiated are less than 50% of that for CM UHMWPE receiving the same radiation doses.

Such low cross-link densities are attributed to the formation of terminal vinyl groups formed at high temperatures. The terminal vinyl group numbers are higher with higher temperatures or longer melting time (Fig. 5.6). Upon irradiation, these vinyl groups are prone to form terminal free radicals that combine with other free radicals to form both H-shaped and Y-shaped cross-links. Consequently, the cross-link density becomes lower. On the other hand, the chain scission may also account for the low cross-link density.

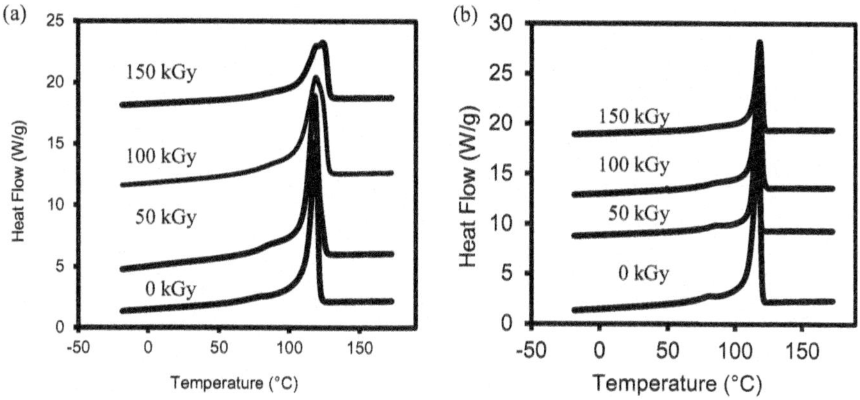

Fig. 5.15 Representative DSC cooling traces of (**a**) CM UHMWPE and (**b**) HTM UHMWPE irradiated to 0, 50, 100, and 150 kGy [29]. (Copyright 2011 Elsevier)

5.3.2 *Crystal Structures of HTM-XL PE*

For compression-molded UHMWPE, radiation cross-linking increases the peak melting point and crystallinity. During e beam irradiation, the temperature could rise close to the T_m of UHMWPE, which leads to partial melting of crystals. Some crystals may grow, while the molten domains are cross-linked into networks. The network may allow for fast recrystallization into small crystallites although the overall chain mobility is hindered.

Figure 5.15 compares representative DSC cooling traces of CM UHMWPE, HTM UHMWPE, and those irradiated to 100 kGy by electron beam. The radiation cross-linked UHMWPE shows a broad melting peak with a distinct shoulder at the low temperature side, suggesting the formation of small crystallites during irradiation. Upon subsequent cooling, the irradiated CM UHMWPEs show broad crystallization peaks at high doses. In contrast, narrow and sharp crystallization peaks are found for the irradiated HTM UHMWPE even at high doses. Such narrow and sharp crystallization peaks suggest the formation of homogeneous cross-linked networks for the irradiated HTM UHMWPE, in contrast to the heterogeneous network for the irradiated CM UHMWPE. Within such networks, the polyethylene lamellae have similar thickness and folding energy, leading to narrow and sharp melting curves upon the second heating tests.

Such uniform network dimension and homogeneous lamellae thickness, together with the low cross-link density, are very important for the strength and toughness of the irradiated HTM UHMWPE.

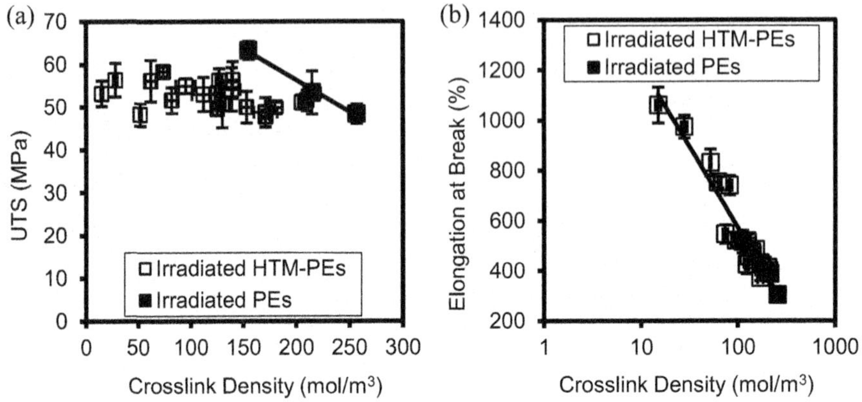

Fig. 5.16 Dependence of (**a**) ultimate tensile strength (UTS) and (**b**) elongation at break of irradiated UHMWPE (irradiated PE) and irradiated HTM UHMWPE [29]. (Copyright 2011 Elsevier)

5.3.3 Mechanical Properties of HTM-XL PE

For the first-generation highly cross-linked UHMWPE, the mechanical strength and toughness are decreased due to the formation of rigid networks and subsequent thermal treatments [8]. Radiation cross-linking reduces the ductility and toughness of UHMWPE. With increasing radiation dose and cross-link density, the ultimate tensile strength (UTS), work to failure (WF), and impact toughness are smaller than those of virgin UHMWPE [8].

After radiation cross-linking, the HTM UHMWPE shows lower cross-link densities than the cross-linked UHMWPE receiving the same doses (Fig. 5.14). Materials with a series of cross-link densities could be prepared by combining the melting temperature and time and radiation doses. The UTS of the irradiated HTM UHMWPE is not sensitive to cross-link density over a broad range. These values (about 47~58 MPa) are very close to those of irradiated CM UHMWPE (Fig. 5.16a) and those of irradiated and remelted polyethylene in literature [30]. These results are encouraging since it suggests a possibility to adjust the cross-link density of UHMWPE over a broad range without a severe loss in mechanical properties. On the other hand, such low cross-link densities mean a high ductility of cross-linked HTM UHMWPE. The elongation at break of such polyethylene, which is proportional to the cross-link density, is significantly higher than those of highly cross-linked UHMWPE in clinical use or in literature (Fig. 5.16b). Meanwhile, the impact toughness of the cross-linked HTM UHMWPE is proportional to cross-link density, following a linear regression of highly cross-linked UHMWPE (Fig. 5.17). Despite the subtle difference in crystallinity and peak melting points of these polymers, it is the cross-linked network that dominates the impact strength of the materials.

Note that the irradiated HTM UHMWPEs are not annealed or remelted after irradiation. However, there is a need to stabilize such materials since there are

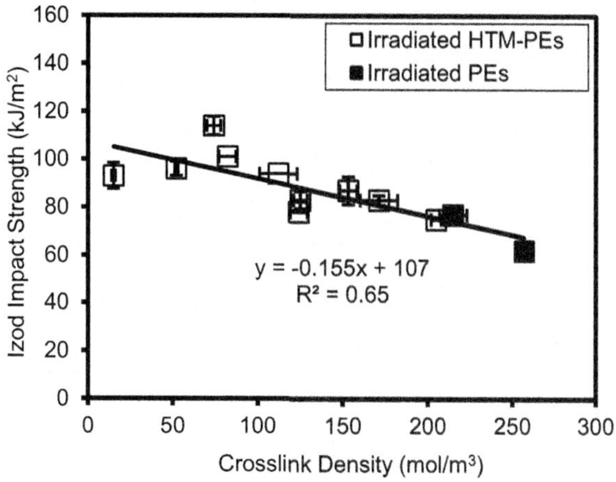

Fig. 5.17 Dependence of double-notched Izod impact strength of high-temperature melted and radiation cross-linked UHMWPE on cross-link density [29]. (Copyright 2011 Elsevier)

numerous residual free radicals in the polymers, particularly in the crystalline phase. This issue will be addressed by blending UHMWPE with vitamin E prior to consolidation and high-temperature melting, which will be discussed later in this chapter.

5.3.4 Wear Properties of Irradiated HTM UHMWPE

It has been well established that high dose irradiation significantly reduces the wear rate of UHMWPE. The POD wear rate is about 10 mg/MC (million cycle) for CM UHMWPE, and gradually reduced down to 0.1 mg/MC as the total dose is increased to 200 or 300 kGy [24]. Post-irradiation melting to eliminate free radicals further increases the cross-link density since the free radicals entrapped the crystal lattice recombine to form new cross-links. Thus, the wear rate is further decreased.

The wear rate of irradiated HTM UHMWPE shows a strong dependence on irradiation dose and the melting temperature and time. The cross-link density of irradiated HTM UHMWPE increases with irradiation dose. As a consequence, the wear rate decreases with increasing cross-link density. On the other hand, since the cross-link density decreases with increasing melting temperature and time (Fig. 5.14), the corresponding wear rate of these irradiated HTM UHMWPE increases. For example, the HTM UHMWPE melted at 280 °C for 5 h after irradiation shows wear rates a little smaller than those for CM UHMWPE at each cross-link density. For UHMWPE melted at higher temperatures (e.g., 300 or 320 °C), the cross-link density is lower after radiation cross-linking than those for irradiated UHMWPE, although the cross-link densities of the irradiated HTM UHMWPE are much lower than those of the irradiated UHMWPE at different doses. Despite the low cross-link densities, it is interesting that, for materials with

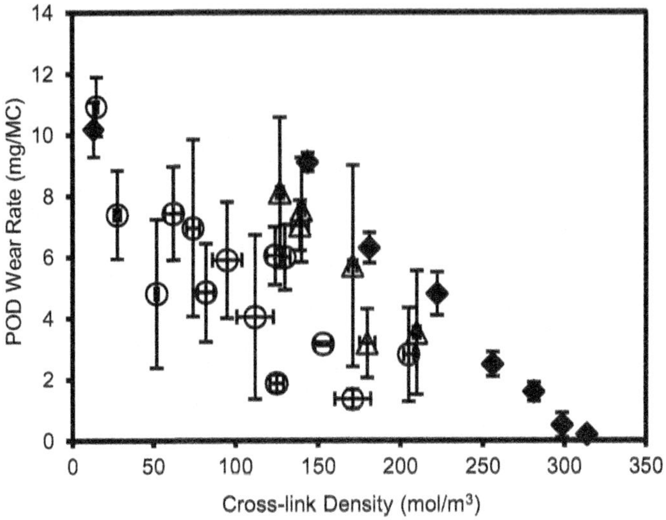

Fig. 5.18 Pin-on-disc wear rates of high-temperature melted and radiation cross-linked UHMWPE (open symbols), in comparison to clinically available irradiated and remelted UHMWPE (solid symbols) [29]. (Copyright 2011 Elsevier)

given cross-link densities, the POD wear rates of the irradiated HTM UHMWPE are always lower than those of irradiated UHMWPE (Fig. 5.18).

5.4 High-Temperature Melted, Radiation Cross-Linked, and Vitamin E-Stabilized UHMWPE

A major challenge for radiation cross-linked UHMWPE is the residual free radicals formed after irradiation that would cause severe oxidation of polyethylene on shelf or in vivo [31]. Post-irradiation thermal treatments, i.e., annealing below T_m [32, 33] or remelting above T_m and subsequent recrystallization [34], have been practiced to eliminate or reduce the free radicals. However, such thermal treatments more or less reduce the mechanical strength and fatigue resistance [7, 35], which are critical for the clinical performance and longevity of polyethylene-based joint implants.

Similar problems would remain for the high-temperature melted and radiation cross-linked UHMWPE, although there are significant gains in mechanical properties due to the elimination of fusion defects and relatively homogeneous lamella dimensions and cross-linked network structures. Severe decreases of the mechanical strength and toughness will occur if these materials were subjected to annealing or remelting after irradiation. Therefore, post-irradiation remelting or annealing should be avoided to retain the advantageous structural and mechanical properties.

In order to stabilize free radicals generated upon irradiation, particularly those entrapped in crystal lamellae, various antioxidants or stabilizers, including vitamin E [36–38], hindered amine light stabilizers [39], or polyphenols [29], have been introduced into UHMWPE before [38] or after [37] irradiation. The presence of these substances efficiently and significantly stabilizes the free radicals and thus improves the oxidation resistance of highly cross-linked UHMWPE. In particular, mixing antioxidants with UHMWPE powders prior to consolidation and irradiation waives thermal treatments [38], which results in materials with improved mechanical strength, toughness, fatigue resistance, and low wear. Alternatively, diffusion of antioxidants into radiation cross-linked UHMWPE at elevated temperatures (below the peak melting point of UHMWPE) yields oxidation-resistant materials with excellent mechanical properties [37]. In order to achieve homogeneous distribution of antioxidants throughout the materials, diffusion at high temperature is usually practiced for tens of hours [40].

Herein, vitamin E is used to stabilize the high-temperature melted and irradiated UHMWPE. Vitamin E is mixed with UHMWPE powders at 0.1 wt% and 0.2 wt% before consolidation. The compression-molded vitamin E/UHMWPE (VE-PE) is irradiated by electron beam at 25 kGy/pass to total doses of 25, 50, 100, and 150 kGy as control. On the other hand, the VE-PE is melted at 300 and 320 °C for 5 h, before being exposed to e beam to receive total doses of 25, 50, 100, and 150 kGy. Despite the presence of vitamin E, the melt viscosity of UHMWPE remains very high so that it does not flow at all at these high temperatures. In order to characterize the concentration of vitamin E in the polymers, FTIR spectra are collected, where the band at 1262 cm^{-1} is used to determine the vitamin E (VE) index. Herein, the VE index is defined as the ratio of the 1262 cm^{-1} band over the 1895 cm^{-1} band. After high-temperature melting, the VE index decreases. For example, the VE index of 0.1 wt% UHMWPE decreases from 0.0092 before melting to 0.0041 after melting at 300 °C for 5 h and 0.0038 after melting at 320 °C for 5 h. It is likely that the phenol groups are partly consumed during high-temperature melting.

The HTM-VE-PE shows excellent mechanical properties. Similar to virgin UHMWPE melted at high temperatures, these VE-PEs are free of fusion defects, which is critical for its tensile and impact properties. With the presence of 0.1 wt% vitamin E, for example, the VE-PE after melting at 300 °C for 5 h shows UTS of 58 MPa, an EAB of 503%, and yield strength of 21 MPa. The double-notched Izod impact strength value is enhanced to 162 kJ/m^2, in comparison to 127 kJ/m^2 for CM UHMWPE and 140 kJ/m^2 for HTM UHMWPE melted at 300 °C for 5 h. With 0.2 wt % vitamin E in HTM UHMWPE, the impact strength is slightly lower (148 kJ/m^2). At higher melting temperature (e.g., 320 °C for 5 h), the impact strength is decreased to about 100 kJ/m^2, close to the VE-PE with 0.1 wt% and 0.2 wt% vitamin E. The decrease in impact strength is related to the high crystallinity and small lamella thickness.

The strength and toughness of semicrystalline polymers are strongly related to the crystalline-amorphous composite structures. High crystallinity and large crystal size usually lead to high strength and modulus, but low toughness. A relatively low crystallinity and small crystal size correspond to low strength but high toughness and ductility. Therefore, in order to create UHMWPE materials with optimal strength and toughness for clinical applications, it is important to balance the crystallinity and lamella size. Herein, this is achieved through optimizing the melting temperature and time. The presence of vitamin E may plasticize the chain mobility and crystallization during and after high-temperature melting.

5.4.1 *Effect of Radiation Cross-Linking on the Mechanical Properties of High-Temperature Melted, Vitamin E-Stabilized UHMWPE*

In order to improve the wear resistance of these high-temperature melted, vitamin E-stabilized UHMWPE, these materials are subjected to irradiation for total doses of 50, 100, and 150 kGy at 25 kGy/pass at room temperature to generate cross-linked structures. During irradiation, free radicals are generated, which may recombine to form cross-links. The residual free radicals are supposed to be stabilized by the vitamin E inside the polyethylene. Previously, it has been demonstrated that vitamin E with concentrations from 0.05 wt% to 0.2 wt% is sufficient to stabilize radiation cross-linked UHMWPE receiving doses up to 150 kGy. Herein, 0.1 wt% and 0.2 wt % vitamin E are used to stabilize the high-temperature melted and radiation cross-linked UHMWPE.

In addition to irradiation doses, the vitamin E concentration and melting conditions have significant influences on the cross-link density. Figure 5.19 compares the cross-link densities of UHMWPE with 0, 0.1, and 0.2 wt% vitamin E melted at 300 °C for 5 h and irradiated with 150 kGy. The cross-link density of irradiated HTM-PE with vitamin E is lower than those free of vitamin E. With increasing vitamin E concentration, the cross-link density is slightly lower. As the melting temperature is increased from 300 °C to 320 °C, the cross-link density decreases,

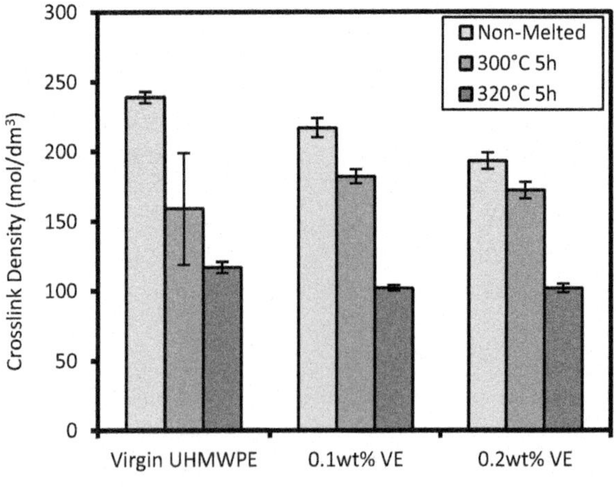

Fig. 5.19 Cross-link densities of UHMWPE with 0.1 wt% and 0.2 wt% vitamin E irradiated by electron beam to 150 kGy. Cross-link density values for polymers before and after melting at 300 °C for 5 h prior to irradiation are compared

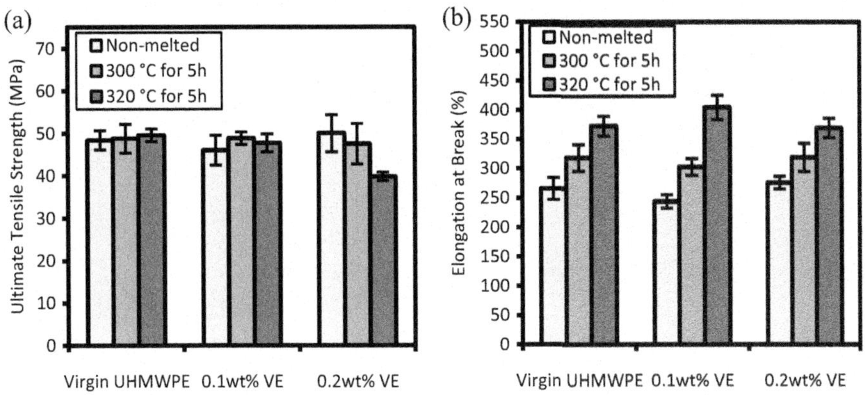

Fig. 5.20 The (**a**) ultimate tensile strength and (**b**) elongation at break of non-melted and high-temperature melted UHMWPE with or without vitamin E irradiated by electron beam to 150 kGy

and, for a given melting temperature, the cross-link density is lower with longer melting time.

The low cross-link densities of high-temperature melted and radiation cross-linked UHMWPE with vitamin E entitle improved tensile toughness and impact strength, in comparison to radiation cross-linked and remelted UHMWPE (XL-RM-PE). Figure 5.20 compares the UTS and EAB of the irradiated polymers. There is no significant difference in the UTS values of these materials (Fig. 5.20a). In contrast, the EAB values are strongly related to the melting history of the materials for polyethylene receiving 150 kGy radiation. The EAB of high-temperature melted and irradiated materials are higher than 300%, while the materials melted at 320 °C show an even higher EAB of about 370% and 400%, much higher than those of 150 kGy irradiated UHMWPE (about 260%). High-temperature melted and irradiated UHMWPEs containing 0.1 wt% and 0.2 wt% show similar EAB values, suggesting that the structures due to HTM and irradiation, rather than vitamin E, dominate the strength and toughness.

These materials show improved impact properties. The double-notched impact strength values of irradiated HTM-VE-PE are compared with irradiated UHMWPE as control (Fig. 5.21). The 150 kGy irradiated UHMWPE shows an impact strength of 62 kJ/m^2, while it is a bit higher (66 and 77 kJ/m^2) for those with 0.1 wt% and 0.2 wt% vitamin E. The presence of vitamin E slightly decreases the cross-link density and thus enhances the impact strength. Interestingly, the 150 kGy irradiated HTM-PE with a history at 300 or 320 °C for 5 h show an impact strength of 83 kJ/m^2, while the values are even higher (94 and 96 kJ/m^2) for irradiated HTM materials with 0.1 wt% and 0.2 wt% vitamin E. These impact strength values are much higher than those of the irradiated and remelted or irradiated and annealed UHMWPE that have been used in joint implants.

It has been established that the tensile and impact toughness are dictated by the cross-link density of the irradiated UHMWPE (Figs. 5.16, 5.17). Herein, the

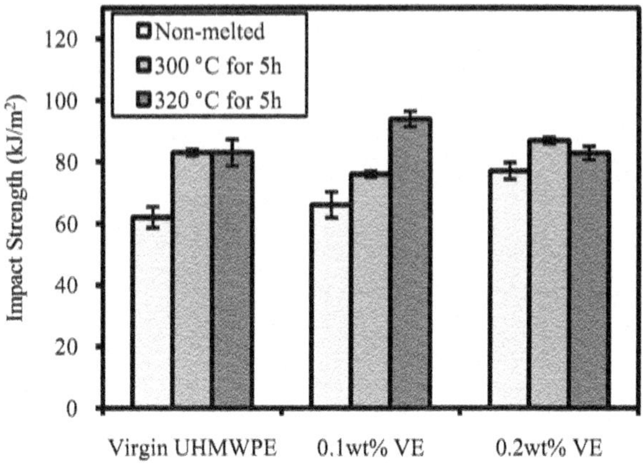

Fig. 5.21 Double-notched impact strength of high-temperature melted and irradiated UHMWPE with or without vitamin E. The irradiation dose is 150 kGy

presence of vitamin E, together with high-temperature melting, decreases the cross-link densities of the irradiated polyethylene (Fig. 5.19). Thus, the vitamin E-blended, high-temperature melted, and irradiated UHMWPEs show improved tensile and impact toughness, in comparison to the first-generation cross-linked UHMWPE.

5.4.2 Pin-on-Disc Wear Rates of High-Temperature Melted and Irradiated Vitamin E/UHMWPE Blends

Wear resistance is another important property of very strong clinical relevance. It has been a long pursuit to create materials with high strength, toughness, and low wear. It's known that the wear rate decreases with increasing cross-link density for the first-generation cross-linked UHMWPE. Herein, the irradiated HTM-UHMWPEs show much lower cross-link densities than their UHMWPE counterparts receiving the same doses from 50 to 100 and 150 kGy. There is a concern on the wear resistance of these materials containing 0.1 wt% and 0.2 wt% vitamin E.

Figure 5.22 shows the pin-on-disc wear rates of irradiated HTM-VE-PE, in comparison to those of first-generation highly cross-linked and remelted UHMWPE. As the cross-link density increases, the POD wear rate decreases from 10.2 mg/MC to 6 mg/MC, 2 mg/MC, 0.5 mg/MC, and lower. The highly cross-linked (about 150–200 mol/dm^3) and medium cross-linked (about 100–150 mol/dm^3) UHMWPEs have been used in hip and knee implants. For the irradiated HTM-VE-PE materials, the cross-link density is between 80 mol/dm^3 and 145 mol/dm^3, which are much lower than those for first-generation cross-linked UHMWPE. Herein, the average wear rates are from 1.6 mg/MC to 4.7 mg/MC. These wear rates are much less sensitive to cross-link density and comparable to those of first-generation highly cross-linked UHMWPE in clinical use.

In comparison to conventional highly cross-linked UHMWPE stabilized by vitamin E, the HTM vitamin E UHMWPEs after irradiation show improved tensile

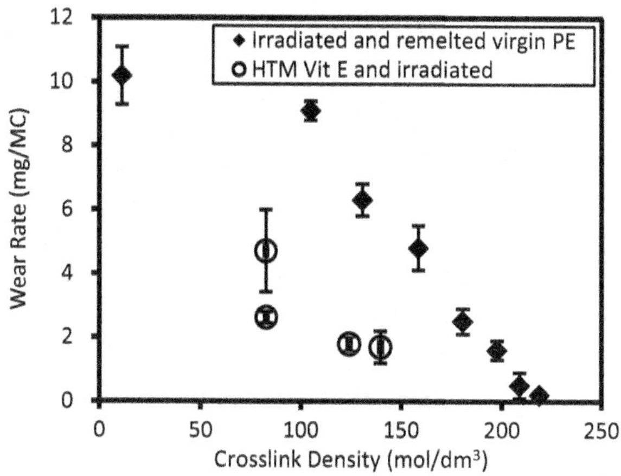

Fig. 5.22 Pin-on-disc wear rates of high-temperature melted, irradiated, and vitamin E-stabilized UHMWPE, in comparison to radiation cross-linked and remelted UHMWPE

and impact strength and toughness. Despite the low cross-link density, the radiation cross-linked HTM-VE-PEs show low wear rates that are comparable to those for clinically used highly cross-linked UHMWPE. In the next section, the oxidative stability of these UHMWPEs will be presented and discussed.

5.4.3 *Oxidation Resistance of High-Temperature Melted, Radiation Cross-Linked, Vitamin E-Stabilized UHMWPE*

The presence of vitamin E is efficient in stabilizing UHMWPE after high dose irradiation, without a need of post-irradiation thermal treatments to reduce or eliminate free radicals. On the other hand, recent studies have reported on in vivo initiated oxidation and oxidation potentials for stabilized UHMWPE components after implantation [41–43]. Unsaturated lipids in synovial fluids are argued to account for such oxidation potentials of UHMWPE since they can diffuse into UHMWPE components, undergo oxidation, and may initiate oxidation of UHMWPE [42, 44]. Oral et al. demonstrated, by using a simulated oxidative challenge to irradiated VE-PE with the presence of squalene, a lipid found in synovial fluids, that vitamin E efficiently prohibited oxidation induced by unsaturated substances [45]. Those encouraging results suggest that vitamin E, after irradiation, is promising for oxidative stabilization of UHMWPE.

In our case, however, it remains open whether or not vitamin E after high-temperature melting and subsequent radiation cross-linking would retain its potency to stabilize UHMWPE against oxidative challenging, particularly in the presence of simulated synovial fluid lipids.

In this section, the high-temperature melted, vitamin E-stabilized, and irradiated UHMWPEs are subjected to accelerated aging at 70 °C in O_2 for 2 weeks to examine

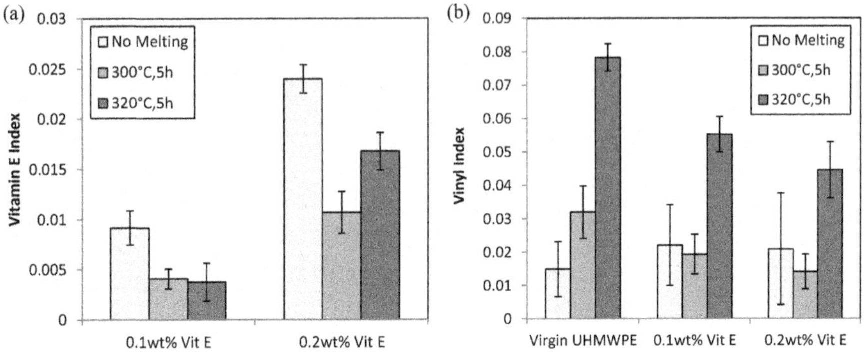

Fig. 5.23 Effect of high-temperature melting on the vitamin E index (**a**) and vinyl index (**b**) of UHMWPE

their oxidative stability. Second, these materials are aged at 120 °C in air and the oxidation process is monitored by real time measurements of the oxidation index. Third, an unsaturated lipid found in synovial fluid, squalene, is diffused into UHMWPE and aged at 70 °C in O_2 to study the oxidation resistance against lipid-initiated oxidation that may occur in vivo.

The vitamin E index (I_{VE}) is characterized by the peak area at band 1262 cm^{-1} against that at 1895 cm^{-1}. The I_{VE} is 0.0092 for 0.1 wt% vitamin E/UHMWPE and 0.024 for 0.2 wt% vitamin E/UHMWPE. After melting at 300 and 320 °C for 5 h, these values decrease to 0.0042 and 0.0038 for 0.1 wt% VE-PE, and 0.011 and 0.017 for 0.2 wt% VE-PE (Fig. 5.23a). High-temperature melting is known to cause chain scission, generating new vinyl end groups, which are characterized by an increase in the band intensity at 909 cm^{-1}. Figure 5.23b shows the vinyl indices of materials with 0.1 wt% and 0.2 wt% vitamin E after melting at different temperatures for 5 h. In comparison to the high-temperature melted virgin UHMWPE, the vinyl indices for the VE-PE are lower, and the vinyl index decreases with increasing vitamin E concentration (Fig. 5.23b). The presence of vitamin E is protective for the polyethylene chains against scissioning at high temperatures.

Upon subsequent radiation cross-linking with 150 kGy, the vitamin E and vinyl indices become undetectable, which suggests the consumption of phenol groups. On the other hand, the vinyl indices become zero after irradiation, suggesting the consumption of vinyl groups, which react with electron beam and recombine to form cross-links.

In order to examine the antioxidation potency of the high-temperature melted, vitamin E-stabilized, and irradiated UHMWPE, cubes of these materials are subjected to accelerated aging at 70 °C in 5 atm oxygen for 2 weeks. After aging, the cubes were cut apart and microtomed into thin (200 μm) slices for microscopic FTIR scanning to characterize the oxidation profiles throughout the samples (Fig. 5.24a). IR spectra are collected at each point along the scan line from one side across the sample to the other side with a step length of 50~100 μm. At each point, the oxidation products show a representative band at 1715 cm^{-1} (C=O

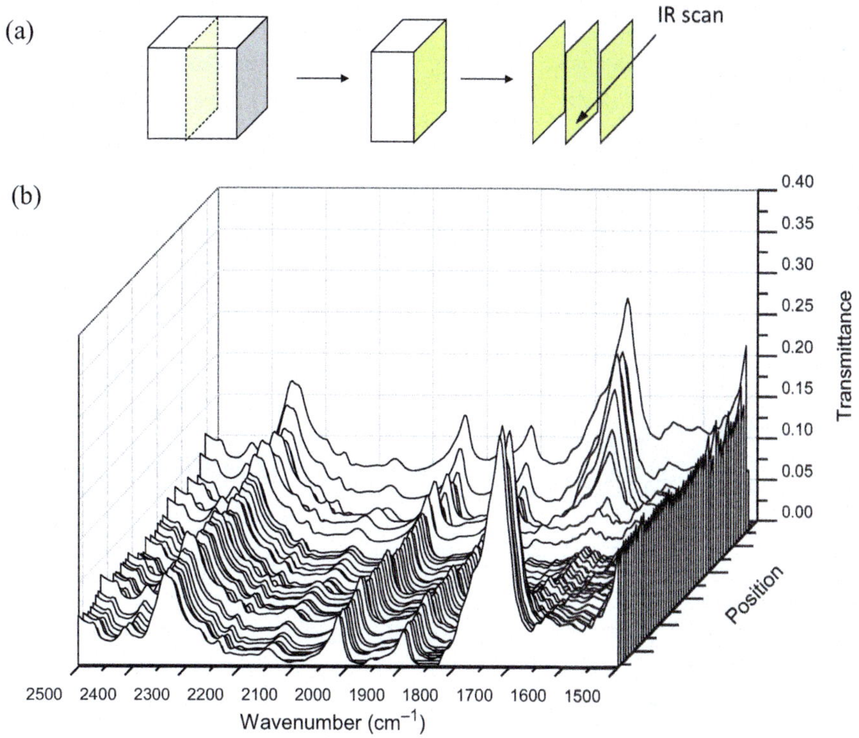

Fig. 5.24 (**a**) Cutting and microtoming of aged irradiated UHMWPE for IR line scan. (**b**) Representative IR spectra collected along the scan line across the slice

stretching). This band is high on the surfaces and decays gradually into the interior of the samples (Fig. 5.24b). The extent of oxidation is characterized by the area of this band normalized against that at 1895 cm^{-1}. Thus, the oxidation profiles along the scan line can be calculated.

Figure 5.25 compares the oxidation profiles of irradiated UHMWPE, vitamin E-stabilized UHMWPE, irradiated vitamin E/UHMWPE, and irradiated HTM-VE-PE after accelerated aging. The irradiated UHMWPE shows severe oxidation. In contrast, UHMWPE containing 0.1 wt% and 0.2 wt% vitamin E show negligible oxidation, even after melting at 320 °C for 5 h. Irradiation at 150 kGy of the vitamin E containing UHMWPE significantly reduces the antioxidation potency of vitamin E. The irradiated UHMWPE with 0.1 wt% and 0.2 wt% vitamin E shows oxidation levels higher than those for unirradiated counterparts. Oxidation occurs throughout the cubic samples, although the maximal oxidation level is less than 0.4. The oxidation level is lower with higher vitamin E. For the vitamin E/UHMWPE melted at 300 or 320 °C for 5 h and irradiated to 150 kGy, their oxidation levels are higher than the 0.2 wt%-150 kGy sample but lower than the 0.1 wt%-150 kGy sample. High-temperature melting and subsequent irradiation reduces the antioxidation potency of vitamin E.

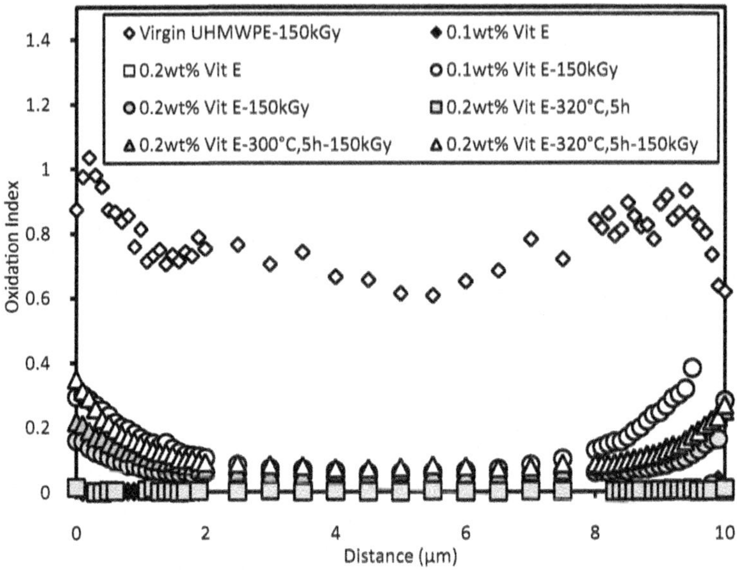

Fig. 5.25 Oxidation profiles of irradiated virgin UHMWPE, vitamin E/UHMWPE, irradiated vitamin E/UHMWPE, and irradiated HTM-VE-PE after accelerated aging for 2 weeks

According to the criterion recommended by ASTM, UHMWPE after accelerated aging with an oxidation level less than 0.5 is regarded as an oxidation-resistant material. Herein, the low vitamin E concentration remains active and efficient after high-temperature melting and irradiation to protect UHMWPE from accelerated oxidation, although vitamin E becomes undetectable for FTIR after the harsh processing.

In order to further examine its potency to protect UHMWPE against lipid-related oxidation, squalene is diffused into cubic samples of high-temperature melted, vitamin E-stabilized, and radiation cross-linked UHMWPE. Squalene is an unsaturated leophilic lipid. When the polyethylene cubes are immersed in squalene at 120 °C for 2 h, squalene gradually diffuses into the polymer, forming a gradient decaying from the surface to the interior. In order to determine the concentration and its distribution of squalene in the materials, the cube samples are cut apart after squalene diffusion and microtomed into slices for FTIR line scans. Figure 5.26 shows typical squalene distribution profiles determined by FTIR for polymers with different cross-link densities and/or thermal histories. The squalene content is the highest throughout the 0.2 wt%/UHMWPE cube, while it becomes a bit lower for that irradiated by 150 kGy. No significant difference is found between those irradiated by 150 kGy, despite the high-temperature melting prior to irradiation. The cross-link density has a remarkable influence on the diffusion of squalene into UHMWPE. A dense network may cause a relatively slow diffusion of squalene into the polymer.

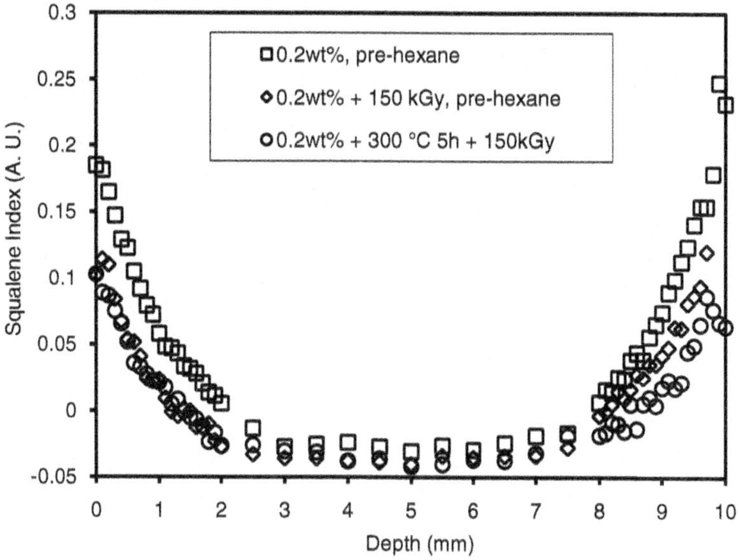

Fig. 5.26 Squalene distribution profiles in 0.2 wt% vitamin E/UHMWPE, 0.2 wt% vitamin E/ UHMWPE irradiated by 150 kGy, and 0.2 wt% vitamin E melted at 300 C for 5 h and subsequently irradiated by 150 kGy

With the presence of squalene, the cubes are aged in oxygen at 70 °C for 2 weeks. Squalene is unsaturated and prone to oxidation, generating free radicals. The squalene free radicals within UHMWPE may initiate cascade oxidation of UHMWPE. After accelerated aging, squalene-diffused UHMWPE samples are cut apart and microtomed for microscopic FTIR line scans across the slices. To reveal the oxidation profiles of these samples, the oxidized thin slices are extracted in hexane to remove any removable oxidation products and squalene.

Figure 5.27 shows representative oxidation profiles of a series of UHMWPE materials. The presence of squalene increases the oxidation levels of irradiated virgin UHMWPE and irradiated HTM-PE. The presence of 0.2 wt% vitamin E effectively suppresses oxidation. The overall maximum oxidation levels are lower than 0.15. The oxidation levels for 150 kGy irradiated materials are slightly higher than those for unirradiated materials. In particular, the high-temperature melted and irradiated VE-PE shows higher oxidation levels than the non-HTM materials. It is likely that high-temperature melting, together with high dose irradiation, may slightly reduce the antioxidation potency. This subtle difference, according to the accelerated aging results, is small. One may conclude that these high-temperature melted and irradiated VE-PEs have excellent oxidative resistance against potential challenges by lipid-related oxidation.

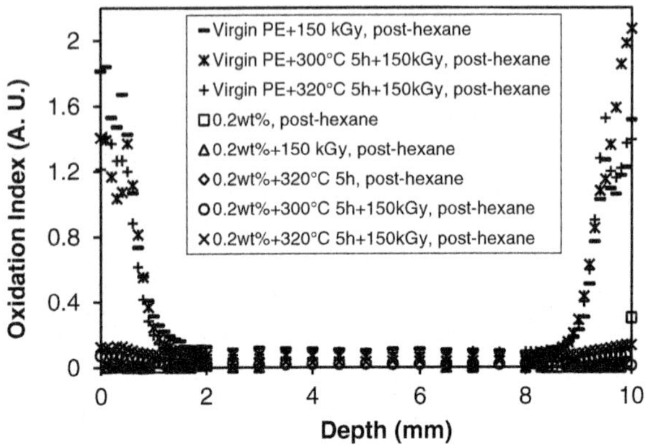

Fig. 5.27 Oxidation profiles of a series of UHMWPE materials after aging at 70C for 2 weeks with the presence of squalene diffused into the polymers

5.5 Clinical Relevance of High-Temperature Melted, Vitamin E-Stabilized, and Irradiated UHMWPE

Low wear, high strength, fatigue resistance, and oxidation stability of UHMWPE have been the most important pursuits for joint implants with high performance and prolonged longevity. Conventional UHMWPE implants have experienced severe problems due to wear [46] and oxidation [47] of the polymer components. Since the middle 1990s, first-generation highly cross-linked UHMWPEs have been developed and showed a wear rate 50–90% lower than conventional UHMWPE [30, 48]. These materials, cross-linked upon exposure to high-energy ionizing radiations, need thermal treatments (i.e., annealing below peak melting point or remelting above peak melting point) to reduce or eliminate residual free radicals that otherwise may cause cascade oxidation upon exposure to oxygen on shelf or in vivo [49]. Since the first approval of highly cross-linked UHMWPE by FDA for clinical applications in 1998, a series of cross-linked UHMWPE materials and tens of low wear artificial joints have been manufactured and applied in patients.

Midterm follow-up studies have showed encouraging results on implants based on highly cross-linked UHMWPE [3]. According to Kurtz's literature review on highly cross-linked polyethylene (HXLPE) implants for total joint arthroplasty, weighted-average analyses of femoral head penetration rates in HXLPE liners result in a mean two-dimensional linear penetration rate of 0.042 mm/year and 0.137 mm/year. The pooled odds ratio for the risk of osteolysis in HXLPE versus conventional liners was 0.13 among studies within a minimum 5-year follow-up. Lower femoral head penetration and an 87% lower risk of osteolysis have been concluded.

Despite the success of first-generation highly cross-linked UHMWPE, demands from younger and more active patients and longer longevity of patients have raised an increasing need in UHMWPE with better strength, toughness, and oxidation resistance while retaining outstanding wear resistance. The use of post-irradiation thermal treatments to stabilize cross-linked UHMWPE is known to compromise the strength and toughness [6, 7]. Alternatively, the introduction of vitamin E, a natural antioxidant, has significantly improved the oxidation resistance, without a need to anneal or remelt the polymer after irradiation. Thus, the mechanical properties are preserved, together with excellent wear and fatigue resistance. The clinical success of vitamin E-stabilized cross-linked UHMWPE has been demonstrated since 2009. Retrievals with short-term (about 18 months) in vivo history show encouraging results that the vitamin E-stabilized, irradiated UHMWPE shows free radical decay over time and no hydroperoxides or other oxidation potentials could be found despite diffusion of lipids into the polyethylene components [50].

In order to further improve the clinical performance and longevity of UHMWPE-based joint implants, the dilemma in mechanical strength, wear resistance, and oxidation stability should be tackled through substantial innovation in UHMWPE technologies to break the bottleneck. In the viewpoint of polymer science, the key problem is the structural defects due to the processing difficulty of UHMWPE with very high viscosity in melt. Medical grade UHMWPE batches are known as porous materials, where the porous structures come from the poor fusion of resin particles. The fusion defects prevent further improvement of strength and toughness, and, worse, upon radiation cross-linking and thermal treatments, such defects may deteriorate the strength and toughness.

High-temperature melting successfully eliminates the fusion defects and significantly improves the mechanical strength, toughness, and ductility. Importantly, the mechanical properties can be manipulated through delicate control of the crystal structures and chain entanglements. Such materials provide a versatile platform for subsequent radiation cross-linking to reduce wear. Inevitably, high dose radiation cross-linking creates three-dimensional networks and thus reduces the ductility of UHMWPE. Nevertheless, the mechanical loss is largely compensated by the mechanical gains in high-temperature melting. The chain scissioning is responsible for the low cross-link density, which, together with the low crystallinity and thin lamellae, entitles relatively high ductility and impact toughness, in comparison to conventional highly cross-linked UHMWPE. It is surprising that the obtained cross-linked HTM-PEs show low wear rates close to those of radiation cross-linked and remelted polyethylenes that are already in clinical use. Therefore, the high-temperature melted and radiation cross-linked UHMWPE is advantageous in both mechanical properties and wear resistance.

For practical applications, certainly, post-irradiation thermal treatments should be avoided for these materials to retain the mechanical properties. Vitamin E has been practically used to stabilize cross-linked UHMWPE in joint implants. In this chapter, vitamin E is further demonstrated to survive the harsh high-temperature melting and high dose irradiation. Accelerated aging, as recommended by ASTM, has demonstrated the excellent potency of vitamin E to protect polyethylene from severe oxidation and even with the presence of unsaturated lipids. Although it is not clear how these tests are related to shelf-aging and in vivo performance, a comparison of these results with those already in clinical use undoubtedly leads to optimistic prediction that these materials are very promising for clinical applications in next-generation joint implants.

5.6 Conclusions

High-temperature melting has been successfully used to eliminate fusion defects and manipulate the chain scission, entanglements, and crystal structures. As a result, the toughness, ductility, and impact strength are highly improved in a controlled manner. Such a freedom to manipulate the structures and properties of UHMWPE enables a manufacturing of polyethylene with tunable and improved toughness and wear resistance by radiation cross-linking. Vitamin E is incorporated into UHMWPE prior to high-temperature melting and radiation cross-linking in order to stabilize the materials. Accelerated aging of these materials with or without the presence of unsaturated lipids shows outstanding oxidation stability, although the vitamin E potency is slightly decreased due to high-temperature melting and radiation cross-linking. Such high-temperature melted, radiation cross-linked, and stabilized UHMWPE are promising for next-generation high-performance joint implants.

References

1. Harris WH (2009) The first 50 years of total hip arthroplasty: lessons learned. Clin Orthop Relat Res 467(1):28–31. https://doi.org/10.1007/s11999-008-0467-1
2. Snir N, Kaye ID, Klifto CS, Hamula MJ, Wolfson TS, Schwarzkopf R, Jaffe FF (2014) 10-year follow-up wear analysis of first-generation highly crosslinked polyethylene in primary total hip arthroplasty. J Arthroplast 29(3):630–633. https://doi.org/10.1016/j.arth.2013.07.034
3. Kurtz SM, Gawel HA, Patel JD (2011) History and systematic review of wear and osteolysis outcomes for first-generation highly crosslinked polyethylene. Clin Orthop Relat Res 469 (8):2262–2277. https://doi.org/10.1007/s11999-011-1872-4
4. Kurtz SM, Medel FJ, MacDonald DW, Parvizi J, Kraay MJ, Rimnac CM (2010) Reasons for revision of first-generation highly cross-linked polyethylenes. J Arthroplast 25(6):67–74
5. Kurtz SM, Lau E, Ong K, Zhao K, Kelly M, Bozic KJ (2009) Future young patient demand for primary and revision joint replacement: national projections from 2010 to 2030. Clin Orthop Relat Res 467(10):2606–2612. https://doi.org/10.1007/s11999-009-0834-6
6. Atwood SA, Van Citters DW, Patten EW, Furmanski J, Ries MD, Pruitt LA (2011) Tradeoffs amongst fatigue, wear, and oxidation resistance of cross-linked ultra-high molecular weight polyethylene. J Mech Behav Biomed Mater 4(7):1033–1045
7. Oral E, Malhi A, Muratoglu O (2006) Mechanisms of decrease in fatigue crack propagation resistance in irradiated and melted UHMWPE. Biomaterials 27:917–925
8. Pruitt LA (2005) Deformation, yielding, fracture and fatigue behavior of conventional and highly cross-linked ultra high molecular weight polyethylene. Biomaterials 26(8):15
9. Kurtz SM (ed) (2009) UHMWPE biomaterials handbook. Ultra-high molecular weight polyethylene in total joint replacement and medical devices, 2nd edn. Elsevier Inc., New York
10. de Gennes P-G (1979) Scaling concepts in polymer physics, 4th edn. Cornell University Press, Ithaca
11. Doi M, Edwards SF (1986) The theory of polymer dynamics. Clarendon, Oxford
12. Rastogi S, Lippits DR, Peters GWM, Graf R, Yao Y, Spiess HW (2005) Heterogeneity in polymer melts from melting of polymer crystals. Nat Mater 4:635
13. Wu JJ, Buckley CP, O'Connor JJ (2002) Mechanical integrity of compression-moulded ultra-high molecular weight polyethylene: effects of varying process conditions. Biomaterials 23:3773–3783

14. Buckley CP, Wu J, Haughie DW (2006) The integrity of welded interfaces in ultra high molecular weight polyethylene: part 1-model. Biomaterials 28:3178–3186
15. Haughie DW, Buckley CP, Wu J (2006) The integrity of welded interfaces in ultra-high molecular weight polyethylene: part 2-interface toughness. Biomaterials 28:3875–3881
16. Poutsma ML (2003) Reexamination of the pyrolysis of polyethylene: data needs, free-radical mechanistic considerations, and thermochemical kinetic simulation of initial product-forming pathways. Macromolecules 36:8931–8957
17. Fu J, Ghali BW, Lozynsky AJ, Oral E, Muratoglu OK (2010) Ultra high molecular weight polyethylene with improved plasticity and toughness by high temperature melting. Polymer 51 (12):2721–2731. https://doi.org/10.1016/j.polymer.2010.04.003
18. Lippits DR, Rastogi S, Talebi S, Bailly C (2006) Formation of entanglements in initially disentangled polymer melts. Macromolecules 39(26):8882–8885. https://doi.org/10.1021/ma062284z
19. Bartczak Z (2010) Effect of chain entanglements on plastic deformation behavior of ultra-high molecular weight polyethylene. J Polym Sci B Polym Phys 48(3):276–285. https://doi.org/10.1002/polb.21873
20. Lippits DR, Rastogi S, Höhne GWH (2006) Melting kinetics in polymers. Phys Rev Lett 96 (21):218303
21. Lotz B, Cheng SZD (2005) A critical assessment of unbalanced surface stresses as the mechanical origin of twisting and scrolling of polymer crystals. Polymer 46(3):577–610. https://doi.org/10.1016/j.polymer.2004.07.042
22. Haward RN (1993) Strain hardening of thermoplastics. Macromolecules 26:5860–5869
23. Na B, Lv R, Xu W, Yu P, Wang K, Fu Q (2007) Inverse temperature dependence of strain hardening in ultrahigh molecular weight polyethylene: role of lamellar coupling and entanglement density. J Phys Chem B 111(46):13206–13210. https://doi.org/10.1021/jp075990q
24. Muratoglu OK, Bragdon CR, O'Connor DO, Jasty M, Harris WH, Gul R, McGarry F (1999) Unified wear model for highly crosslinked ultra-high molecular weight polyethylenes (UHMWPE). Biomaterials 20(16):1463–1470
25. Wang A (2001) A unified theory of wear for ultra-high molecular weight polyethylene in multi-directional sliding. Wear 248:38–47
26. Bragdon CR, O'Connor DO, Lowenstein JD, Jasty M, Biggs SA, Harris WH (2001) A new pin-on-disk wear testing method for simulating wear of polyethylene on cobalt-chrome alloy in total hip arthroplasty. J Arthroplast 16(5):658–665
27. Furmanski J, Gupta S, Chawan A, Kohm A, Lannutti J, Jewett B, Pruitt LA, Ries MD (2007) Aspherical femoral head with highly cross-linked ultra-high molecular weight polyethylene surface cracking: a case report. J Bone Joint Surg Ser A 89(10):2266–2270
28. Muratoglu OK, Harris WH (2001) Identification and quantification of irradiation in UHMWPE through *trans*-vinylene yield. J Biomed Mater Res 56(4):584–592
29. Fu J, Shen J, Gao G, Hou R, Xu Y, Cong Y, Cheng Y (2013) Natural polyphenol-stabilised highly crosslinked UHMWPE with high mechanical properties and low wear for joint implants. J Mater Chem B 1:4727–4735. https://doi.org/10.1039/c3tb20707b
30. Kurtz SM, Muratoglu OK, Evans M, Edidin AA (1999) Advances in the processing, sterilization, and crosslinking of ultra-high molecular weight polyethylene for total joint arthroplasty. Biomaterials 20(18):1659–1688
31. Wannomae K, Bhattacharyya S, Freiberg A, Estok D, Harris W, Muratoglu O (2006) In vivo oxidation of retrieved crosslinked UHMWPE acetabular components with residual free radicals. J Arthroplast 21(7):1005–1011
32. Oral E, Rowell S, Muratoglu O (2008) Near-melt annealing of irradiated UHMWPE under pressure. In: Transactions, 54th annual meeting of the orthopaedic research society, San Francisco, CA. p 1680
33. Sobieraj MC, Kurtz SM, Wang A, Manley MM, Rimnac CM (2008) Notched stress-strain behavior of a conventional and a sequentially annealed highly crosslinked UHMWPE. Biomaterials 29(35):4575–4583. https://doi.org/10.1016/j.biomaterials.2008.08.010

34. McKellop H (2002) Moderate crosslinking and remelting of UHMW polyethylene for reduced Wear in hip prostheses. In
35. Medel FJ, Pena P, Cegonino J, Gomez-Barrena E, Puertolas JA (2007) Comparative fatigue behavior and toughness of remelted and annealed highly crosslinked polyethylenes. J Biomed Mater Res B Appl Biomater 83(2):380–390
36. Wolf C, Krivec T, Blassing J, Lederer K, Schneider W (2002) Examination of the suitability of α-tocopherol as a stabilizer for ultra-high molecular weight polyethylene used for articulating surfaces in joint endoprostheses. J Mater Sci Mater Med 13:185–189
37. Oral E, Wannomae KK, Hawkins NE, Harris WH, Muratoglu OK (2004) α-Tocopherol doped irradiated UHMWPE for high fatigue resistance and low wear. Biomaterials 25(24):5515–5522
38. Oral E, Greenbaum ES, Malhi AS, Harris WH, Muratoglu OK (2005) Characterization of irradiated blends of alpha-tocopherol and UHMWPE. Biomaterials 26(33):6657–6663. https://doi.org/10.1016/j.biomaterials.2005.04.026
39. Gijsman P, Smelt H, Schumann D (2010) Hindered amine light stabilizers: an alternative for radiation cross-linked UHMwPE implants. Biomaterials 31:6685–6691
40. Oral E, Wannomae K, Muratoglu O (2005) The effect of doping conditions on α-tocopherol stabilized UHMWPE. In: Transactions, 51st annual meeting of the orthopaedic research society, Washington, DC
41. Currier BH, Van Citters DW, Currier JH, Collier JP (2010) In vivo oxidation in remelted highly cross-linked retrievals. J Bone Joint Surg 92(14):2409–2418
42. Muratoglu OK, Wannomae KK, Rowell SL, Micheli BR, Malchau H (2010) Ex vivo stability loss of irradiated and melted ultra-high molecular weight polyethylene. J Bone Joint Surg 92 (17):2809–2816
43. Currier BH, Van Citters DW, Currier JH, Carlson EM, Tibbo ME, Collier JP (2013) In vivo oxidation in retrieved highly crosslinked tibial inserts. J Biomed Mater Res B Appl Biomater 101B(3):441–448. https://doi.org/10.1002/jbm.b.32805
44. Costa L, Bracco P, Brach del Prever EM, Luda MP, Trossarelli L (2001) Analysis of products diffused into UHMWPE prosthetic components in vivo. Biomaterials 22(4):307–315
45. Wannomae K, Oral E, Neils A, Rowell S, Muratoglu O (2012) Trace amounts of grafted vitamin E protect polyethylene against squalene-initiated oxidation. J Bone Joint Surg. British Volume 94-B (SUPP XXXVII):130
46. Amstutz H, Campbell P, Kossovsky N, Clarke I (1992) Mechanism and clinical significance of wear debris-induced osteolysis. Clin Orthop 276:7–18
47. Costa L, Jacobson K, Bracco P, Brach del Prever EM (2002) Oxidation of orthopaedic UHMWPE. Biomaterials 23:1613–1624
48. Wang A, Zeng H, Yau S-S, Essner A, Manely M, Dumbleton J (2006) Wear, oxidation and mechanical properties of a sequentially irradiated and annealed UHMWPE in total joint replacement. J Phys D Appl Phys 39:3213–3219
49. Muratoglu OK, Bragdon CR, O'Connor DO, Jasty M, Harris WH (2001) A novel method of cross-linking ultra-high-molecular-weight polyethylene to improve wear, reduce oxidation, and retain mechanical properties. Recipient of the 1999 HAP Paul Award. J Arthroplast 16(2):149–160. https://doi.org/10.1054/arth.2001.20540
50. Rowell SL, Muratoglu OK (2016) Investigation of surgically retrieved, vitamin E-stabilized, crosslinked UHMWPE implants after short-term in vivo service. J Biomed Mater Res B Appl Biomater 104(6):1132–1140. https://doi.org/10.1002/jbm.b.33465

Chapter 6
Drug-Loaded UHMWPE to Inhibit Wear Particle-Induced Osteolysis: Processing, Characterization, and Biological Evaluation

Shuxin Qu, Yumei Liu, and Kemeng Gong

Abstract Osteolysis induced by ultrahigh molecular weight polyethylene (UHMWPE) wear particles is currently recognized as a major cause to the aseptic loosening of joint prosthesis. Improving the wear resistance of UHMWPE significantly reduces wear particle-related osteolysis. However, the current wear-resistant UHMWPE is still difficult to satisfy the clinical requirement for the increasing needs of young and active patients. In this chapter, UHMWPE loaded with 17β-estradiol (E2) or alendronate sodium (ALN), potential drugs to prevent wear particle-induced osteolysis, was processed and characterized. Furthermore, in vitro release of drugs and cell responses to drug-loaded UHMWPE wear particles were investigated. Results showed that the mechanical strength of UHMWPE-E2 and UHMWPE-ALN decreased with the addition of drugs in a dose-dependent manner. However, there was no significant difference in mechanical strength of UHMWPE-ALN (1 wt. % ALN) and UHMWPE due to the improved distribution of ALN in UHMWPE, in which surfactant Pluronic F68 and mechanical activation were used. Correspondingly, the wear rate and coefficient of friction of UHMWPE-ALN were confirmed similar to those of UHMWPE. The release of E2 and ALN from wear particles of drug-loaded UHMWPE in phosphate buffered solution (PBS) in vitro both includes three stages: the initial burst release, the following rapid release, and the final slow release. The ALN release rate in wear process was higher than that of non-wear test. Furthermore, the released drugs increase the proliferation and alkaline phosphatase activity of osteoblasts, while inhibiting the proliferation and cytokines of macrophages. As a result, drug-loaded UHMWPE might have potential clinical application to prevent the wear particle-induced osteolysis in artificial joint replacements.

Keywords Ultra-high molecular weight polyethylene · 17β-Estradiol · Alendronate sodium · Pluronic F68 · Mechanical activation · Mechanical properties · Tribological properties · Drug release · Biological evaluation

---

S. Qu (✉) · Y. Liu · K. Gong
Key Laboratory of Advanced Technologies of Materials, Ministry of Education, School of Materials Science and Engineering, Southwest Jiaotong University, Chengdu, China
e-mail: qushuxin@swjtu.edu.cn

J. Fu et al. (eds.), *UHMWPE Biomaterials for Joint Implants*, Springer Series in Biomaterials Science and Engineering 13,
https://doi.org/10.1007/978-981-13-6924-7_6

6.1 Introduction

Ultrahigh molecular weight polyethylene (UHMWPE) has been extensively used as a bearing material in the prosthetic joint for more than half century due to its excellent wear resistance, stable chemistry performance, and good biocompatibility [1]. Nevertheless, UHMWPE wear particles have been recognized as one of the major causes of osteolysis and aseptic loosening of prosthetic joint because UHMWPE is the weakest component in prosthetic joint [2–5]. It is believed that macrophages became activated after phagocytosing those wear particles and released a series of cytokines and other mediators of inflammation. Those could trigger the bone resorption ability of osteoclasts, which was so-called wear particle-induced osteolysis [2].

Significant efforts have been devoted to reducing the wear particles to prevent osteolysis. The main approaches include altering articulation bearing pairs in terms of tribology and material science [6, 7]; developing improved bearing materials, e.g., highly cross-linked, ion-implanted, and filler-modified UHMWPE [8, 9]; optimizing the biomechanical design of artificial articulations [10]; and improving surgery technology [11]. These endeavors have led to considerable improvements in reducing wear; however, the osteolysis remains an issue in clinical performance of artificial joint prostheses.

In addition, pharmacological interventions have become available because of advances in the understanding of the molecular mechanisms of osteolysis [12, 13]. In clinic, various drugs have been used to treat the osteolysis, e.g., simvastatin [14], erythromycin [15], and bisphosphonates [16–19]. Bisphosphonates are currently the most effective inhibitors of bone resorption [20]. However, oral bisphosphonate therapies are associated with disadvantages such as esophageal erosions, low bioavailability, and long periods of therapy [21]. As reported in literatures, 17β-Estradiol (E2), which is thought to prevent postmenopausal bone loss, could impede the differentiation and activity of osteoclasts in many ways [22–28]. E2 is also capable to directly enhance osteoblastic activity and bone formation [25, 26] and could further suppress the production of interleukin-6 (IL-6) [27, 28], which is one of the most potent cytokines that induce osteoclasts maturation and thereby stimulate bone resorption [29]. Furthermore, more researchers preferred topical administration to oral administration, which could give enough doses at the focus to overcome the side effect of drug resulting from the oral administration [30–33].

In this chapter, UHMWPE, which is the most extensively used as the bearing materials of artificial joint in clinic, has been loaded with drugs. It is hypothesized that the released drug from the inevitable UHMWPE wear particles could decrease the biological response to wear particles [34]. The processing, characterization, tribological behavior, drug release in vitro, and the biological response of drug-loaded UHMWPE are studied.

6.2 Processing

Drug-loaded UHMWPE is firstly proposed to explore the possibility to treat particle-induced osteolysis [35]. Two kinds of drug-loaded UHMWPE are studied and are described as below.

6.2.1 UHMWPE Loaded with 17β-Estradiol (UHMWPE–E2)

UHMWPE loaded with E2 was firstly investigated in the study of Liu et al. [35]. In their study, the drug, 1.2 g E2, was dissolved in 220 mL acetone and then mixed with 180 g UHMWPE (the Second Factory of Auxiliary Agents, Beijing, China) powder. The mixture was dried under vacuum at room temperature (RT). The amount of E2 in the UHMWPE–E2 mixture was 0.67 wt%, and this primary mixture rate was chosen according to the previous literature [36, 37]. The control UHMWPE powder without E2 was also dried. Then the UHMWPE–E2 mixture and the control UHMWPE powder were both compression-molded under 7 MPa at 180 °C in the Manual Hydraulic Press (Carver 3925, USA).

6.2.2 UHMWPE Loaded with Alendronate Sodium (UHMWPE–ALN)

6.2.2.1 UHMWPE–ALN

As an estrogen hormone, however, E2 is only effective on females. ALN, one of the third generation bisphosphonates, is not limited by the gender of patients [38]. Liu et al. [39] followed this method to prepare consolidated UHMWPE–ALN by solution blending and compression molding. Therein, the molding temperature was from 180 to 240 °C because of the high melt viscosity of UHMWPE. However, severe oxidation of UHMWPE–ALN molded at 240 °C was found in the following studies. On the other hand, higher mechanical and tribological properties were obtained while elevating the molding pressure. Thus, 180 °C of molding temperature and 15 MPa of molding pressure with annealing at 140 °C and 20 MPa were adopted in the following studies in their study [34, 39–41].

As described by Yang et al. [40], the ALN is water soluble, while UHMWPE powder is hydrophobic, which would cause the nonuniform distribution of ALN and poor interfacial adhesion between ALN and UHMWPE. With the increasing content of ALN in UHMWPE, the mechanical and tribological properties of UHMWPE–ALN decreased.

6.2.2.2 UHMWPE–ALN Doping with Surfactant Pluronic F68

An optimization to improve the dispersion of ALN in UHMWPE using orthogonal design was studied by Zhang et al. [42]. This design combined with statistical analyses revealed an optimal content of 1 wt% Pluronic F68 in UHMWPE to assist the processing of ALN in UHMWPE. The presence of F68 led to a uniform distribution of ALN in UHMWPE, and the material is denoted as UHMWPE–ALN–F68.

6.2.2.3 UHMWPE–ALN after Mechanical Activation

Another method without additives was developed by Gong et al. [41]. In their study, mechanical activation of UHMWPE–ALN was performed in a planetary ball mill (QM-1SP4, Instrument plant of Nanjing University, China) equipped with polyamides jars (400 mL volume) and 200 agate balls (4 mm diameter). UHMWPE–ALN powder (50 g) was put into a polyamide jar with 200 agate balls and 100 mL absolute ethanol. The jar was mounted vertically in the ball mill carrier and subsequently rotated at 400 rpm for 60 min. Then the mixture was totally dried into powder (i.e., UHMWPE–ALN–ma) at 70 °C. The mechanical properties and wear of mechanical activated UHMWPE–ALN were significantly increased compared to those without mechanical activation.

6.3 Characterization of Physical, Chemical, Mechanical, and Tribological Properties

There are many critical properties to be considered when developing a new UHMWPE material for implant applications. This section summarizes the test techniques used to characterize various properties of drug-loaded UHMWPE, including of physical, chemical, mechanical, and tribological properties.

6.3.1 UHMWPE–E2

6.3.1.1 Crystallinity

Differential scanning calorimetry (DSC) was employed [35] to determine the crystallinity of UHMWPE–E2 composite and the control UHMWPE. Firstly, the samples were heated from RT to 200 °C with a heat rate of 10 °C/min under nitrogen atmosphere. It was kept at 200 °C for 3 min to eliminate the thermal history [43]. Then the temperature was reduced to RT with a cooling rate of 10 °C/min. Finally, another heating cycle was conducted from RT to 200 °C. The heat of fusion

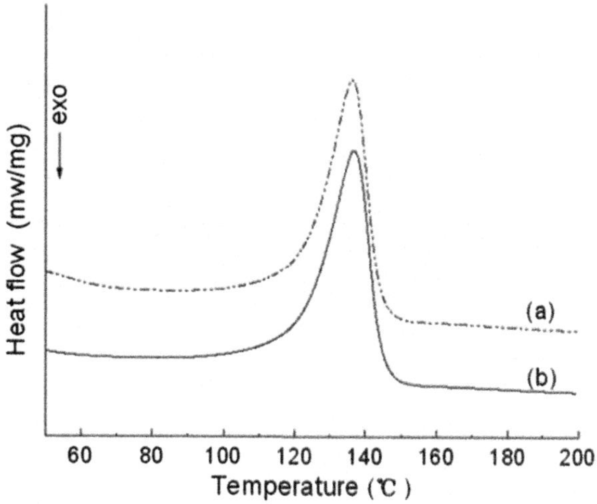

Fig. 6.1 DSC curves of (**a**) UHMWPE–E2, (**b**) UHMWPE. (Reprinted from Ref. [35], Copyright 2008, with permission from Wiley)

(ΔH) was obtained by integrating the area under the endothermic peak from about 80 °C to 160 °C. The melting temperature was registered as the maximum of the endothermic melting peak. The degree of crystallinity was calculated by Eq. 6.1:

$$X_c(\%) = \frac{\Delta H}{\Delta H_{100}} \times 100\% \tag{6.1}$$

where ΔH is the heat of fusion of the sample, and ΔH_{100} is the heat of fusion for 100% crystallinity UHMWPE, or 291 J/g [44].

Figure 6.1 presents the DSC curves of UHMWPE–E2 and the control UHMWPE. The melting temperatures (T_m) of UHMWPE–E2 and the control UHMWPE are 136.8 and 137.1 °C, respectively. According to Eq. 6.1, the crystallinities of UHMWPE–E2 and UHMWPE were calculated as 65.8 and 65.9%. There were no significant differences in the crystallinity between UHMWPE–E2 and the control UHMWPE ($p > 0.05$).

6.3.1.2 Mechanical Properties

The tensile properties of UHMWPE–E2 and UHMWPE were tested [35] by using a universal material testing machine (5567, Instron Corp., USA) at a crosshead speed of 50 mm/min in accordance with ASTM D638-03 [45]. The dimension of a dog-bone-shaped specimen was 100 mm × 6 mm × 1 mm with the gauge length of 25 mm. The tensile strength values of UHMWPE–E2 and the control UHMWPE were 47.2 ± 3.6 MPa and 52.1 ± 1.0 MPa, respectively. The yield strength of UHMWPE–E2 (25.0 ± 0.6 MPa) is similar to that of the control UHMWPE (25.6 ± 0.4 MPa). The elongations at break of UHMWPE–E2 and UHMWPE were 320 ± 25.5% and 327.3 ± 13.8%, respectively. The differences in the tensile strength, yield strength, and elongation between UHMWPE–E2 and UHMWPE were not statistically significant ($p > 0.05$).

6.3.1.3 Tribological Properties

Two-body in vitro wear test was conducted on UHMWPE-E2 in a ball-on-flat configuration at RT with the relative humidity of 65 ± 5% under dry condition [35]. Taking into account the principle that the contact stress between the counterfaces should be larger than the yield strength of UHMWPE (21–28 MPa), the wear tests were conducted with normal loads of 10, 30, and 50 N for 10,000 cycles at 5 Hz. The maximum contact stress for the above normal loads were approximately 22, 32, and 38 MPa according to Hertzian contact stress analysis [46]. The titanium–aluminum–vanadium alloy (Ti–6Al–4V) ball with a diameter of 40 mm was used as thc sphere counterpart, which was cleaned with alcohol before the wear test. The coefficients of friction were measured during the whole test.

Stylus profilometer (XP-2, Ambios Technology Inc., USA) was used to measure the roughness of UHMWPE–E2 and UHMWPE to describe wear grooves after the wear test. The wear grooves of UHMWPE–E2 had profiles similar to those of the control UHMWPE at the same normal load under dry friction, as shown in Fig. 6.2a, b. The variations of wear depth and width as a function of normal load were shown in Fig. 6.2c, d. It was observed that the wear depth and width of UHMWPE–E2 were similar with those of UHMWPE at the same normal load. Furthermore, the wear depth and width of UHMWPE–E2 and the control UHMWPE increased with the increasing of normal load. One-way analysis of variance (ANOVA) revealed that there were no significant differences ($p > 0.05$) in the wear depth and width between UHMWPE–E2 and UHMWPE at the same normal loads.

The wear surface of UHMWPE–E2 and UHMWPE was observed by Scanning Electron Microscopy (SEM, Quanta 200, FEI, USA) [35]. The morphology of wear surfaces of UHMWPE–E2 and UHMWPE were similar at the same normal loads (Fig. 6.3). Typical micrographs of the elliptic wear surface at the load of 50 N were presented in Fig. 6.3a, e. Figure 6.3b–d, and f–h showed the center of the wear surface of UHMWPE–E2 and UHMWPE at different loads. Various shapes of wear particles were observed on the wear surfaces for both UHMWPE–E2 and UHMWPE. However, there were different features between wear surfaces at the same normal loads. At 10 N, obvious deformation flow, ploughing, and micro-cracks were observed on the wear surface of both samples (Fig. 6.3b, f). With the increasing normal loads, the wear surface became smooth with less and narrower micro-cracks. The ploughing and deformation flow were not obviously observed at 50 N (Fig. 6.3d, h).

The wear test results showed similar coefficients of friction curves and similar wear depth and width between UHMWPE–E2 and UHMWPE under the same normal load. Moreover, SEM observation showed that there was no obvious difference in the wear surface between UHMWPE–E2 and UHMWPE (Fig. 6.3). The

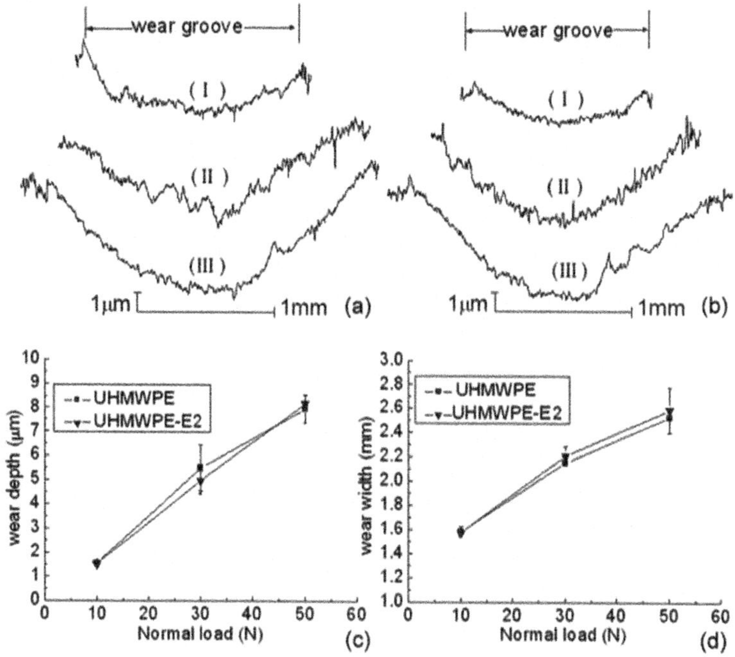

Fig. 6.2 Surface profile measurements on wear grooves: (**a**) UHMWPE–E2 (I) $F_n = 10$ N, (II) $F_n = 30$ N, (III) $F_n = 50$ N. (**b**) UHMWPE (I) $F_n = 10$ N, (II) $F_n = 30$ N, (III) $F_n = 50$ N. Variations of wear depth and width as the function of normal load: (**c**) wear depth and (**d**) wear width. (Reprinted from Ref. [35], Copyright 2008, with permission from Wiley)

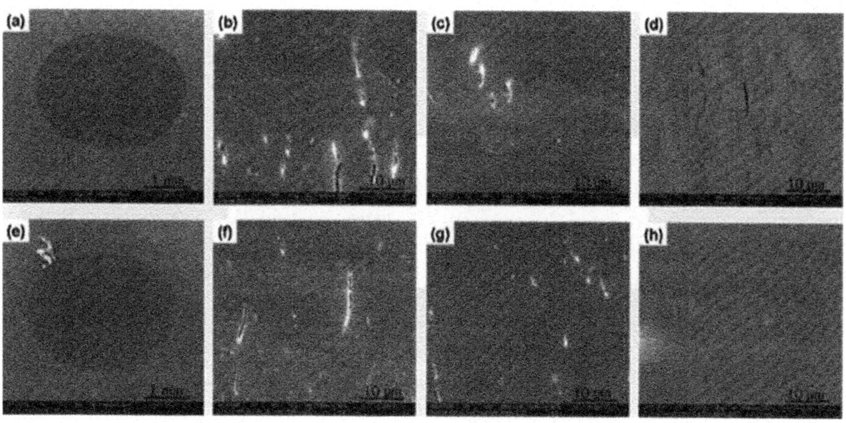

Fig. 6.3 Typical SEM micrographs in the wear surfaces subjected to 10,000 cycles at different loads: (**a–d**) UHMWPE, (**e–h**) UHMWPE–E2. (**a**, **e**) $F_n = 50$ N, (**b**, **f**) $F_n = 10$ N, (**c**, **g**) $F_n = 30$ N, and (**d**, **h**) $F_n = 50$ N. (Reprinted from Ref. [35], Copyright 2008, with permission from Wiley)

results indicated that UHMWPE–E2 and UHMWPE had similar friction and wear performance.

The coefficient of friction is inversely proportional to the normal load, which is common in engineering contacts and could be elucidated by the Hertzian theory [46]. There were some differences among the morphologies of wear surfaces under three different normal loads either UHMWPE–E2 or UHMWPE. The most notable characteristics of the wear surfaces of the two samples were the deformation flow, ploughing and micro-cracks on the wear surface at 10 N (Fig. 6.3b, f). These characterizations suggested that abrasive wear was the main wear mechanism for both the UHMWPE–E2 and UHMWPE under dry condition. The micro-cracks running transverse to the wear track was typical on the wear surface of UHMWPE [47]. However, the micro-cracks became less and limited with the increasing of load. The wear surface at 50 N was smoother than that of 10 N (Fig. 6.3d, h). It likely resulted from the high friction heat under high normal load, which would improve the fluidity of the materials. It was deduced that the micro-cracks on the wear surface did not have sufficient time to diffuse before they were removed to the edge of the wear surface. On the other hand, some of the micro-cracks might be filled during the flowing of the materials. Therefore, only slight micro-cracks were observed on the exposed fresh wear surface under high load.

6.3.2 *UHMWPE–ALN*

6.3.2.1 Mechanical Properties

The tensile properties of UHMWPE–ALN and UHMWPE were measured by the universal material testing machine at a crosshead speed of 50 mm/min [40]. The dimensions of dog-bone-shaped samples were 75 mm × 12 mm × 2 mm, with the gauge length of 25 mm. Table 6.1 shows the mechanical properties of UHMWPE–ALN. The elastic modulus of UHMWPE–ALN 1.0, 0.5 wt% and the control UHMWPE were not significantly different ($p > 0.05$). The average elongation at break was decreased by 3.2% and 20.3% after loading ALN 0.5 and 1.0 wt%, respectively. The elongation at break of UHMWPE–ALN 0.5 wt% was not significantly different ($p > 0.05$), while those of UHMWPE–ALN 1.0 wt% were significantly reduced by about 20% compared with that of the control UHMWPE ($p < 0.05$).

Table 6.1 Mechanical properties of UHMWPE–ALN and UHMWPE (Modified from ref. 40)

Sample	Elastic modulus (GPa)	Elongation at break (%)
UHMWPE	0.554 ± 0.016	439 ± 33
UHMWPE–ALN 0.5 wt. %	0.544 ± 0.023	425 ± 22
UHMWPE–ALN 1.0 wt. %	0.540 ± 0.019	350 ± 25

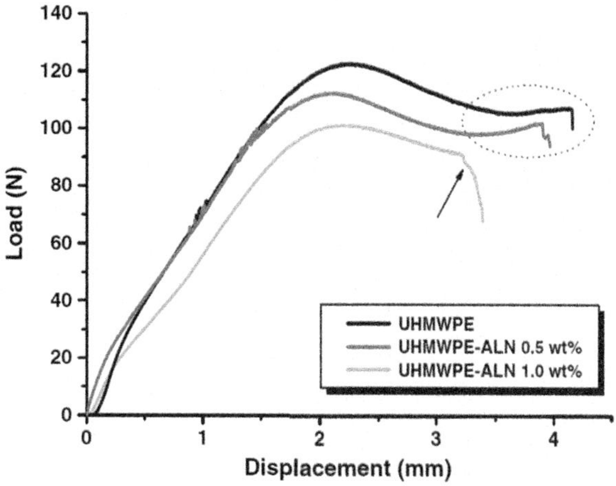

Fig. 6.4 The typical load-displacement curve of small punch testing for UHMWPE–ALN 1.0 wt. % and UHMWPE; the ellipse, which represents strain-hardening, was replaced by a roll-off (indicated by arrow) for UHMWPE–ALN 1.0 wt. %. (Reprinted from Ref. [40], Copyright 2011, with permission from Elsevier)

Table 6.2 Work-to-failure of UHMWPE–ALN and UHMWPE (Modified from Ref. 40)

Sample	Work-to-failure (mJ)
UHMWPE	363 ± 33
UHMWPE–ALN 0.5 wt. %	351 ± 28
UHMWPE–ALN 1.0 wt. %	280 ± 42

Small punch test [40] was carried out by a self-made apparatus, which was dimensioned according to ASTM F2183-02 [48]. Disk-shaped samples of 6.4 mm in diameter and 1.0 mm in thickness were tested by the universal material testing machine at a constant punch displacement rate of 0.5 mm/min at RT. During the testing, the punch load and displacement were digitally recorded. The work to failure was calculated by integrating the area under the load–displacement curve, which provides additional quantitative measures of the sample ductility and fracture resistance [49, 50]. Typical small punch test curves were shown in Fig. 6.4, while the work-to-failure of UHMWPE–ALN and UHMWPE–ALN were shown in Table 6.2. The strain hardening region (marked in ellipse) on the curve of control UHMWPE was replaced by a roll-off (indicated by arrow) for UHMWPE–ALN 1.0 wt%, which reduces the work to failure significantly. According to small punch test results, the presence of ALN weakens the strain hardening at large deformation, which leads to the roll-off of load–displacement curve and the reduced work to failure for UHMWPE–ALN 1.0 wt%. The presence of ALN may form microphase separation from UHMWPE and consequently reduces the fracture resistance of UHMWPE.

Tensile test results show that both UHMWPE–ALN and UHMWPE have similar elastic modulus, while the elongation at break of UHMWPE–ALN 1.0 wt% decreased by 20% compared with that of UHMWPE. The decrease in crystallinity might be the primary cause to the decrease in mechanical properties. On the other hand, for polymer composites, the dispersion of filled phase and the interfacial adhesion among composite components play important roles in determining their mechanical properties [44]. The hydrophilic ALN and hydrophobic UHMWPE might phase separate to form nonuniform distribution of ALN with poor interfacial adhesion between ALN and UHMWPE. The elastic modulus of UHMWPE was measured at relatively low deformation, and thus insensitive to the ALN content [51]. After yielding, the large deformation induces cold drawing and hardening in the semicrystalline polymer, where the chains in amorphous regions disentangle, followed by the unfolding of lamellar structures in crystalline regions [52]. The poor interfacial adhesion between ALN and UHMWPE would introduce voids or crazes at the interfaces, which would accelerate the rupture of UHMWPE–ALN.

6.3.2.2 Tribological Properties

In vitro friction-wear tests were performed [40] by using a unidirectional rotating ball-slide-on-plate apparatus (UMT II, CETR, USA). A ZrO_2 ball with the diameter of 9.8 mm (Shanghai Research Institute of Materials, China) was used as the counterface with 25 vol% calf serum as lubricant. One complete revolution was a ring of 3.0 mm in diameter and the sliding speed was 15.7 mm/s. A normal load of 10 N was applied, which gave rise to the maximum contact pressure of 31.2 MPa [46]. To investigate the evolution of volumetric loss and wear surface morphology, tests of 5000, 15,000 and 30,000 cycles were applied for UHMWPE–ALN 0.5, 1.0 wt% and UHMWPE. Variation of coefficient of friction was recorded automatically as a function of cycles. Three samples of UHMWPE–ALN 0.5, 1.0 wt% and the control UHMWPE were performed at RT for each test condition.

The cross-section profiles of the wear tracks of UHMWPE–ALN and UHMWPE were scanned along the radial direction by using a 3D profiler (NanoMap-D, AEP Tech., USA), which was repeated 3 times per sample at different directions (Fig. 6.5).

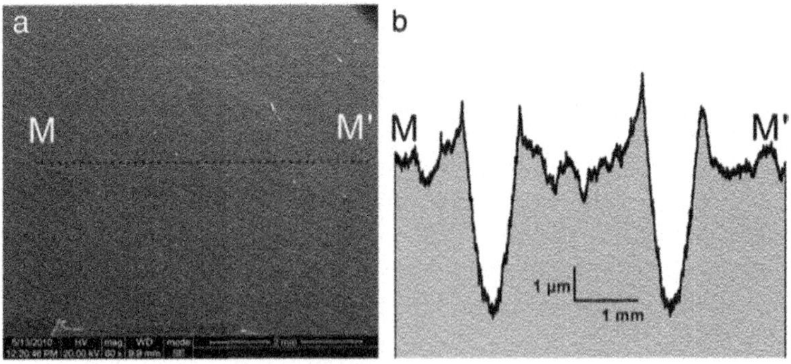

Fig. 6.5 (**a**) SEM micrograph of the wear surface on UHMWPE 1.0 wt. % after 30,000 circulating cycles. (**b**) The cross-sectional profile of (**a**) along M–M′. (Reprinted from Ref. [40], Copyright 2011, with permission from Elsevier)

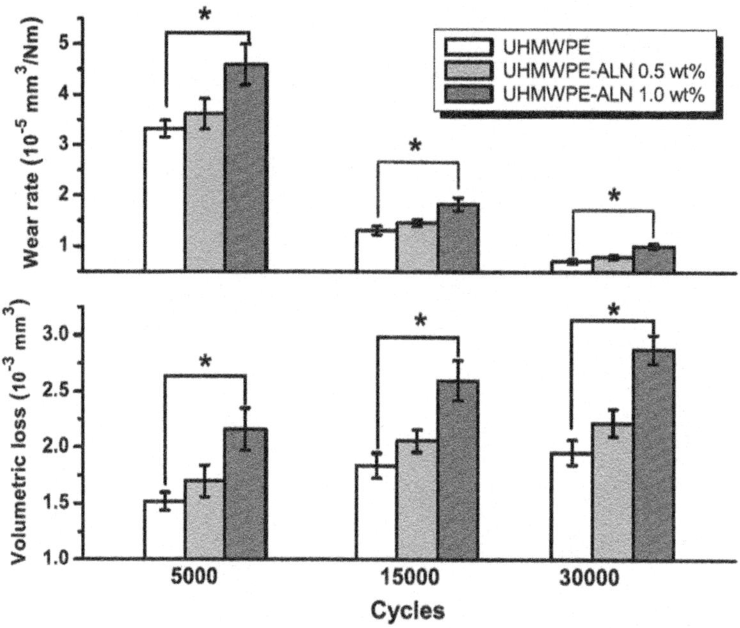

Fig. 6.6 Volumetric losses and wear rates under different circulating cycles. $^*p < 0.05$. (Reprinted from Ref. [40], Copyright 2011, with permission from Elsevier)

The volumetric loss was calculated by the average cross-section wear area multiplying the circumference of the wear track. The wear property of UHMWPE–ALN or UHMWPE with increasing testing duration was evaluated by the wear rate K, as defined by Eq. 6.2 [53]:

$$K = \Delta V/(N \cdot D) \tag{6.2}$$

where ΔV, N and D represent the volumetric loss, normal load and sliding distance, respectively.

The wear performances of UHMWPE–ALN and the control UHMWPE were evaluated by volumetric loss and wear rate (Fig. 6.6). The volumetric loss was introduced for comparing the wear resistance between different samples during the same period. For each testing period, i.e., 5000, 15,000 or 30,000 cycles, the average volumetric losses of UHMWPE–ALN 1 wt% were 42.1, 41.3 and 46.9% greater than those of UHMWPE ($p < 0.05$), whereas no significant difference in volumetric loss was observed between UHMWPE–ALN 0.5 wt% and the control UHMWPE ($p > 0.05$). During the wear process of both UHMWPE–ALN and UHMWPE, the wear rates prior to 5000 circulating cycles were 3–4 times higher than the total wear rate that was measured after 30,000 circulating cycles. It implied that the most severe wear occurred during the initial 5000 cycles, which is consistent with literatures [54, 55].

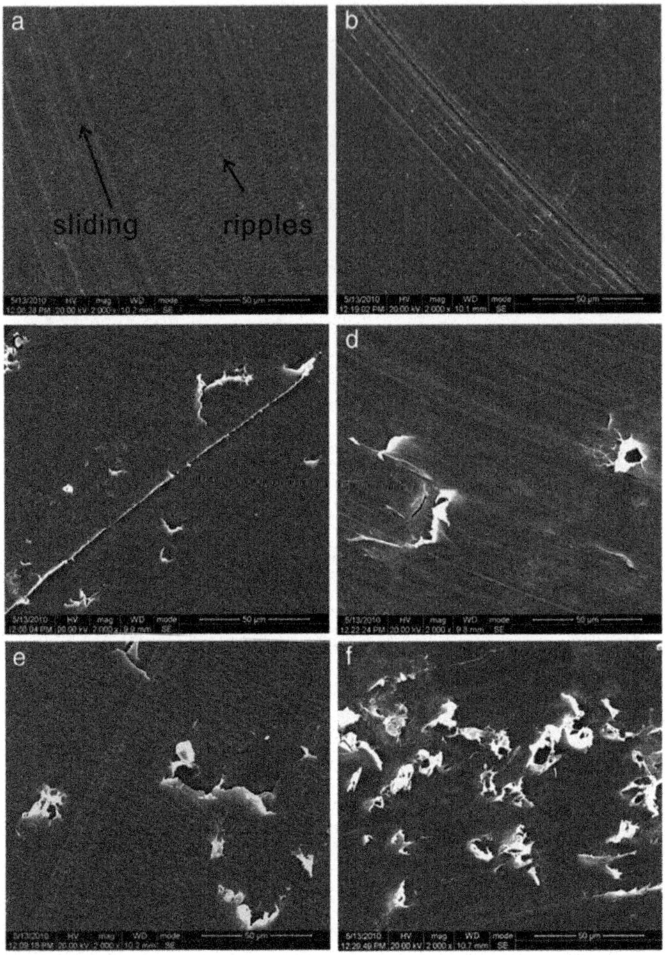

Fig. 6.7 SEM micrographs of control UHMWPE (**a**, **c**, **e**) and UHMWPE–ALN 1.0 wt. % (**b**, **d**, **f**) wear surfaces under different circulating cycles: (**a**, **b**) 5000 cycles, (**c**, **d**) 15,000 cycles, (**e**, **f**) 30,000 cycles. (Reprinted from Ref. [40], Copyright 2011, with permission from Elsevier)

There was a significant difference in the wear rate between UHMWPE–ALN 1.0 wt% and UHMWPE ($p < 0.05$), while there was no significant difference in the wear rate between UHMWPE–ALN 0.5 wt% and UHMWPE ($p > 0.05$). The wear surfaces of UHMWPE–ALN 1.0 wt% and UHMWPE were examined using SEM [40]. There were distinct differences in the morphology of wear surface between UHMWPE and UHMWPE–ALN 1.0 wt% (Fig. 6.7). After 5000 cycles, shallow ploughing and regular arrays of ripples perpendicular to the sliding direction of ZrO_2 appeared on the wear surface of the control UHMWPE (Fig. 6.7a). The obvious ploughing was observed on the wear surface of UHMWPE–ALN 1.0 wt% (Fig. 6.7b). After 15,000 cycles, few as-detached fibrils were found on the wear

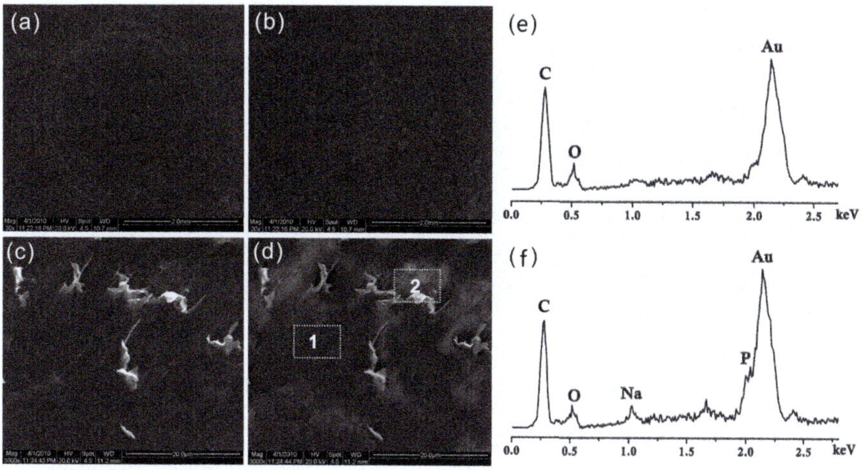

Fig. 6.8 SEM-BSE micrographs of the wear surfaces of UHMWPE–ALN 1.0 wt%: (**a**, **c**) SEM micrographs of wear track and bright spots observed, (**b**, **d**) BSE micrographs of the same site on the wear surface of UHMWPE–ALN 1.0 wt% ALN. EDS of the wear surfaces of UHMWPE–ALN 1.0 wt% which corresponds to the rectangular regions in (**d**): (**e**) region 1; (**f**) region 2. EDS determined the chemical elements, Na and P (characteristic elements of ALN) existing on the UHMWPE–ALN 1.0 wt%. (Reprinted from Ref. [40], Copyright 2011, with permission from Elsevier)

surfaces of UHMWPE and UHMWPE–ALN 1 wt% (Fig. 6.7c, d). Subsequently, the ploughing was still the dominant feature on the UHMWPE wear surface, accompanied by certain sheet-like peelings and fibrils (Fig. 6.7e) after 30,000 cycles. The wear surface of UHMWPE–ALN 1.0 wt% accumulated a lot of non-oriented loose fibrils, where non-oriented loose fibrils were smaller than those sheetlike peelings and fibrils (Fig. 6.7f).

6.3.2.3 Dispersion of ALN in UHMWPE–ALN

The wear surfaces of UHMWPE–ALN 1.0 wt% and UHMWPE were examined using SEM, Back Scattered Electron (BSE) and Energy Dispersive X-ray spectrum (EDS) to reveal the distribution of ALN and investigate the effect of ALN on the wear performance of UHMWPE–ALN [39].

Figure 6.8a–d show SEM and BSE images at the same site on the wear surface of UHMWPE–ALN 1.0 wt% with different magnifications. For UHMWPE–ALN, the characteristic elements of ALN, i.e., Na and P, showed higher intensities than those of C, H, and O in BSE images. Thus, the darker regions corresponded to UHMWPE matrix, whereas the brighter regions represented the presence of Na and P [56, 57].

Fig. 6.9 SEM-BSM micrographs of UHMWPE–ALN 0.5 wt%: (**a**) SEM mode (**b**) BSE mode. (Reprinted from Ref. [40], Copyright 2011, with permission from Elsevier)

The bright spots on the dark background in Fig. 6.8b and bright regions in Fig. 6.8d indicated the presence of ALN. Certain as-detached fibrils appeared on the wear surface of UHMWPE–ALN 1.0 wt% (Fig. 6.8c). Correspondingly, Fig. 6.8d showed some bright regions around the as-detached fibrils in the BSE image of UHMWPE–ALN 1.0 wt%. Figure 6.8e and f show the EDS results of different regions 1 and 2 in Fig. 6.8d. P and Na were detected in region 2, which corresponded to the bright regions. Taking into account the results of EDS spectrum, SEM-BSE micrographs suggested that the loose fibrils from the wear surface of UHMWPE–ALN 1.0 wt% were associated with the presence of agglomerated ALN. Figure 6.8 showed SEM and BSE image at the same site on wear surface of UHMWPE–ALN 0.5 wt%. It was mainly characterized by the ploughing instead of loose fibrils which appeared on UHMWPE–ALN 1.0 wt%. No agglomeration of ALN was observed on BSE images (Fig. 6.8b).

BSE and EDS helped us to understand the relationship between loose fibrils and the presence of ALN. It was hypothesized that the primary reason for the distinct wear morphology on UHMWPE–ALN was the agglomeration of ALN in UHMWPE–ALN 1.0 wt%. Figure 6.8d showed that the agglomerated ALN was the weak phase to be worn out. On the other hand, the worn region contained higher content of ALN compared with the rest region of UHMWPE–ALN 1.0 wt%. With the progressive generation of wear particles from UHMWPE–ALN, ALN could be released into the peri-implants, which was probably beneficial for preventing osteolysis.

By using solvent evaporation method to load UHMWPE with ALN, no obvious agglomeration of ALN was found at low content (e.g., 0.5 wt%) (Fig. 6.9) [39]. However, further addition of ALN would probably lead to agglomeration of ALN (Fig. 6.9). This problem occurred as well for UHMWPE composites reinforced with bovine bone hydroxyapatite (BHA) [58]. The consolidation of UHMWPE–ALN and UHMWPE was realized by compression molding. Thus, the UHMWPE chains or chain segments in adjacent particles intermingled at a molecular level [1]. The kinetics of intergranular diffusion was promoted by close proximity of the interfaces and thermally activated mobility of the polymer chains [1]. The loaded

ALN, however, slightly hindered the intergranular diffusion during consolidation, and hence decreased the crystallinity of UHMWPE. The decrease in crystallinity was about 4% and 8% by loading 0.5 wt% and 1.0 wt% ALN into UHMWPE, respectively [40]. This phenomenon was similar to that happened in highly cross-linked UHMWPE. When the high-density cross-linking networks restrict the motion, fold or crystallization of molecular chains, the crystallinity of UHMWPE decreases significantly [59].

6.3.3 *UHMWPE–ALN–F68*

F68 as surfactant was used to assist the dispersion of ALN in UHMWPE. Tensile test, X-ray diffraction (XRD) and DSC were used to characterize and optimize the processing parameters of UHMWPE–ALN. With the presence of F68, the tensile strength and elongation at break values were improved in comparison to those free of F68 (UHMWPE–ALN). The crystallinity of UHMWPE–ALN–F68 was slightly higher than that of UHMWPE–ALN. The addition of F68 led to the more uniform distribution of ALN in UHMWPE.

6.3.3.1 Crystallinity

XRD and DSC were conducted to characterize crystallinities of UHMWPE–ALN–F68, UHMWPE–ALN and UHMWPE [42]. The crystallinities according to XRD of UHMWPE–ALN–F68, UHMWPE–ALN and UHMWPE were 54.9%, 52.0% and 59.7%, respectively, while the crystallinities from DSC were 54.5%, 53.7% and 60.2%, respectively. These crystallinities changes indicated that the surfactant F68 contributed to the recrystallization of UHMWPE on the cooling stage of compression molding and led to a higher crystallinity for ALN loaded UHMWPE.

6.3.3.2 Mechanical Properties

Figure 6.10 shows the tensile strength and elongation at break for UHMWPE–ALN–F68, UHMWPE–ALN and UHMWPE. The tensile strength of UHMWPE–ALN–F68 was significantly higher than that of UHMWPE, while the elongation at break of UHMWPE–ALN–F68 decreased significantly than that of UHMWPE.

The surfactant F68 assists the dispersion of ALN in UHMWPE [60]. In solution, the terminal hydrophilic groups of F68 absorb on the terminal hydrophilic groups of ALN, while the terminal hydrophobic groups of F68 absorb on UHMWPE, which leads to the formation of bilayer structure on the surface of UHMWPE particle. As a result, ALN could homogeneously disperse into the consolidated UHMWPE after consolidation.

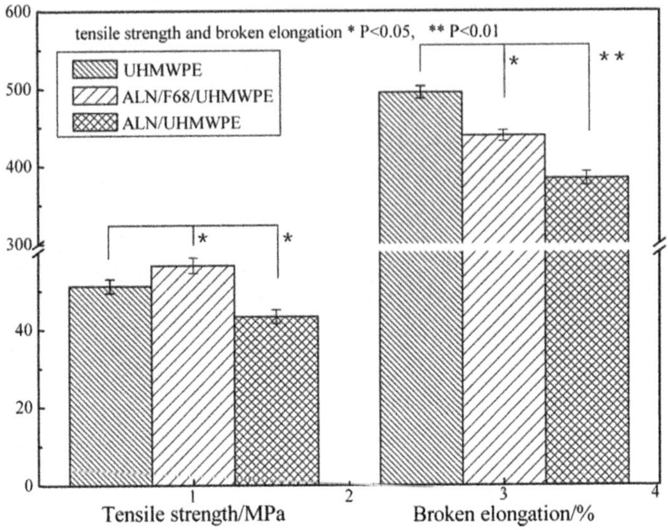

Fig. 6.10 Tensile strength and broken elongation of UHMWPE–ALN–F68, UHMWPE–ALN and UHMWPE (Modified from ref. [42])

6.3.4 UHMWPE–ALN–ma

In order to explore a way to improve the mechanical and tribological properties of UHMWPE–ALN without any new addition, Gong et al. [41] concentrated on the treatment of mechanical activation to UHMWPE–ALN. The mechanical properties and tribological properties were measured by using tensile test, small punch test, and reciprocating sliding wear test on UHMWPE–ALN–ma. SEM and EDS were employed to characterize the fusion defects and the dispersion of ALN in UHMWPE–ALN–ma.

6.3.4.1 Mechanical Properties

Figure 6.11 shows various mechanical values of UHMWPE–ALN–ma and UHMWPE–ALN. UHMWPE–ALN–ma showed higher Young's modulus, tensile strength, and elongation at break than those of UHMWPE–ALN. The yield strength of UHMWPE–ALN–ma, however, was close to that of UHMWPE–ALN. Figure 6.11 illustrated the significant increasing of Young's modulus and tensile strength of UHMWPE–ALN–ma by 5.9% and 5.5% compared to those of UHMWPE–ALN with statistical significance. However, the yield strength and elongation at break of UHMWPE–ALN–ma showed no statistically significant difference from those of UHMWPE–ALN.

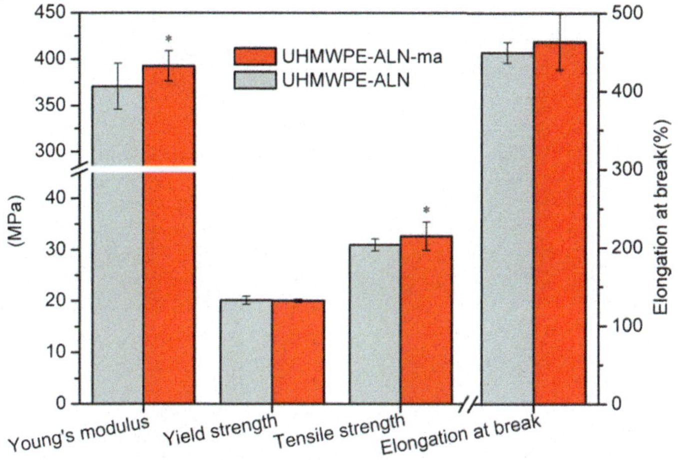

Fig. 6.11 Young's modulus, yield strength, tensile strength and elongation at break of UHMWPE–ALN–ma and UHMWPE–ALN. (Reprinted from Ref. [41], Copyright 2016, with permission from Elsevier)

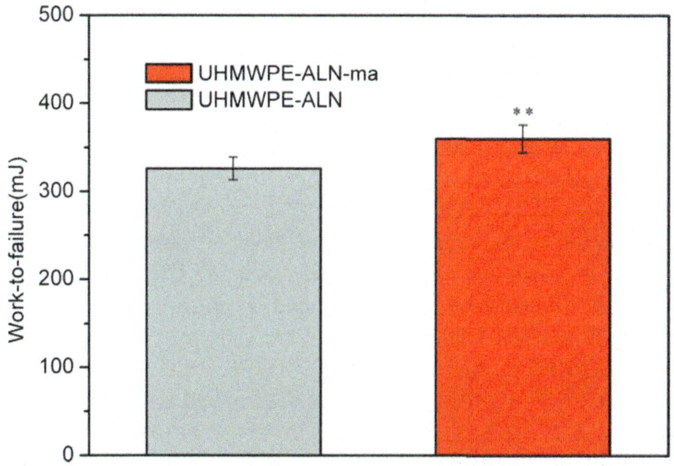

Fig. 6.12 Work-to-failure of UHMWPE–ALN–ma and UHMWPE–ALN. (Reprinted from Ref. [41], Copyright 2016, with permission from Elsevier)

Small punch test on UHMWPE–ALN–ma showed a higher fracture load than that of UHMWPE–ALN [41]. The load-displacement curve of UHMWPE–ALN–ma had a higher integrated area than that of UHMWPE–ALN, which represented a higher work-to-failure. The average work-to-failure of UHMWPE–ALN–ma increased significantly by 10.4% compared to that of UHMWPE–ALN (Fig. 6.12).

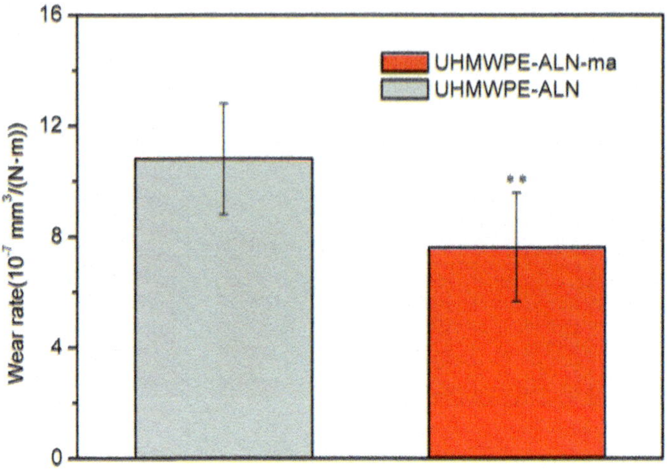

Fig. 6.13 Wear rates of UHMWPE–ALN–ma and UHMWPE–ALN. **: $p < 0.01$. (Reprinted from Ref. [41], Copyright 2016, with permission from Elsevier)

6.3.4.2 Tribological Properties

The tribological properties of UHMWPE-ALN-ma were evaluated [41] in 25 vol% calf serum according to ASTM F732–00 by ball-on-plate method with normal load of 30 N. The counter body was a 28 mm diameter 316 L stainless steel ball and the sliding velocity was 20 mm/s at 25 °C. A complete reciprocating sliding process of 10 mm sliding distance was expressed as a cycle. The whole process was 72,000 cycles in total. The wear of three specimens of each group was tested.

This study concentrated on the wear rate of UHMWPE–ALN–ma and UHMWPE–ALN. Wear rate K was defined as Eq. 6.2. The wear rates of UHMWPE–ALN–ma and UHMWPE–ALN are shown in Fig. 6.13. The average wear rates of UHMWPE–ALN–ma and UHMWPE–ALN were 7.61×10^{-7} and 10.81×10^{-7} mm^3/ (N·m), respectively. The mechanical activation could result in a significant ($p < 0.05$) reduction of 29.6% in wear rate of UHMWPE–ALN–ma compared to that of UHMWPE–ALN.

The wear surface of UHMWPE–ALN–ma and UHMWPE–ALN were observed using SEM [41]. Figure 6.14 shows the typical morphology of the wear surfaces of UHMWPE–ALN–ma and UHMWPE–ALN after a 72,000 cycle wear test. Indistinct wear surface boundary was observed on the overall wear surface of UHMWPE–ALN–ma (Fig. 6.14a). In contrast, wear surface boundary and numerous voids were observed on the wear surface of UHMWPE–ALN (Fig. 6.14d). At a higher magnification, ripples of plastic deformation of UHMWPE were observed on the wear surface of UHMWPE–ALN–ma (Fig. 6.14b and c), while curly UHMWPE-ALN fibers were observed on UHMWPE–ALN (Fig. 6.14e and f).

Reciprocating sliding of steel ball under normal load was likely to have removed the UHMWPE-ALN wear particles from the subsurface fracture of the fusion defects

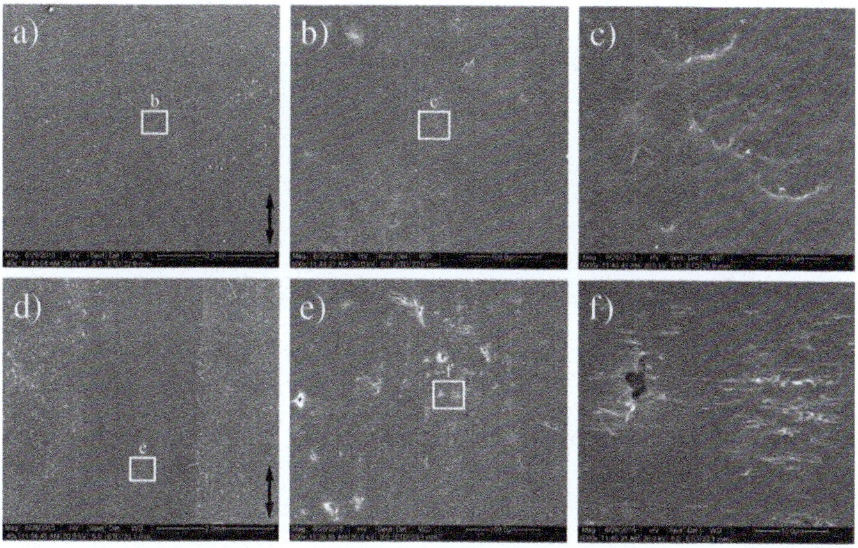

Fig. 6.14 Typical morphologies of wear surfaces of (**a-c**) UHMWPE–ALN–ma and (**d-f**) UHMWPE–ALN. (Reprinted from Ref. [41], Copyright 2016, with permission from Elsevier)

of UHMWPE-ALN asperity contacts [61], leaving curly tensile fibers (Fig. 6.14c). Subsequently, the wear regime turned into three-body abrasion, which turned out to be an increasing of friction coefficient [41]. On the contrary, ripples of plastic deformation of UHMWPE, perpendicular to the sliding direction of steel ball, were observed on the wear surface of UHMWPE–ALN–ma (Fig. 6.14b).

6.3.4.3 Fusion Defects

UHMWPE does not flow like lower molecular weight polyethylene when heated above its melting temperature. For this reason, the ultrahigh molecular weight of UHMWPE is difficult to achieve complete consolidation and self-diffusion. Poor consolidation and self-diffusion resulted in two distinct types of fusion defects in compression molded UHMWPE [62]. Type 1 defects are residual interparticle voids arising from incomplete powder compaction, while type 2 defects are caused by poor self-diffusion of UHMWPE chains at grain boundaries [62].

Both UHMWPE–ALN–ma and UHMWPE–ALN (approximately $100 \times 10 \times 10$ mm^3) were precracked and deeply frozen in liquid nitrogen for 2 h before they were fractured [41]. The fracture surface was etched in an etching solution to remove the amorphous phase [63].

Typical SEM images of freeze-fractured surfaces after etching were shown in Fig. 6.15. The surface of UHMWPE–ALN–ma showed few small voids (Fig. 6.15a, black arrows). In contrast, the surface of UHMWPE–ALN (Fig. 6.15b) showed

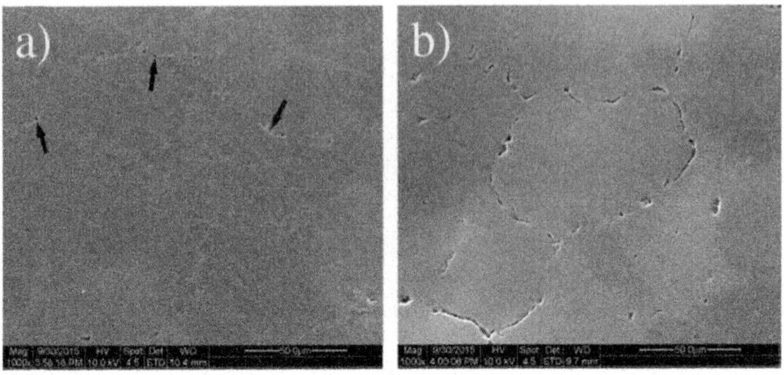

Fig. 6.15 Typical SEM micrographs of freeze-fractured surfaces after etching of (**a**) UHMWPE–ALN–ma and (**b**) UHMWPE–ALN. (Reprinted from Ref. [41], Copyright 2016, with permission from Elsevier)

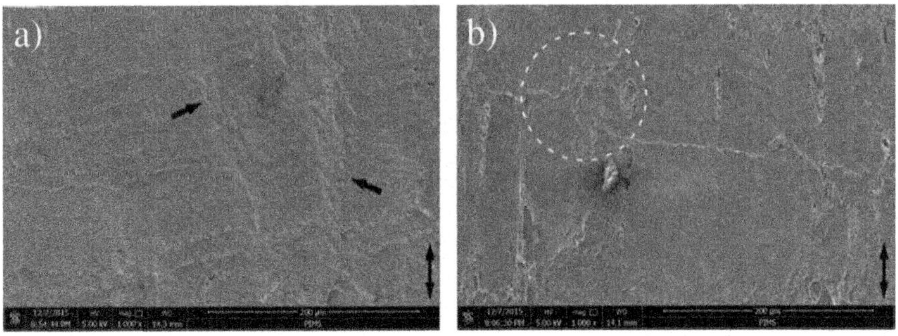

Fig. 6.16 Typical SEM micrographs of lateral surfaces of (**a**) UHMWPE–ALN–ma and (**b**) UHMWPE–ALN specimens after extension to fracture. (Reprinted from Ref. [41], Copyright 2016, with permission from Elsevier)

distinct circular voids at the boundaries of UHMWPE grains, which represented as the type 1 defects [62].

The consolidation conditions, including the molding temperature, dwell time, and annealing were critical for the fusion and self-diffusion of UHMWPE. Numerous voids (i.e., type 1 defects) were observed in UHMWPE–ALN (Fig. 6.15b). However, the UHMWPE–ALN–ma showed few small voids, which illustrated an obvious diminishing of interparticle voids (Fig. 6.15a).

The lateral surfaces of UHMWPE–ALN–ma and UHMWPE–ALN specimens after tensile fracture were observed by SEM [64]. Figure 6.16 shows the typical lateral surface morphologies of UHMWPE–ALN–ma and UHMWPE–ALN tensile specimens. Vertical ripples indicated by black arrows and horizontal ripples of machining marks were observed on UHMWPE–ALN–ma (Fig. 6.16a), while the disconnected and dislocated machining marks caused by stamping for dog-bone shaped specimens were observed on UHMWPE–ALN, as shown in the dotted circle area in Fig. 6.16b.

Type 2 defects as the grain boundaries with poor self-diffusion, were difficult to be observed for the UHMWPE. Nevertheless, indirect features revealing the inhomogeneity were observed beneath the surface of UHMWPE as straight or closed-path lines [62]. Similar phenomenon was found in Fig. 6.16 [41]. The machining marks caused by stamping for dog-bone-shaped specimens disconnected and dislocated in UHMWPE–ALN (Fig. 6.16b), while those marks appeared relatively straight in UHMWPE–ALN–ma (Fig. 6.16a). This observation suggested that a slip process occurred at the original particle boundaries, revealing that a memory of the interparticle interfaces existed in UHMWPE–ALN after compression molding [62]. The vertical ripples observed on the surface of UHMWPE–ALN–ma (Fig. 6.16a), however, revealed a diminishing of the memory of the interfaces after mechanical activation. Short molecular chains of UHMWPE–ALN–ma, as a flow accelerator, were likely to promote self-diffusion of molecular chains of UHMWPE at particle boundaries [65], which explored the space of adjacent particles during compression molding, and hence established molecular connectivity between them [62]. As a result, mechanical activation increased the mobility of molecular chains of UHMWPE–ALN–ma during compression molding, and consequently, reduced type 2 defects.

6.3.4.4 Dispersion of ALN in UHMWPE

The distribution of sodium on the surfaces of both UHMWPE–ALN–ma and UHMWPE–ALN are detected by EDS (Fig. 6.17). Homogeneous light dots in UHMWPE–ALN–ma were observed in comparison to that of UHMWPE–ALN. These results demonstrated a homogeneous distribution of ALN in UHMWPE–ALN–ma due to mechanical activation. For UHMWPE–ALN composite, the dispersion of filled phase and the interfacial adhesion among different composite components play important roles in determining the mechanical and tribological properties [43].

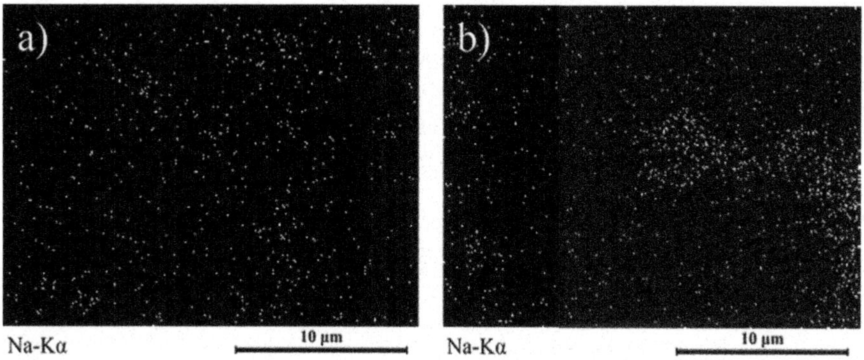

Fig. 6.17 Typical EDS images of sodium on the surfaces of (**a**) UHMWPE–ALN–ma and (**b**) UHMWPE–ALN. (Reprinted from Ref. [41], Copyright 2016, with permission from Elsevier)

6.4 Drug Release Behavior

Drug-loaded UHMWPE was compression molded according to the method mentioned in Sect. 6.2.2.1. It was hypothesized that the drug could be released from the inevitable UHMWPE wear particles in vivo. In this section, the drug release behavior was investigated under two different conditions: drug release from the drug-loaded UHMWPE wear particles under agitation and during the wear process of drug-loaded UHMWPE.

6.4.1 Drug Release of Drug-Loaded Wear Particles

6.4.1.1 Drug Release of UHMWPE–E2 Wear Particles

UHMWPE–E2 wear particles were generated by polishing the consolidated UHMWPE–E2 with a grinder [66]. The concentration of E2 in UHMWPE–E2 wear particles was 0.67 wt%. The mean diameter of UHMWPE–E2 wear particles was 71.95 μm.

In vitro drug release was performed by immersing 0.5 g UHMWPE–E2 wear particles in test tubes containing 20 mL of PBS with 0.05 wt% Pluronic F68 (Sigma–Aldrich, USA) at 37 °C under agitation at 100 rpm. A volume of 1 mL supernatant was collected and replaced with fresh PBS at various time points up to 60 days. The drug release was determined using Fluorescence Spectrophotometry (F-7000, Hitachi, Japan). The excitation and emission wavelength were 260 and 314 nm, respectively [67]. The fluorescence peak intensities were detected and the concentrations of E2 in the collected supernatants were calculated according to the established calibration curve.

The profile of E2 release from UHMWPE–E2 wear particles in vitro (Fig. 6.18) included three remarkable stages: initial burst release, the following aggressive release, and final steady release. About 14% of E2 was released from UHMWPE–E2 wear particles in the first 3 days at initial burst stage, which is due to the desorption of E2 from the surface of UHMWPE–E2 wear particles. The following aggressive release stage lasted nearly 40 days to release 33% of E2. In this stage, E2 was released from the amorphous region of UHMWPE–E2 wear particles. In addition, the mesopores on the surface of wear particles and the presence of Pluronic F68 would synergistically improve the hydrophilicity of UHMWPE–E2 wear particles and accelerate the release of E2. In the final steady release stage, E2 release was hypothesized to be associated with the drug embedded in UHMWPE–E2 wear particles. With respect to UHMWPE, a typical linear semicrystalline polymer possessed a higher chemical stability [68]. The release of the remaining 55% of E2 at this stage would slow down and need a much longer term.

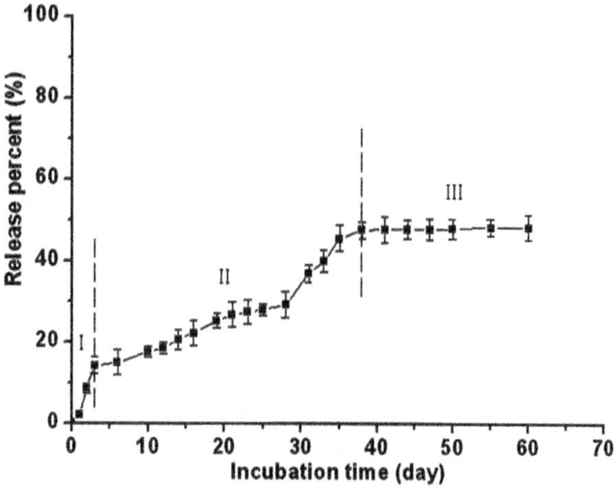

Fig. 6.18 In vitro E2 release profile from UHMWPE–E2 wear particles. There were three remarkable stages including initial burst release (I), the following celerity release (II), and final steady release (III) during the release process. (Reprinted from Ref. [66], Copyright 2012, with permission from Elsevier)

6.4.1.2 Drug Release of UHMWPE–ALN Wear Particles

UHMWPE–ALN wear particles were generated by grinding UHMWPE–ALN 0.5 wt% or UHMWPE–ALN 1.0 wt% with light table emery wheel [34]. The generated wear particles were dispersed in ethanol and filtered sequentially through 75 and 45 μm nylon sieve to obtain wear particles in the size ranges of <45 μm and 45–75 μm, respectively.

In vitro ALN release was performed by immersing the above UHMWPE–ALN wear particles (1 g) in test tubes containing 5 mL of PBS as description in Sect. 6.4.1.1. The ninhydrin reaction was employed to determine the concentration of ALN in PBS [69]. The reactive solution formed the Ruhemann's purple reaction product with a maximum absorbance at 568 nm was measured by UV–vis spectrophotometry (Shimadzu UV-250, Tokyo, Japan) against the reagent blank [69]. The absorbance was transformed into the ALN concentration using an established calibration curve. Accumulative release profiles were correspondingly calculated.

The profile (Fig. 6.19) of in vitro ALN release from UHMWPE–ALN wear particles also showed three stages: the initial burst release, the following rapid release, and the final slow release.

The initial burst release occurred during the first 24 h, the accumulative release of ALN from 45 to 75 μm groups of UHMWPE–ALN 0.5 wt%, and UHMWPE–ALN 1.0 wt% wear particles were 13.4% and 15.6%, respectively [34]. However, the accumulative release of ALN from UHMWPE–ALN 0.5 wt% and UHMWPE–ALN 1.0 wt% wear particles <45 μm increased to 20.2% and 23.8% in initial burst release. It indicates that the release of ALN from UHMWPE–ALN wear particles was mainly controlled by particle size at this stage. The ALN release in this stage might result from desorption of ALN from the surface of UHMWPE–ALN wear particles, in which the larger specific surface area of smaller wear particles would be beneficial to the release of ALN (Fig. 6.20 stage 1) [70]. Subsequently, a rapid release started from 2 days and lasted until 60 days in Fig. 6.19. In this stage, the accumulative

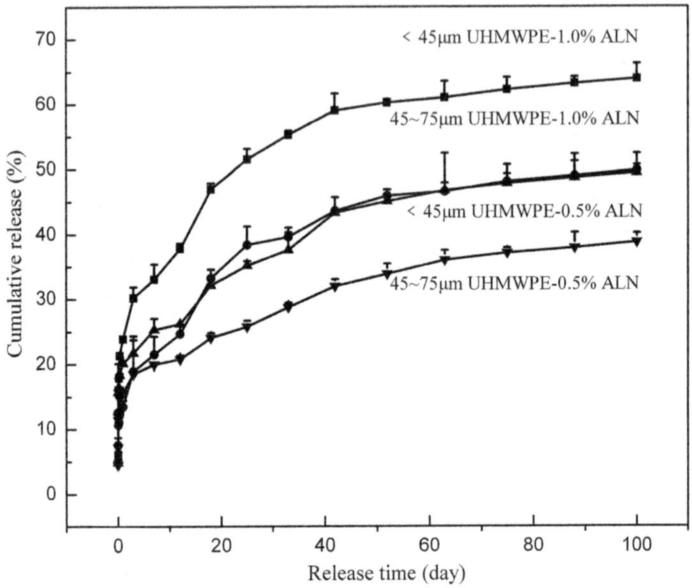

Fig. 6.19 In vitro ALN cumulative release from UHMWPE–ALN 0.5 wt. % and UHMWPE–ALN 1.0 wt. % wear particles (<45 μm size-ranged groups and 45–75 μm size-ranged groups). (Reprinted from Ref. [34], Copyright 2012, with permission from Wiley)

release of ALN from UHMWPE–ALN 0.5 wt% wear particles ranged in 45–75 μm and < 45 μm groups were 20.4% and 26.5%.

However, the accumulative release percentages, excluding the release during burst release, of ALN from UHMWPE–ALN 1.0 wt% wear particles with size ranges of 45–75 μm and < 45 μm groups markedly increased to 33.1% and 37.2%, respectively. It reflected that the ALN content in UHMWPE–ALN wear particles played a key role in the release of ALN in this stage. It was suggested that a continuous diffusion release occurred after the swelling of the amorphous region of UHMWPE–ALN wear particles and subsequent dislocation of UHMWPE chains (Fig. 6.20 stage 2) [71].

The final slow release stage started from 60 days and lasted until 100 days, although the ALN release continued thereafter. Hypothetically, the ALN release in this stage might be attributed to ALN embedded in UHMWPE–ALN wear particles (Fig. 6.20 stage 3). Figure 6.20 showed the hypothesized mechanism of the in vitro ALN release from the UHMWPE–ALN wear particles.

6.4.2 *Drug Release Behavior During Wear Process*

The drug release of drug-loaded UHMWPE during wear process was further investigated under simulated in vivo conditions. Yang et al. [72] investigated the drug

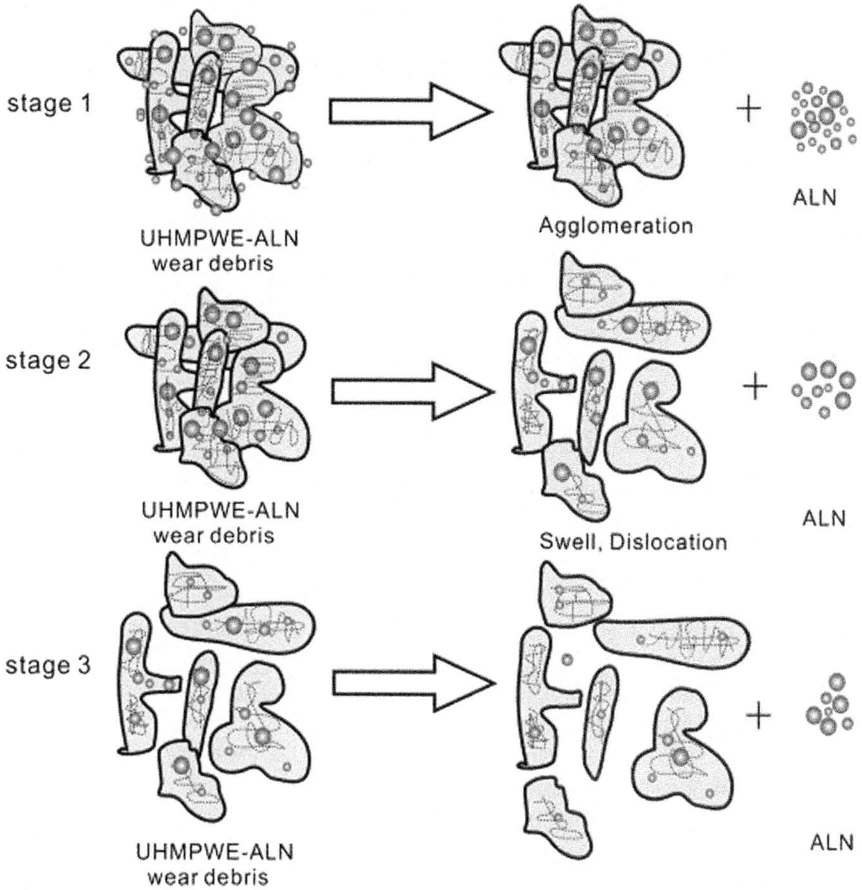

Fig. 6.20 Schematic diagrams of three stages of the in vitro ALN release from UHMWPE–ALN wear particles. (Reprinted from Ref. [34], Copyright 2012, with permission from Wiley)

release from UHMWPE–ALN 1.0 wt% during wear on reciprocating ball-sliding-on-block wear test apparatus at RT in deionized water (20 mL). As the control group, another identical UHMWPE–ALN sample was immersed in deionized water (20 mL) for 310 h. 0.2% NaN_3 was added into the deionized water of two groups to prevent deterioration of lubricants.

The concentration and accumulative mass of ALN in the lubricants [69] during wear process were shown in Fig. 6.21. It revealed two noticeable stages of ALN release. The concentrations of ALN were lower than 5 μg/mL at the initial stage (before 200 h) and the accumulative mass of ALN increased linearly with the duration of wear process [72]. At the second stage, the ALN concentration increased sharply from 7 to 12 μg/mL, and the accumulative mass increased more quickly compared with that before 200 h, which indicates that the speed of ALN release was accelerated. Compared with the conventional drug release pattern, the ALN release

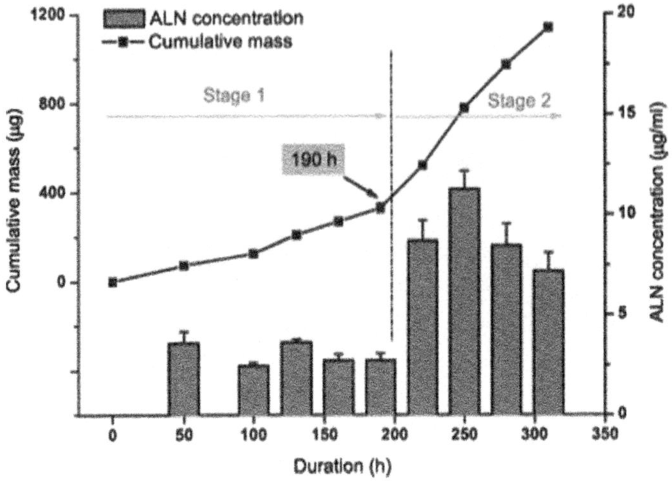

Fig. 6.21 The ALN concentration and the cumulative mass of the ALN during wear process. (Reprinted from Ref. [72], Copyright 2011, with permission from Elsevier)

rate of UHMWPE–ALN was relatively low at the onset and then kept increasing during the whole wear process, with no stable release rate reached till the end of test. In contrast, the amount of released ALN was very limited and not detectable without wear process. Sliding against the counterpart under wear process, the UHMWPE–ALN was subjected to different stresses in different regions [71]. These stresses would destroy the crystalline structure of the polymer and the amorphous region, which increase the possibility for the contact of water molecule with ALN. The wear particles accumulated during wear process had higher specific surface area than the matrix, which also improved the diffusion of ALN into deionized water. The morphological change of the wear surface and the wear particles generated by the friction and wear contributed to the promotion of ALN diffusion in lubricants.

6.5 Biological Evaluation

It has been shown that the size, concentration, and morphology of UHMWPE wear particles are the major factors affecting the release of a variety of cytokines from macrophages, the stimulation of osteoclast, and final development of osteolysis [59, 62]. Furthermore, the response of osteoblasts to wear particles has been critical to both initial osteointegration of implant and ongoing regeneration of the periprosthetic bed [73]. Therefore, it is essential to investigate the responses of macrophages and osteoblasts to drug-loaded UHMWPE wear particles with different sizes. Two different co-culture systems are described in this section.

6.5.1 Conventional Cell Culture System

Qu et al. [34] investigated the response of murine RAW 264.7 macrophages and murine MC3T3-E1 osteoblasts to UHMWPE–ALN wear particles. UHMWPE–0.5 wt% ALN, UHMWPE–1.0 wt% ALN and UHMWPE wear particles in the size ranges of <45 μm and 45–75 μm were generated as described in 4.1.2. UHMWPE–ALN wear particles were co-cultured with cells following Huang et al. [74]. Briefly, sterilized UHMWPE–ALN wear particles were coated on culture plates. Then 1 mL of macrophage or osteoblast cell suspension was dispensed into each well of this wear particle-coated 24-well culture plates at 1×10^4 cells/mL, respectively. Cells cultured with medium alone served as the control.

6.5.1.1 Cell Morphology

SEM was used to study the morphologies of macrophages co-cultured with UHMWPE–ALN wear particles after fixation, dehydration, and drying [34]. The macrophages grew well with the presence or absence of wear particles (Fig. 6.22). The macrophages appeared in round, long, or flattened irregular cell patterns. Some cells had long cytoplasm with a few filopodia, which indicates the cellular chemotaxis induced by wear particles. In particular, for groups with wear particles, some macrophages showed long filopodia and even connected with each other via filopodia. Some macrophages had obvious phagocytotic vesicles in cytoplasm due to the phagocytosis of small wear particles, as indicated by arrows in Fig. 6.22b.

6.5.1.2 Cell Proliferation

The proliferation of osteoblasts and macrophages was evaluated [34] by MTT [3-(4, 5, 5-dimethylthiazol-2-yl)-2, 5-diphenyl tetrazolium bromide] assay after co-cultured with wear particles (Fig. 6.23). The concentrations of ALN in cell culture medium were estimated to range from 1.1×10 to 6 mol/L for UHMWPE–0.5 wt% ALN and 3.7×10 to 6 mol/l for UHMWPE–1.0 wt% ALN groups according to the result of in vitro ALN release study in Sect. 6.4.1.2. There was no significant difference in the MTT values after co-culture with UHMWPE–ALN wear particles containing different contents of ALN. On the other hand, the small size of UHMWPE–ALN wear particles decreased the MTT values of macrophages (Fig. 6.23). Evans [75] found that the number of macrophages was reduced after the addition of bisphosphonates (10^{-9} to 10^{-5} M). In this study, the ALN concentrations in the culture media containing UHMWPE–0.5 wt% ALN and UHMWPE–1.0 wt% ALN wear particles were both in the range of effective ALN concentrations reported previously [75]. The MTT values of macrophages treated with UHMWPE–ALN groups were lower than those of UHMWPE groups.

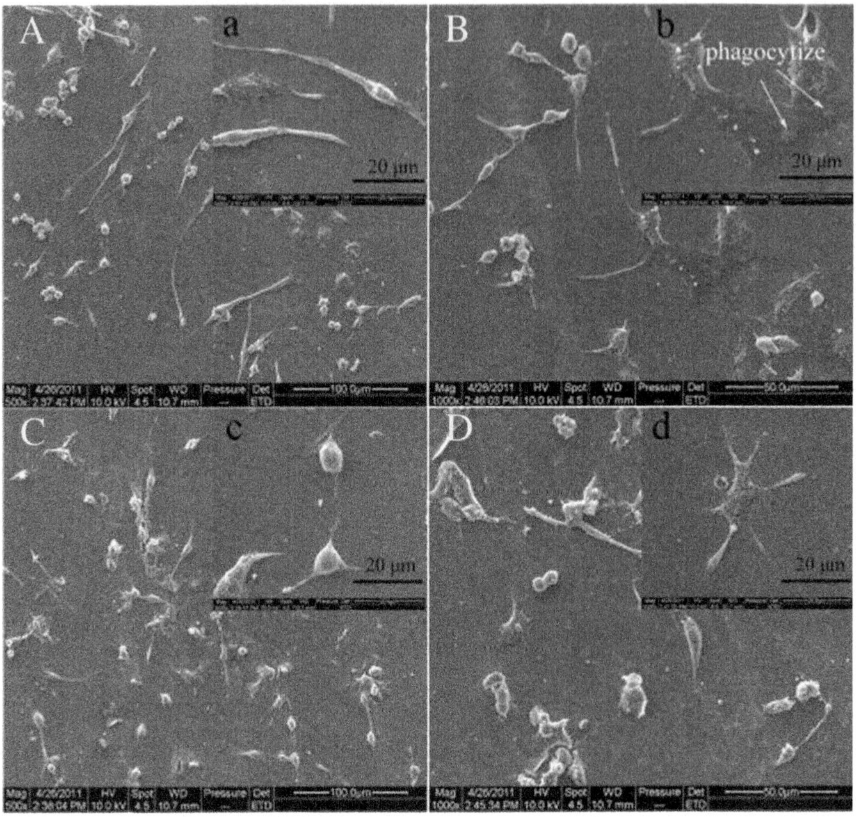

Fig. 6.22 SEM morphologies of macrophages co-cultured with UHMWPE–0.5 wt. % ALN (A and a), UHMWPE–1 wt. % ALN (B and b) and UHMWPE (C and c) wear particles and control macrophages (D and d) culture in medium alone. (Reprinted from Ref. [34], Copyright 2012, with permission from Wiley)

Figure 6.24 showed the effect of particle size and ALN concentrations on the osteoblast proliferation after co-culture for 72 h. The proliferations of osteoblasts co-cultured with UHMWPE–ALN or UHMWPE wear particles were suppressed compared with the control free of particles. With increasing particle size and ALN concentration, the proliferations of osteoblasts co-cultured with UHMWPE–ALN particles were slightly increased, without significant difference. Wear particles smaller than 5 μm might undergo phagocytosis by mature osteoblasts, with potential adverse effects on cellular proliferation and function [73]. Some studies find that the effective concentration of ALN to inhibit the growth of osteoblast cells is 2.5×10^{-5} mol/L but has influence on osteoblast growth in the range of 2.5×10^{-8} mol/L to 2.5×10^{-6} mol/l [76]. The ALN concentrations in this study were estimated ranged at 1.1×10^{-6} mol/L for UHMWPE–0.5 wt% ALN and 3.7×10^{-6} mol/L for UHMWPE–1.0 wt% ALN wear particles. Therefore, despite the inhibition of osteoblast proliferation by wear particles, released ALN from

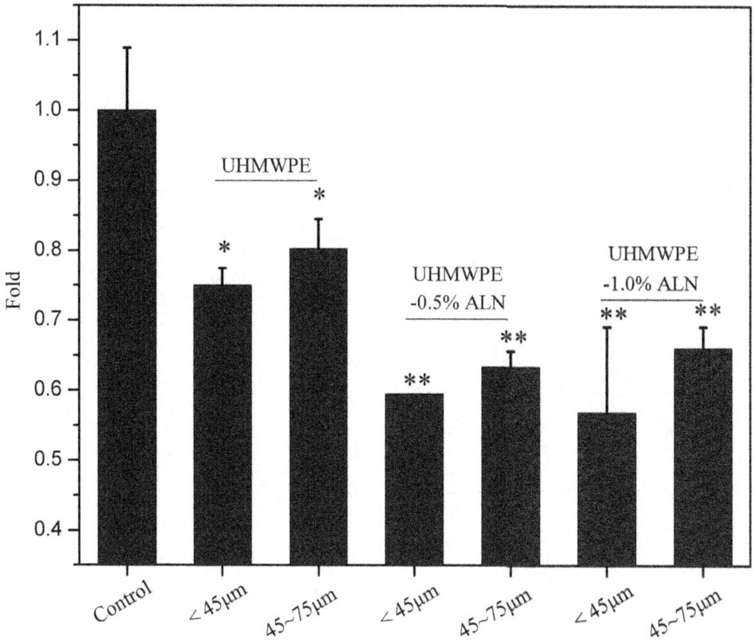

Fig. 6.23 MTT values of macrophages co-cultured with UHMWPE–0.5 wt. % ALN, UHMWPE–1.0 wt. % ALN and UHMWPE wear particles. *$p < 0.05$ and **$p < 0.001$. (Reprinted from Ref. [34], Copyright 2012, with permission from Wiley)

UHMWPE-ALN wear particles might counteract the adverse effect of wear particles. As a result, the proliferations of osteoblasts were promoted [34].

6.5.1.3 Alkaline Phosphatase (ALP) Activity

The ALP activity was determined as an indicator of osteoblastic differentiation [34]. Results in Fig. 6.25 show that the ALP activities of osteoblasts after co-culture with UHMWPE–ALN wear particles for 72 h are increased with the increase in particle size and ALN content, without significant differences [34]. Furthermore, ALN incorporation in UHMWPE–ALN wear particles up-regulated the ALP of osteoblasts in size- and concentration-dependent manners.

6.5.1.4 Secretion of Cytokines

The culture supernatants from macrophages co-cultured with UHMWPE–ALN or UHMWPE particles were assayed for the secretion of tumor necrosis factor-alpha (TNF-α) and IL-6 by enzyme-linked immunosorbent assay (ELISA) kits (Biosource) [34]. After co-culture for 24 h, UHMWPE and UHMWPE–ALN wear particles

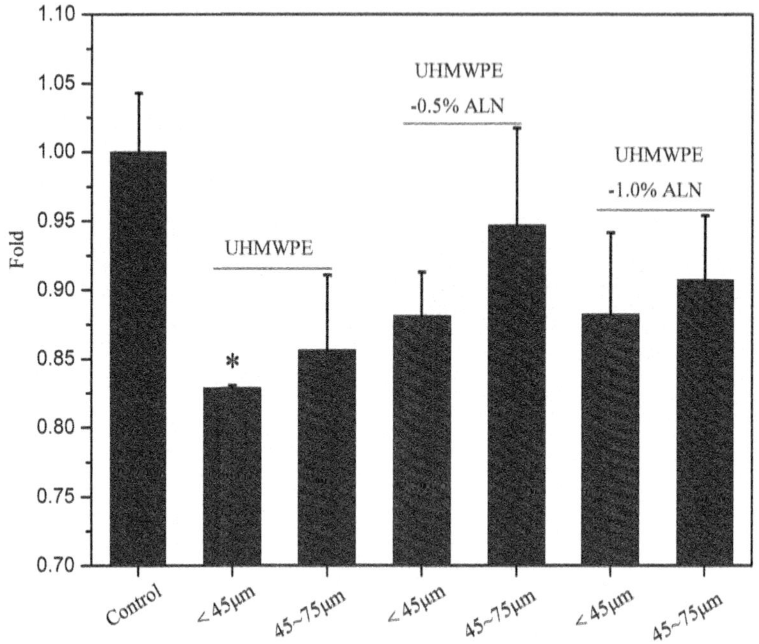

Fig. 6.24 The effect of UHMWPE–0.5 wt% ALN, UHMWPE–1.0 wt% ALN and UHMWPE wear particles on the proliferations of osteoblasts. *$p < 0.05$. (Reprinted from Ref. [34], Copyright 2012, with permission from Wiley)

caused the obvious elevations in the secretion of TNF-α from macrophages in comparison with that of the control. But it showed no obvious effect on the secretion of IL-6, (Fig. 6.26). For UHMWPE–0.5% ALN wear particles, the secretion of TNF-α had no significant difference either in <45 μm or in 45–75 μm size-ranged groups compared to that of the control. It indicated that the stimulation effect of UHMWPE–0.5% ALN wear particles on the secretion of TNF-α could be reduced by ALN released from UHMWPE–0.5% ALN wear particles. However, for UHMWPE–1.0% ALN wear particles, the secretion of TNF-α in both of <45 μm or 45–75 μm size-ranged groups were significantly elevated compared to that of the control. For UHMWPE wear particles, the secretion of TNF-α was significantly elevated in <45 μm size-ranged group, while there was no significant difference between 45 and 75 μm size-ranged group and the control. There were some wear particles less than 10 μm in <45 μm size-ranged group due to the filtration by 45 μm filter membrane, while there almost was no wear particles less than 10 μm in 45–75 μm size-ranged group. Therefore, it was deduced that the elevation of the TNF-α secretion from macrophage was attributed to the UHMWPE wear particles less than 10 μm by induced-phagocytization as in previous studies [76]. Additionally, it was found that there was no significant difference in the secretion of TNF-α

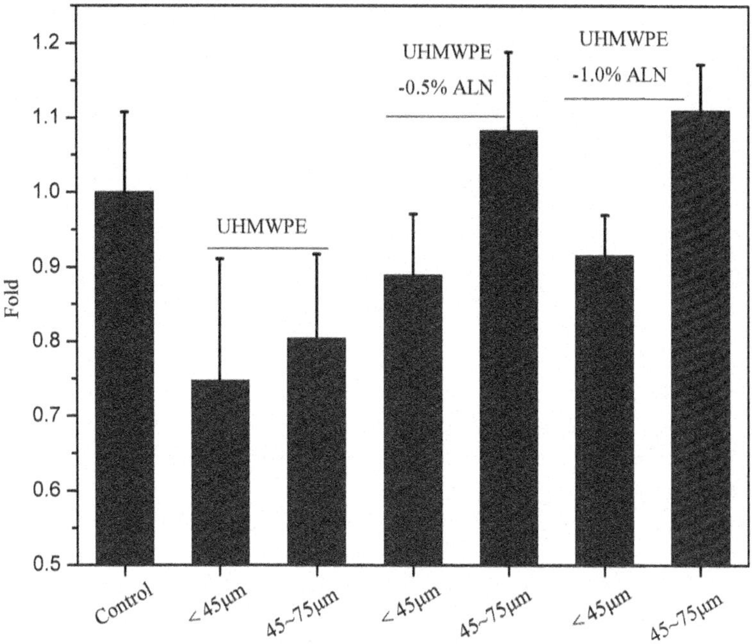

Fig. 6.25 The effect of UHMWPE–0.5 wt. % ALN, UHMWPE–1.0 wt. % ALN and UHMWPE wear particles (<45 μm and 45–75 μm size-ranged groups) on the alkaline phosphatase (ALP) activity of osteoblasts. (Reprinted from Ref. [34], Copyright 2012, with permission from Wiley)

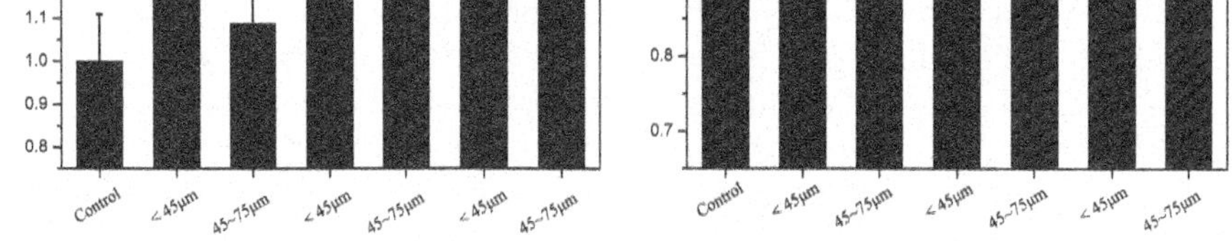

Fig. 6.26 The TNF-α and IL-6 secretion of macrophages after culture with UHMWPE–0.5 wt% ALN, UHMWPE–1.0 wt% ALN and UHMWPE wear particles (<45 μm and 45–75 μm size-ranged groups). $^{*}p < 0.05$. (Reprinted from Ref. [34], Copyright 2012, with permission from Wiley)

between the <45 μm and 45–75 μm size-ranged groups for UHMWPE-ALN wear particles, while the significant elevation of TNF-α secretion appeared in <45 μm size-ranged group compared to that of 45–75 μm size-ranged group for UHMWPE wear particles [76]. As described above, there were some wear particles less than

10 μm in <45 μm size-ranged group, while there almost was no wear particles less than 10 μm in 45–75 μm size-ranged group. Thus, there should be the stronger response of macrophage induced in <45 μm size-ranged group. However, this stimulating effect of smaller wear particles on the TNF-α secretion of macrophage in vitro could be reduced by ALN released in UHMWPE–ALN wear particles. The present results had implications for the treatment of implant loosening induced by macrophage-derived cytokines.

6.5.2 Inverted Cell Culture System

UHMWPE wear particles float on the culture medium due to their low density, which results in insufficient contact between particles and cells. Fang et al. [77] developed an inverted cell culture system to enable the interactions between UHMWPE wear particles and cells. Qu et al. [66] adopted this system to investigate the response of macrophages and osteoblasts to UHMWPE–E2 wear particles in vitro. The cell suspension was seeded into each well of 24-well culture block at 1×10^5 cell/mL for 24 h. UHMWPE–E2 and UHMWPE wear particles were suspended in culture medium after they were γ-irradiation (14.5 kGy) sterilized. In brief, 1 mL particle suspension stocks (0.2, 1.0, 3.0, 5.0, 6.0, 8.0, 10.0, and 15.0 mg/mL) were added into each well after the original medium was drained out. Then the wells were fully filled with fresh medium and covered with a sterilized flat polyvinyl chloride film with some ventilation holes. Finally, the block was covered and inverted quickly.

6.5.2.1 Cell Morphology

Inverted Phase Contrast Microscopy (DMIL, Leica, Germany) was used to observe the morphologies of osteoblasts after co-cultured with UHMWPE–E2 and UHMWPE wear particles for 96 h (Fig. 6.27) [66]. The morphologies of osteoblasts co-cultured with UHMWPE–E2 wear particles were similar with those co-cultured with UHMWPE wear particles (about 71 μm diameter) at low concentration, 1.0 mg per well (Fig. 6.27a and b). The typical polygonal, fusiform, and triangle morphologies indicated that the cells spread very well. Wear particles were dispersed among the cells. The morphologies of cells co-cultured with 15.0 mg per well of UHMWPE wear particles became round and irregular (Fig. 6.27d). However, cells co-cultured with 15.0 mg per well of UHMWPE–E2 wear particles grew as well as those co-cultured with 1.0 mg per well of UHMWPE–E2 wear particles.

The morphologies of macrophages co-cultured with 15.0 mg per well of UHMWPE–E2 and UHMWPE wear particles for 96 h were shown in Fig. 6.28. Macrophages co-cultured with UHMWPE–E2 and UHMWPE wear particles were about 100 μm in diameter, while cells in the control group were 20 μm. There was no obvious difference in morphology and size between cells co-cultured with

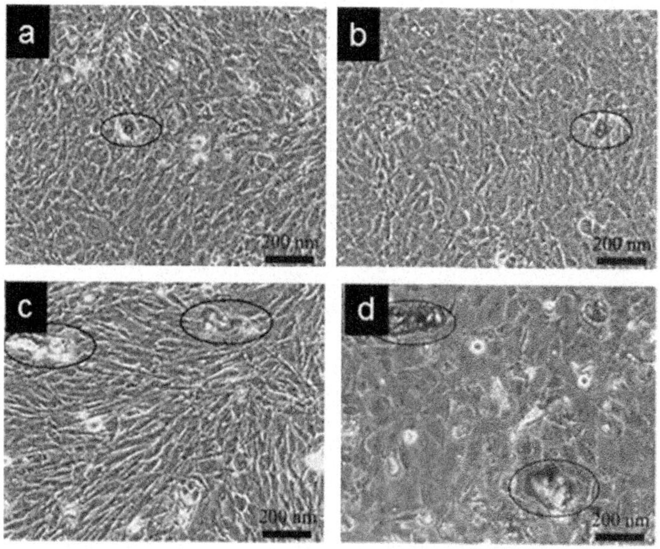

Fig. 6.27 Microscopic images of osteoblasts co-cultured with UHMWPE–E2 and UHMWPE wear particles at different concentrations: (**a**) 1 mg of UHMWPE–E2 wear particles; (**b**) 1 mg of UHMWPE wear particles; (**c**) 15 mg of UHMWPE–E2 wear particles and (**d**) 15 mg of UHMWPE wear particles. Wear particles were indicated in the black circles. (Reprinted from Ref. [66], Copyright 2012, with permission from Elsevier)

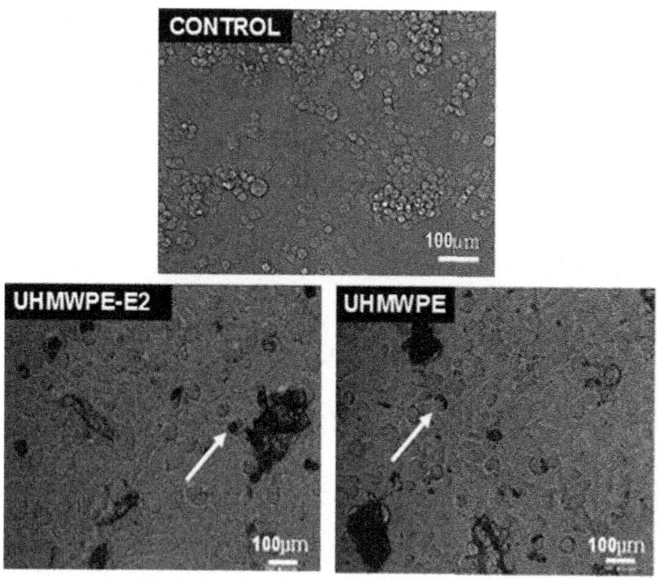

Fig. 6.28 Microscopic images of macrophages co-cultured with 15 mg of UHMWPE–E2 and UHMWPE wear particles. Cells cultured with medium only were used as control. (Reprinted from ref. [66], Copyright 2012, with permission from Elsevier)

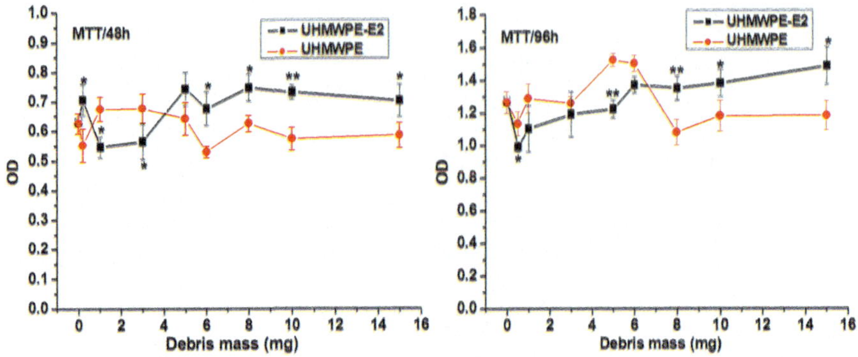

Fig. 6.29 MTT results of osteoblasts co-cultured with UHMWPE–E2 and UHMWPE wear particles for 48 and 96 h. $*p < 0.05$ and $**p < 0.001$. (Reprinted from Ref. [66], Copyright 2012, with permission from Elsevier)

UHMWPE–E2 and UHMWPE wear particles. Small wear particles were found inside the cells, which indicated that the wear particles activated the phagocytosis of macrophages [66].

6.5.2.2 Osteoblastic Proliferation

Figure 6.29 shows that the osteoblastic proliferation with the presence of UHMWPE wear particles was not significantly affected with the dosages of 1.0, 3.0, 5.0, 8.0, 10.0, and 15.0 mg wear particles per well for 48 h, in comparison to that of the wear particles-free group. After 96 h, osteoblasts co-cultured with UHMWPE wear particles showed the higher proliferation at the dosages of 5.0 and 6.0 mg per well. Compared to UHMWPE wear particles group, the proliferations of osteoblasts with UHMWPE–E2 wear particles of 8.0–15.0 mg per well were statistically higher than those with UHMWPE wear particles ($p < 0.05$) [66].

6.5.2.3 Alkaline Phosphatase (ALP) Activity

The ALP levels of osteoblasts co-cultured with UHMWPE–E2 and UHMWPE wear particles at 48 h and 96 h were shown in Fig. 6.30. After co-culture for 48 h, there was no significant difference between the ALP values of osteoblasts in both of UHMWPE–E2 and UHMWPE wear particles at all dosages [66]. Compared with UHMWPE group, UHMWPE–E2 wear particles with 8.0, 10.0 and 15.0 mg per well caused significantly higher levels of ALP, which was consistent with the proliferation profile ($p < 0.05$).

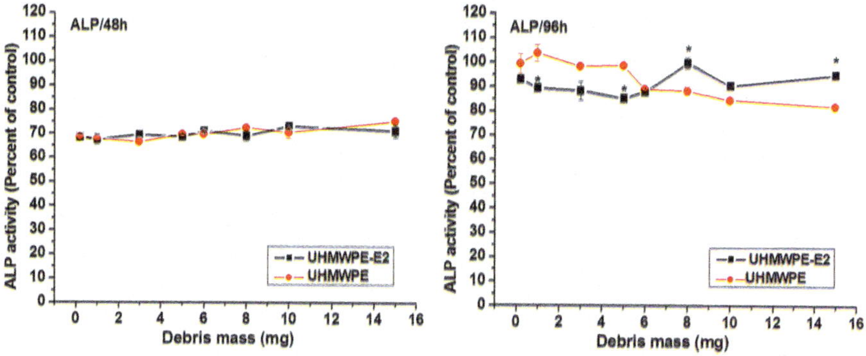

Fig. 6.30 ALP results of osteoblasts co-cultured with UHMWPE–E2 and UHMWPE wear particles for 48 and 96 h. $*p < 0.05$. (Reprinted from Ref. [66], Copyright 2012, with permission from Elsevier)

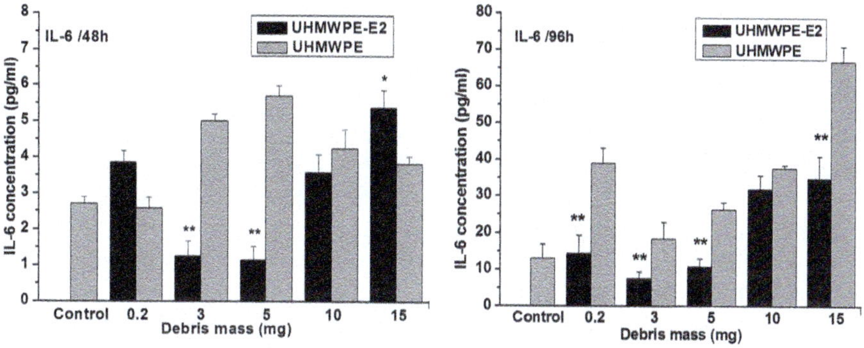

Fig. 6.31 IL-6 secretion of osteoblasts co-cultured with UHMWPE–E2 and UHMWPE wear particles for 48 and 96 h. $*p < 0.05$ and $**p < 0.01$, * A statistically significant IL-6 value of cells treated with UHMWPE–E2 wear particles compared to the cells treated with UHMWPE wear particles. (Reprinted from Ref. [66], Copyright 2012, with permission from Elsevier)

6.5.2.4 Interleukin-6 (IL-6) Secretion

IL-6 secretion of osteoblasts co-cultured with UHMWPE–E2 and UHMWPE wear particles at 48 and 96 h were shown in Fig. 6.31. After cultured for 48 h, UHMWPE wear particles caused the significant elevation of IL-6 secretion at all dosages, except that with 0.2 mg particles, in comparison with the control [66]. The levels of IL-6 in UHMWPE–E2 group were significantly lower than those in UHMWPE group at the dosages of 3.0 and 5.0 mg per well. The cells in UHMWPE wear particles groups yielded high levels of IL-6 after culture for 96 h. The levels of IL-6 were increased with the increasing amount of particles. The IL-6 levels with UHMWPE–E2 wear particles at the dosages of 0.2, 3.0, 5.0 and 15.0 mg per well were lower than those in UHMWPE wear particles group after 96 h. It was probably attributed to the release of E2 from UHMWPE–E2 wear particles [27].

6.6 Concluding Remarks and Future Directions

Drug-loaded UHMWPE was successfully prepared by liquid/solid blending and compression molding. The present results showed that drug-loaded UHMWPE possessed excellent performances, e.g. crystallinity, mechanical and tribological properties, similar to those of the control UHMWPE when the content of drug was low. The distribution of drug in UHMWPE played an important role in determining the properties of drug-loaded UHMWPE, in particular, mechanical strength and tribological behavior. Multiple factors, e.g. the content of drug, the properties of drug, and the consolidation conditions, would affect the distribution of drug in UHMWPE. The heterogeneous distribution of drug in UHMWPE would deteriorate the mechanical strength and accelerate the wear of drug-loaded UHMWPE. The use of surfactant and mechanical activation treatment of drug-loaded UHMWPE powders prior to compression molding effectively assisted the uniform distribution of drugs in UHMWPE.

The release of drug in vitro was studied from the drug-loaded UHMWPE wear particles and during wear tests, respectively. The release behavior of drug-loaded UHMWPE wear particles was similar to that of traditional polymer drug delivery system, including the initial burst release, slow release, and final steady release. However, the final accumulated release amount of drug from particles appeared relatively lower because UHMWPE is a stable polymer with limited swelling and degradation. Therefore, a certain amount of drug remains in the particles, which might be released intracellularly if they were phagocytized. Interestingly, drug released from drug-loaded UHMWPE during friction was firstly investigated in this study. It was found that friction and wear would influence the release of drug. The serious wear would promote the release of drug from the drug-loaded UHMWPE during friction.

Cell culture results showed that TNF-α secreted from macrophages, one of the inflammatory factors, could be reduced significantly by co-culturing macrophages and UHMWPE-ALN wear particles compared to that of UHMWPE. Moreover, the proliferations and alkaline phosphatase activities of osteoblasts increased moderately with the increasing of particle sizes and ALN concentrations after co-cultured osteoblasts and drug loaded UHMWPE wear particles. These results suggested that incorporation of drug in UHMWPE might be a potential approach to prevent or reduce particle-induced osteolysis.

However, it is necessary to optimize the fabrication of drug loading in UHMWPE to get the optimum tribological and mechanical properties firstly. Secondly, the UHMWPE wear particles floated at the top of the culture medium due to its lower density, which resulted in insufficient contact between UHMWPE wear particles and cells adhering to the bottom of the culture plate. Therefore, it is necessary to develop a new strategy to co-culture cells and the drug-loaded UHMWPE wear particles to provide sufficient contact between cells and wear particles for a longer term. Additionally, in vivo experiment should be in proceeding to verify the role of drug loading in UHMWPE.

References

1. Kurtz SM, Muratoglu OK, Evans M et al (1999) Advances in the processing, sterilization, and crosslinking of ultra-high molecular weight polyethylene for total joint arthroplasty. Biomaterials 20:1659–1688
2. Ingham E, Fisher J (2005) The role of macrophages in osteolysis of total joint replacement. Biomaterials 26:1271–1286
3. Amstutz HC, Campbell P, Kossovsky N et al (1992) Mechanism and clinical significance of wear particles-induced osteolysis. Clin Orthop Res 276:7–18
4. Harris WH (1995) The problem is osteolysis. Clin Orthop Res 311:46–53
5. Willert HG, Bertram H, Buchhorn GH (1990) Osteolysis in alloarthroplasty of the hip. The role of ultra-high molecular weight polyethylene wear particles. Clin Orthop Relat Res 258:95–107
6. Learmonth ID, Young C, Rorabeck C (2007) The operation of century total hip replacement. Lancet 370:1508–1519
7. Pruitt LA (2005) Deformation, yielding, fracture and fatigue behavior of conventional and highly cross-linked ultra high molecular weight polyethylene. Biomaterials 26:905–915
8. Chiesa R, Tanzi MC, Alfonsi S et al (2000) Enhanced wear performance of highly crosslinked UHMWPE for artificial joints. J Biomed Mater Res 50:381–387
9. Shi W, Dong H, Bell T (2003) Wear performance of ion implanted ultra high molecular weight polyethylene. Surf Eng 19:279–283
10. Sariali E, Veysi V, Stewart T (2008) Biomechanics of the human hip–consequences for total hip replacement. Curr Orthop 22:371–375
11. Seon JK, Song EK (2006) Navigation-assisted less invasive total knee arthroplasty compared with conventional total knee arthroplasty. J Arthroplast 21:777–782
12. Agarwal S (2004) Osteolysis—basic science, incidence and diagnosis. Curr Orthop 18:220–231
13. Wang ML, Sharkey PF, Tuan RS (2004) Particle bioreactivity and wear-mediated osteolysis. J Arthroplast 19:1028–1038
14. von Knoch F, Heckelei A, Wedemeyer C et al (2005) The effect of simvastatin on polyethylene particle-induced osteolysis. Biomaterials 26:3549–3555
15. Ren WP, Li XH, Chen BD et al (2004) Erythromycin inhibits wear particles-induced osteoclastogenesis by modulation of murine macrophage NF-kB activity. J Orthopaed Res 22:21–29
16. Millett PJ, Allen MJ, Bostrom MP (2002) Effects of alendronate on particle-induced osteolysis in a rat model. J Bone Joint Surg Am 84:236–249
17. Thadani PJ, Waxman B, Sladek E et al (2002) Inhibition of particulate particles-induced osteolysis by alendronate in a rat model. Orthopedics 25:59–63
18. Iwase M, Kim KJ, Kobayashi Y et al (2002) A novel bisphosphonate inhibits inflammatory bone resorption in a rat osteolysis model with continuous infusion of polyethylene particles. J Orthopaed Res 20:499–505
19. Shanbhag AS, Hasselman CT, Rubash HE (1997) Inhibition of wear particles mediated osteolysis in a canine total hip arthroplasty model. Clin Orthopaed Relat Res 344:33–43
20. Im GI, Qureshi SA, Kenney J et al (2004) Osteoblast proliferation and maturation by bisphosphonates. Biomaterials 25:4105–4115
21. Akesson K (2003) New approaches to pharmacological treatment of osteoporosis. Bull World Health Organ 81:657–664
22. Harris SA, Tau KR, Turner RT et al (1996) Estrogen and progestins. In: Bilezikian JP, Raitz LG, Rodan GA (eds) Principles of bone biology, 1st edn. Academic, San Diego, pp 507–520
23. Kameda T, Mano H, Yuasa T et al (1997) Estrogen inhibits bone resorption by directly inducing apoptosis of the bone-resorbing osteoclasts. J Exp Med 186:489–495
24. Holzer G, Einhorn TA, Majeska RJ (2002) Estrogen regulation of growth and alkaline phosphatase expression by cultured human bone marrow stromal cells. J Orthopaed Res 20:281–288
25. Disilvio L, Jameson J, Gamie Z et al (2006) In vitro evaluation of the direct effect of estradiol on human osteoblasts (HOB) and human mesenchymal stem cells (h-MSCs). Injury 37:S33–S42

26. Qu Q, Perälä-Heape M, Kapanen A et al (1998) Estrogen enhances differentiation of osteoblasts in mouse bone marrow culture. Bone 22:201–209
27. Girasole G, Jilka RL, Passeri G et al (1992) 17β-estradiol-inhibits interleukin-6 production by bone marrow-derived stromal cells and osteoblasts in vitro: a potential mechanism for the antisteoporotic effect of estrogens. J Clin Invest 89:883–891
28. Jilka RL, Hangoc G, Girasole G et al (1992) Increased osteoclast development after estrogen loss: mediation by interleukin-6. Science 257:88–91
29. Steeve KT, Marc P, Sandrine T et al (2004) IL-6, RANKL, TNF-alpha/IL-1: interrelations in bone resorption pathophysiology. Cytokine Growth Factor Rev 15:49–60
30. Astrand J, Aspenberg P (2004) Topical, single dose bisphosphonate treatment reduced bone resorption in a rat model for prosthetic loosening. J Orthopaed Res 22:244–249
31. Peter B, Pioletti DP, Laib S et al (2005) Calcium phosphate drug delivery system: influence of local zoledro zoledronate release on bone implant osteointegration. Bone 36:52–60
32. Peter B, Gauthier O, Laib S et al (2006) Local delivery of bisphosphonate from coated orthopedic implants increases implants mechanical stability in osteoporotic rats. J Biomed Mater Res A 76:133–143
33. Duan K, Fan YW, Wang RZ (2005) Electrolytic deposition of calcium etidronate drug coating on titanium substrate. J Biomed Mater Res B 72:43–51
34. Qu SX, Bai YL, Liu XM et al (2013) Study on in vitro release and cell response to alendronate sodium-loaded ultrahigh molecular weight polyethylene loaded with alendronate sodium wear particles to treat the particles-induced osteolysis. J Biomed Mater Res A 101:394–403
35. Liu AQ, Qu SX, Chao MM et al (2008) UHMWPE carrying estradiol to treat the particle-induced osteolysis—processing and characterizing. J Biomed Mater Res A 90:496–505
36. Yang HL, Xu YZ, Zhu M et al (2016) Inhibition of titanium-particle-induced inflammatory osteolysis after local administration of dopamine and suppression of osteoclastogenesis via D2-like receptor signaling pathway. Biomaterials 80:1–10
37. Lombardi A (1999) Treatment of paget's disease of bone with alendronate. Bone 24:S59–S61
38. Otsuka M, Matsuda Y, Baig AA et al (2000) Calcium-level responsive controlled drug delivery from implant dosage forms to treat osteoporosis in an animal model. Adv Drug Deliv Rev 42:249–258
39. Liu XM, Qu SX, Lu X et al (2009) Time-of-flight secondary ion mass spectrometry study on the distribution of alendronate sodium in drug-loaded ultra-high molecular weight polyethylene. Biomed Mater 4:065008
40. Yang D, Qu SX, Huang J et al (2012) Characterization of alendronate sodium-loaded UHMWPE for anti-osteolysis in orthopedic applications. Mater Sci Eng C 32:83–91
41. Gong KM, Qu SX, Liu YM et al (2016) The mechanical and tribological properties of UHMWPE loaded ALN after mechanical activation for joint replacements. J Mech Behav Biomed Mater 61:334–344
42. Zhang CZ, Qu SX, Liao Y et al (2011) Optimization on the dispersion of alendronate sodium in drug-loaded UHMWPE by orthogonal design. Acta Mater Compos Sin 28:109–117
43. Xie XL, Tang CY, Chan KYY et al (2003) Wear performance of ultrahigh molecular weight polyethylene/quartz composites. Biomaterials 24:1889–1896
44. Spiegelberg S (2004) Thermal transitions. In: Kurtz SM (ed) The UHMWPE handbook. Ultra-high molecular weight polyethylene in total joint replacement, 1st edn. Academic, New York, p 266
45. ASTM D638-03 (2003) Standard test method for tensile properties of plastics. American Society for Testing and Materials, New York
46. Johnson KL (1987) Normal contact of elastic solids-hertz theory. In: Johnson KL (ed) Contact mechanics, 1st edn. Cambridge University, Britain, p 90
47. Nusbaum HJ, Rose RM, Paul IL et al (1979) Wear mechanisms for ultrahigh molecular weight polyethylene in the total hip prosthesis. J Appl Polym Sci 23:777–789
48. ASTM F2183-02 (2002) Standard test method for small punch testing of ultra-high molecular weight polyethylene used in surgical implants. American Society for Testing and Materials, New York

49. Kurtz SM, Foulds JR, Jewett CW et al (1997) Validation of a small punch testing technique to characterize the mechanical behavior of ultra-high molecular weight polyethylene. Biomaterials 18:1659–1663
50. Edidin AA, Kurtz SM (2000) Influence of mechanical behavior on the wear of 4 clinically relevant polymeric biomaterials in a hip simulator. J Arthroplast 15:321–331
51. Fu SY, Feng XQ, Lauke B et al (2008) Effects of particle size, particle/matrix interface adhesion and particle loading on mechanical properties of particulate–polymer composites. Compos Part B Eng 39:933–961
52. Kanaga Karuppiah KS, Bruck AL, Sundararajan S et al (2008) Friction and wear behavior of ultra-high molecular weight polyethylene as a function of polymer crystallinity. Acta Biomater 4:1401–1410
53. Jin ZM, Stone M, Ingham E et al (2006) (v) Biotribology. Curr Orthop 20:32–40
54. Wang A, Sun DC, Stark C et al (1995) Wear mechanisms of UHMWPE in total joint replacements. Wear 181–183:241–249
55. El-Domiaty A, El-Fadaly M, Nassef AE (2002) Wear characteristics of ultrahigh molecular weight polyethylene (UHMWPE). J Mater Eng Perform 11:577–583
56. Angker L, Nockolds C, Swain MV et al (2004) Quantitative analysis of the mineral content of sound and carious primary dentine using BSE imaging. Arch Oral Biol 49:99–107
57. Ho SP, Carpick RW, Boland T et al (2002) Nanotribology of CoCr–UHMWPE TJR prosthesis using atomic force microscopy. Wear 253:1145–1155
58. Wang QL, Liu JL, Ge SR (2009) Study on biotribological behavior of the combined joint of CoCrMo and UHMWPE/BHA composite in a hip joint simulator. J Bionic Eng 6:378–386
59. Muratoglu OK, Bragdon CR, O'connor DO et al (1999) Unified wear model for highly crosslinked ultra-high molecular weight polyethylenes (UHMWPE). Biomaterials 20:1463–1470
60. Sauter C, Emin MA, Schuchmann HP et al (2008) Influence of hydrostatic pressure and sound amplitude on the ultrasound induced dispersion and de-agglomeration of nanoparticles. Ultrason Sonochem 15:517–523
61. Barbour PSM, Stone MH, Fisher J (1999) A study of the wear resistance of three types of clinically applied UHMWPE for total replacement hip prostheses. Biomaterials 20:2101–2106
62. Wu JJ, Buckley CP, O'Connor JJ (2002) Mechanical integrity of compression-moulded ultra-high molecular weight polyethylene: effects of varying process conditions. Biomaterials 23:3773–3783
63. Fu J, Ghali BW, Lozynsky AJ et al (2010) Ultra high molecular weight polyethylene with improved plasticity and toughness by high temperature melting. Polymer 51:2721–2731
64. Butler JH, Joy DC, Bradley GF et al (1995) Low-voltage scanning electron microscopy of polymers. Polymer 36:1781–1790
65. Huang YF, Xu JZ, Li JS et al (2014) Mechanical properties and biocompatibility of melt processed, self-reinforced ultrahigh molecular weight polyethylene. Biomaterials 35:6687–6697
66. Qu SX, Liu AQ, Liu XM et al (2012) Study on drug release of and biological response to UHMWPE wear particles carrying estradiol. Appl Surf Sci 262:168–175
67. Goncalves S, Santos NC, Martins-Silva J et al (2007) Fluorescence spectroscopy evaluation of fibrinogen-β-estradiol binding. J Photoch Photobiol B 86:170–176
68. Edidin AA (2009) Development and application of the small punch test to UHMWPE. In: Kurtz SM (ed) UHMWPE biomaterials handbook. Ultra-high molecular weight polyethylene in total joint replacement, 2nd edn. Academic, New York, p 490
69. Taha EA, Youssef NF (2003) Spectrophotometric determination of some drugs for osteoporosis. Chem Pharm Bull 51:1444–1447
70. Berkland C, King M, Cox A et al (2002) Precise control of PLG microsphere size provides enhanced control of drug release rate. J Control Release 82:137–147
71. Galetz MC, Glatzel U (2010) Molecular deformation mechanisms in UHMWPE during tribological loading in artificial joints. Tribol Lett 38:1–13
72. Yang D, Qu SX, Lin SZ et al (2012) Preliminary study on the effect of wear process on drug release of ALN-loaded UHMWPE. Appl Surf Sci 262:207–211

73. Goodman SB, Ma T, Chiu R et al (2006) Effects of orthopaedic wear particles on osteoprogenitor cells. Biomaterials 27:6096–6101
74. Huang ZN, Ma T, Ren PG et al (2010) Effects of orthopedic polymer particles on chemotaxis of macrophages and mesenchymal stem cells. J Biomed Mater Res A 94:1264–1269
75. Evans CE (2002) Bisphosphonates modulate the effect of macrophage-like cells on osteoblast. Int J Biochem Cell B 34:554–556
76. Igarashi K, Hirafuji M, Adachi H et al (1997) Effects of bisphosphonates on alkaline phosphatase activity, mineralization, and prostaglandin E2 synthesis in the clonal osteoblast-like cell line MC3T3-E1. Prostaglandins Leukot Essent Fatty Acids 56:121–125
77. Fang HW, Ho YC, Yang CB et al (2006) Preparation of UHMWPE particles and establishment of inverted macrophage cell model to investigate wear particles induced bioactivities. J Biochem Biophys Methods 68:175–187

Chapter 7
Biomechanics and Tribology of Artificial Knee Joint

Zhenxian Chen, Jing Zhang, Xiao Zhang, and Zhong-Min Jin

Abstract Ultrahigh molecular weight polyethylene (UHMWPE) remains the gold standard polymeric bearing material for use in total knee arthroplasty (TKA). However, the wear of UHMWPE inserts, the prosthetic aseptic loosening, and the osteolysis induced by wear particles limit the in vivo performance and lifetime of TKA. The loading and wear performance of UHMWPE components in TKA largely depend on the in vivo biomechanics of artificial knee joint. In turn, the wear of the bearing surfaces of UHMWPE components influences the biomechanics of artificial knee joint. Moreover, the biomechanics and wear of artificial knee joint are influenced by the design, surgical, and patient factors. The development of patient-specific musculoskeletal multibody dynamics simulation provides a reliable virtual platform for the investigation and evaluation of the coupled wear and biomechanical performance of TKA. In the future, in addition to the improvements of the UHMWPE material performance, patient-specific prosthetic design, surgery, and rehabilitation guidance will be the research directions for improving the in vivo wear performance of the UHMWPE components in TKA. In this chapter, both the biomechanics and tribology of knee implants are reviewed. The first section introduces briefly TKA and the current clinical problems. The second section describes the biomechanics of artificial knee joint, including the knee contact forces, the

Z. Chen · J. Zhang
Key Laboratory of Road Construction Technology and Equipment of MOE, Chang'an University, Xi'an, Shaanxi, China

X. Zhang
State Key Laboratory for Manufacturing System Engineering, School of Mechanical Engineering, Xi'an Jiaotong University, Xi'an, Shaanxi, China

Z.-M. Jin (✉)
Tribology Research Institute, School of Mechanical Engineering, Southwest Jiaotong University, Chengdu, Sichuan, China

State Key Laboratory of Manufacturing System Engineering, School of Mechanical Engineering, Xi'an Jiaotong University, Xi'an, Shaanxi, China

Institute of Medical and Biological Engineering, School of Mechanical Engineering, University of Leeds, Leeds, UK
e-mail: zmjin@mail.xjtu.edu.cn

J. Fu et al. (eds.), *UHMWPE Biomaterials for Joint Implants*, Springer Series in Biomaterials Science and Engineering 13,
https://doi.org/10.1007/978-981-13-6924-7_7

kinematics, and the stress and strain of the UHMWPE insert. The third section reviews the tribology of artificial knee joint, including wear mechanism, wear tests, and prediction of the UHMWPE insert. The fourth section mainly introduces the interaction between the biomechanics and wear of artificial knee joint. The fifth section reviews the effects of prosthetic design and material, surgical and patient factors on the biomechanics, and wear of artificial knee joint. In the final section, the major points are summarized.

Keyword Ultrahigh molecular weight polyethylene · Artificial knee joint · Wear failure · Total knee arthroplasty · Biomechanics · Patient-specific · Wear testing · Computational simulation

7.1 Artificial Knee Joint

Degenerative joint diseases (osteoarthritis) caused by wear and tear, trauma, infection, and disease damage the articular cartilage and hence cause knee pain and limited knee function. TKA surgery is a successful treatment approach for severe knee joint diseases. An increasing number of TKA are carried out every year over the world. Artificial knee joints are designed to replace the damaged biological tissues to help the patient to perform the daily normal activities. Section 7.1.1 mainly introduces the components and material of artificial knee joint. Wear of the polyethylene component of artificial knee joint is a major reason for TKA revision. Section 7.1.2 presents the wear failure of the UHMWPE inserts in TKA.

7.1.1 Components and Materials

In general, the artificial knee joint mainly consists of three main components: femoral, tibial, and patellar components (Fig. 7.1). The tibial component is comprised of a tibial tray and a tibial insert. The femoral component and the tibial tray are made of stainless steel, titanium alloys or cobalt chromium molybdenum (CoCrMo) alloy, while the tibial insert and the patellar components are typically made of UHMWPE [1]. The femoral component is used to replace the lower end of the femoral cartilage and bone [2]. The tibial tray is used to attach the UHMWPE tibial insert to the upper end of the tibial bone [2]. The UHMWPE insert provides a tibiofemoral (TF) articular load bearing surface with low friction during the motion of the knee joint [2], and the UHMWPE patellar component provides a patellofemoral (PF) articular load bearing surface.

Typical designs of artificial knee joint in TKA including the fixed-bearing type, the mobile-bearing type, the unicompartmental type, and the rotating-hinge type are shown in Fig. 7.2. The fixed-bearing type and mobile-bearing type further include posterior cruciate substitution (PS) and posterior cruciate retention (CR). These

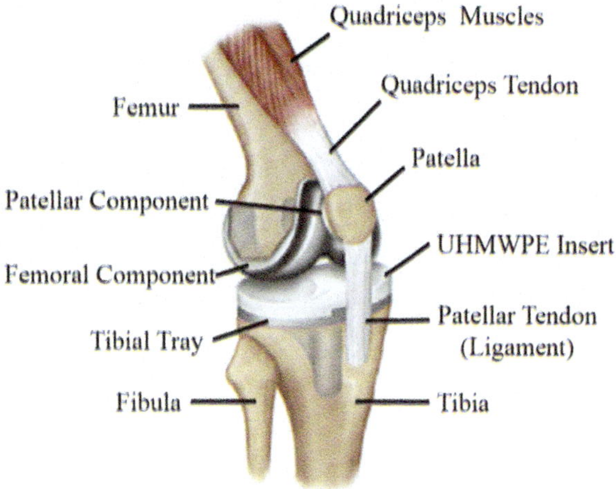

Fig. 7.1 Total knee arthroplasty

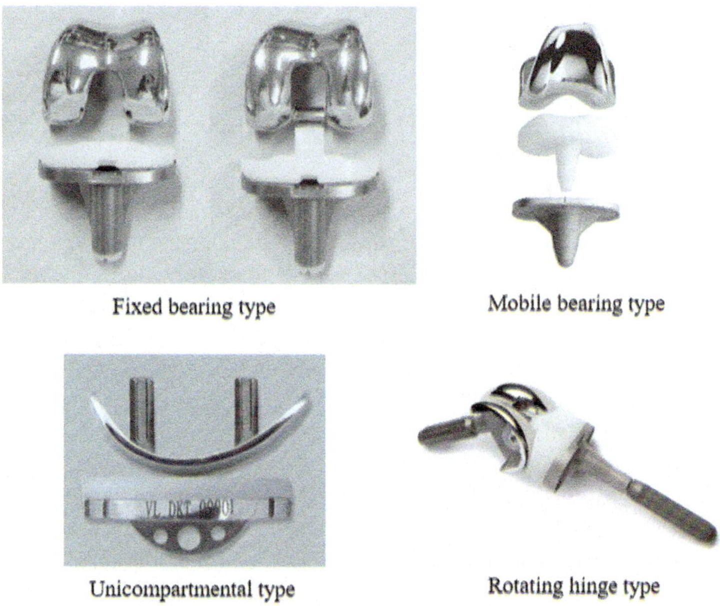

Fig. 7.2 Some typical implant design

designs are available for surgeons to implant depending on the sclerotin, the state of the joint disease, and the expected activity of the patient. The mobile-bearing knee prostheses allow greater unconstrained internal–external rotation of tibial insert, whereas the tibial insert in fixed bearings is fixed by interference fit between insert backside with tibial tray and cannot move relative to the tibial tray. The mobile-bearing knee prostheses may be implanted in order to obtain high conforming bearing with free rotation. For the fixed-bearing type, the anterior cruciate ligament (ACL) is always sacrificed in TKA, but the posterior cruciate ligament (PCL) may be

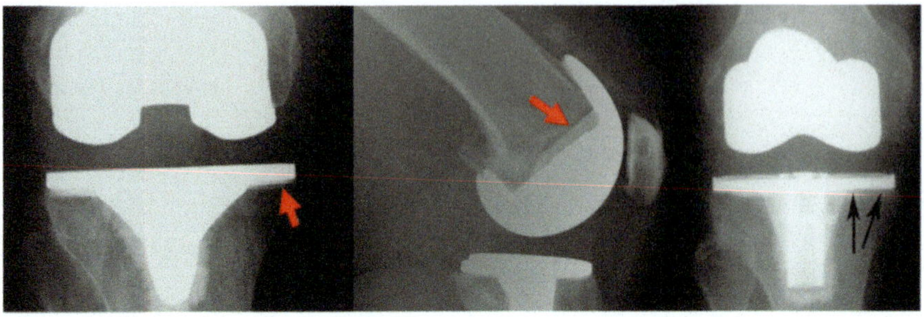

Fig. 7.3 Typical aseptic loosening in TKAs (http://kctlegal.com/zimmer-persona-recall/)

substituted or retained. If the PCL is removed in some case, a PS knee prosthesis with post-cam design should be implanted in order to simulate the function of the PCL. For the unicompartmental type, if only the medial or lateral compartment of the knee is damaged by the arthritis, a unicompartmental knee arthroplasty may be performed.

7.1.2 Wear Failure of the UHMWPE Insert

Though TKA is widely used to restore normal knee joint function and to relieve pain, the failure of TKA caused by wear of the UHMWPE, breakage, dislocation, loosening within the bone, or fatigue failure of the tibial tray limits the long-term survivorship of artificial knee joint and causes secondary pain and surgery costs. Especially, the wear and aseptic loosening are the two most common reasons for knee revision according to annual report from the Swedish Knee Arthroplasty Registery [3]. Whereas the aseptic loosening mainly results from the polyethylene and metallic wear particles, the wear particles result in the osteolytic lesions around the artificial knee joint (Fig. 7.3) [4]. Kurtz et al. [5] suggested that the demand for primary TKA is projected to grow by 673% to 3.48 million procedures in the United States from 2005 to 2030, while TKA revisions are projected to grow by 601%. An increasing number of young patients require TKA, and thus the expected lifetime and function of artificial knee joint is considerably increasing. So, TKA revisions should be improved by reducing the wear failure of artificial knee joint.

UHMWPE was first used in TKA in the early 1960s [1]. UHMWPE is used as the gold standard material in TKA due to its good mechanical and physical properties. However, there are seven damage patterns of UHMWPE in TKAs, including abrasion, delamination, pitting, scratching, burnishing, deformation, and embedded debris [6]. Pitting, delamination, and abrasive–adhesive modes are three basic wear patterns for generating the primary UHMWPE wear debris (Fig. 7.4) [7]. The generation of UHMPWE wear debris from bearing surfaces is a major problem for the lifetime of artificial knee joint. The possible reasons leading to wear include defective implant design, material selection, surgical procedure, and patient's gait

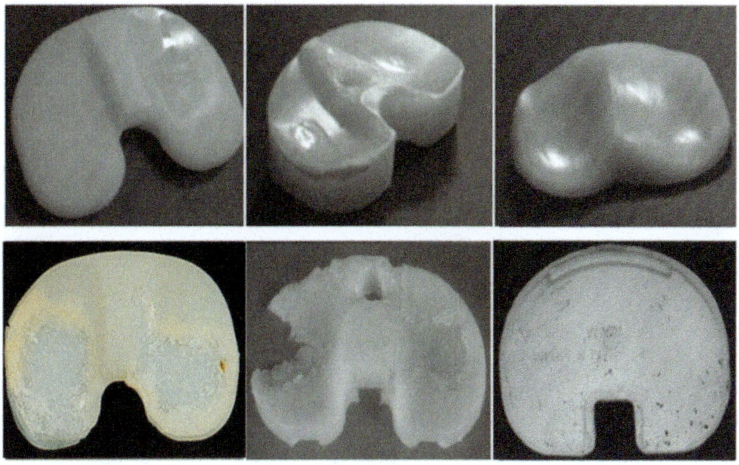

Fig. 7.4 Typical UHMWPE wear patterns in TKA implants

pattern. The performances of current artificial knee joint are substantially dependent on the contact mechanics, joint kinematics, and wear of joint components [8]. The joint dynamic loading and motion of the knee joint as well as the stability from muscles and ligaments may contribute to the wear failures of TKA as well [8].

7.2 Biomechanics of Artificial Knee Joint

The knowledge of in vivo biomechanics of artificial knee joint is essential to understand the wear failure mechanism of the UHMWPE insert in TKA. Section 7.2.1 presents the experimental measurements and prediction by patient-specific musculoskeletal multibody dynamics simulation of knee contact forces and kinematics. Based on both experimental measurement and finite element analysis, Sect. 7.2.2 mainly discusses on the stress and strain of the UHMWPE insert under knee joint loading and kinematics.

7.2.1 Knee Contact Forces and Kinematics

Knee contact forces and kinematics are important factors for in vivo working condition of artificial knee joint. Knee kinematics includes the primary and secondary knee kinematics, the contact track and position of femoral condyle relative to the insert, etc. In vivo knee joint forces and kinematics are generally determined by experimental or computational means. However, it is difficult to measure the knee contact forces and kinematics, especially for natural knee joint.

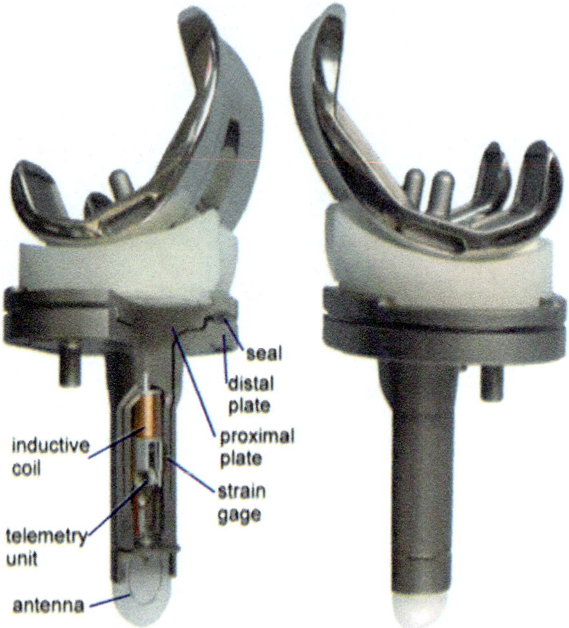

Fig. 7.5 Instrumented tibial tray. (Reprinted from Ref. [9], Copyright 2009, with permission from Elsevier)

Currently, the knee contact forces are only measured from a limited number of patients with instrumented knee prostheses during dynamic activities. Although the primary kinematics can be obtained based on motion capture analysis, the secondary knee kinematics is measured by fluoroscopy from a limited number of patients. However, these measurements of in vivo knee contact forces and kinematics are indispensable for the validation of corresponding prediction methods.

7.2.1.1 Experimental Measurements

Instrumented knee implants with force-sensing transducers are developed [9, 10] to measure in vivo knee joint forces and moments during different gait patterns or activities, which provides a feasible real-time quantitative evaluation of the joint contact forces, especially the TF medial and lateral contact forces. An example of instrumented tibial tray developed by Kutzner et al. [9] is shown in Fig. 7.5. The three forces and three moments acting in the implant can be measured using the custom-made telemetric tibial tray. Furthermore, the medial and lateral contact forces of the TF joints could be determined using the obtained data with regression equations. The total TF forces transmitted across the knee joint are 2–3 times the body weight (×BW) for walking, approximately 3 × BW for stair ascent, and 3.5 × BW for stair descent.

Motion capture analysis (Fig. 7.6) is currently the most widely used non-invasive method for assessing the primary function movement. Primary knee kinematics is

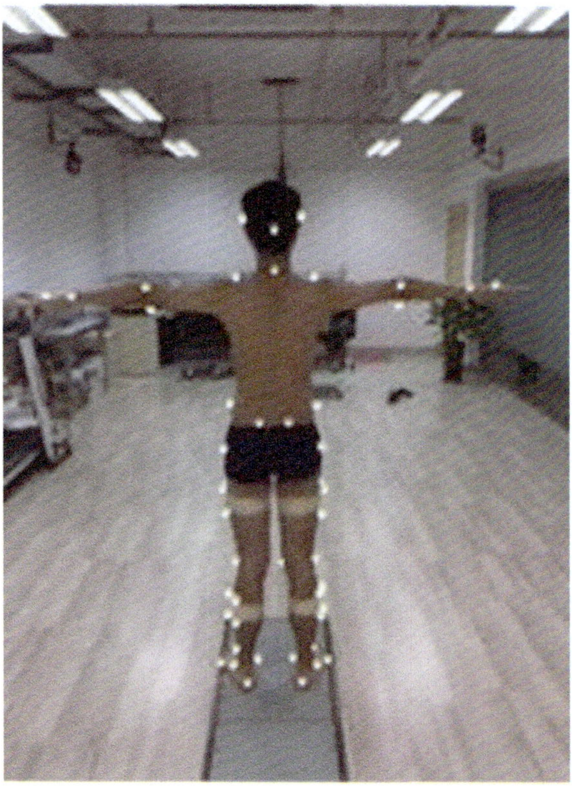

Fig. 7.6 Motion capture experiment

usually obtained from markers in a gait laboratory. Different activities, such as gait, sit–stand–sit, step–descent, lunge and squat activities, etc. can be performed in the motion capture experiments. This method gives an insight into the knee kinematics and kinetics. Motion capture and force plate data can be used as input condition for musculoskeletal modeling system. However, only the primary knee flexion–extension motion can be obtained based on the maker trajectory analysis; the secondary knee kinematics, including the in vivo internal–external and varus–valgus rotation and anterior–posterior and medial–lateral translations, cannot be determined accurately.

Fluoroscopic measurement is an important method to obtain the in vivo joint kinematics of TKA, especially secondary knee kinematics (Fig. 7.7). The dual orthogonal fluoroscopic imaging system developed by Li et al. [11] provides an easy and powerful tool for accurately determining six degree of freedom (DOF) positions of the knee. The motions of femoral component relative to tibial insert can be quantified using fluoroscopic data. Based on the fluoroscopy imaging analysis, Shandiz et al. [12] found that there are differences between the preoperative and postoperative knee kinematics for the specific patients, while the joint kinematics of TKA are different between different patients. The measured results cannot necessarily be extrapolated to other patients. In previous studies [12–14], the knee kinematics

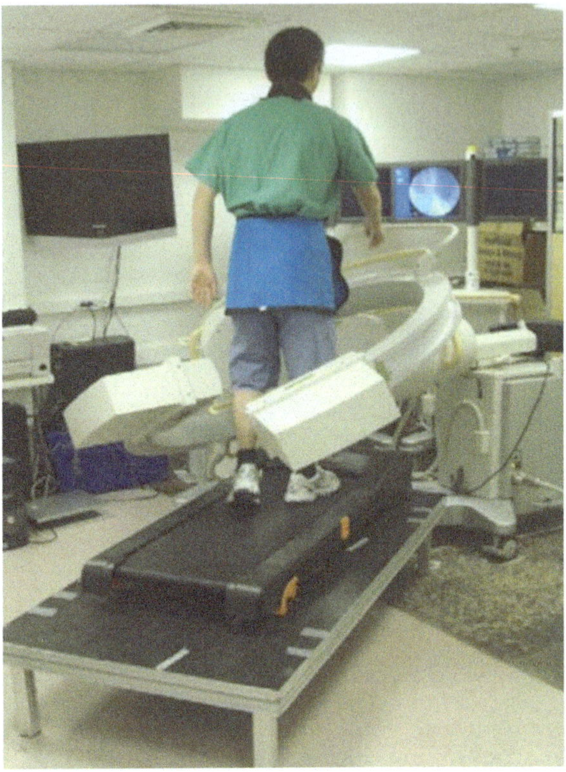

Fig. 7.7 Fluoroscopic measurement. Reprinted from Ref. [11], Copyright 2009, with permission from Elsevier

during a non-weight-bearing or weight-bearing deep knee bend or lunge were reported. Few studies were performed to investigate the in vivo knee kinematics during the walking gait cycle.

Using the data obtained from instrumented knee implants and fluoroscopic measurement, the knee joint mechanical or wear simulators can be programmed according to realistic load profiles, and musculoskeletal multibody dynamics models of TKA can be validated. However, in vivo knee contact forces and kinematics are measured in a limited number of patients; the measured results can not necessarily be transferred to other patients. Besides, the fluoroscopy measurements were difficult to obtain joint kinematics data for different over-ground gait trails like jumping, running, turning-right gait, etc. [15]. Moreover, the finding by Kozanek et al. [11] demonstrates that the knee kinematics is activity-dependent and the motion patterns of one activity cannot be generalized to interpret the motion patterns of other activities. On the other hand, if one wants to know the in vivo knee loading and kinematics under different implant designs or surgical malalignments, it is impossible via measurement method for the same patient. The performance of multiparameter analyses is inhibited due to the laboratory experimental costs, time costs, etc.

7.2.1.2 Patient-Specific Musculoskeletal Multibody Dynamics Simulation

Most recently, many available musculoskeletal (MSK) multibody dynamics (MBD) software packages have been introduced for the purpose to measure in vivo joint forces and kinematics. For example, LifeMod (Lifemodeler Inc., CA, USA) a musculoskeletal-specific plugin to ADAMS (MSC Software, CA, USA), and AnyBody (AnyBody Technology, Aalborg, Denmark), OpenSim (Delp et al. 2007 [16]) are popularly used to model the human MSK system. Particularly, a novel force-dependent kinematics (FDK) approach was developed by Andersen and Rasmussen [17] in AnyBody, which can be potentially used to predict the muscle, joint, and ligament forces and internal joint kinematics. To obtain more realistic and accurate prediction, patient-specific modeling should be considered, including patient-specific CT data, gait pattern, and TKA report.

Patient-specific MSK MBD model of TKA is an attractive platform to obtain in vivo knee joint forces and kinematics during movement [18]. The knee contact forces and secondary knee kinematics of TKA have been quantitized with a reasonable accuracy by using a patient-specific MSK modeling framework based on the FDK approach [18, 19]. The recently available datasets [20] "Grand Challenge Competition to Predict In Vivo Knee Loads", which are obtained from patients with instrumented knee prostheses, offer a unique opportunity to evaluate the patient-specific MSK MBD models of TKA by comparing the predicted results with the experimental measurements.

Using the public datasets [20], a patient-specific MSK MBD modeling framework of TKA using FDK in AnyBody [17, 18] is shown in Fig. 7.8. To mitigate many challenges of the development of a full MSK model, the generic MSK model of software model repository is always modified to establish new MSK model. Here, the generic MSK model extracted from AnyBody Managed (version 1.6) was modified for patient-specific MSK modeling. In order to obtain a patient-specific MSK model, the bone geometries of the generic model in Anybody are morphed to patient's bone geometries. The preoperative and postoperative bone geometries of patient's knee are created and exported from Mimics (version 15, Materialise NV, Leuven, Belgium) based on the CT image data [20]. Because the preoperative CT images only include the distal femur and the proximal tibia, the partial knee preoperative bone geometries are registered with the postoperative full-leg bone geometries to obtain the complete preoperative femur and tibia models in Geomagic studio (version 12, Geomagic, USA). A lot of patient-specific landmark points on the surface of the complete preoperative bones are selected and recorded in Geomagic studio. The corresponding point locations for the corresponding bones of the generic model are obtained by the same operation. Based on the selected patient-specific landmark points, an advanced bone morphing method [21] is adopted to match the topology of the generic model to the corresponding patient-specific preoperative bones. A radial basis function (RBF) interpolation scheme reported by Marra et al. [19] is used to interpolate the points and create the scaling transformation. During bone morphing, the muscle attachment sites are adjusted automatically according to the changes in bone geometries.

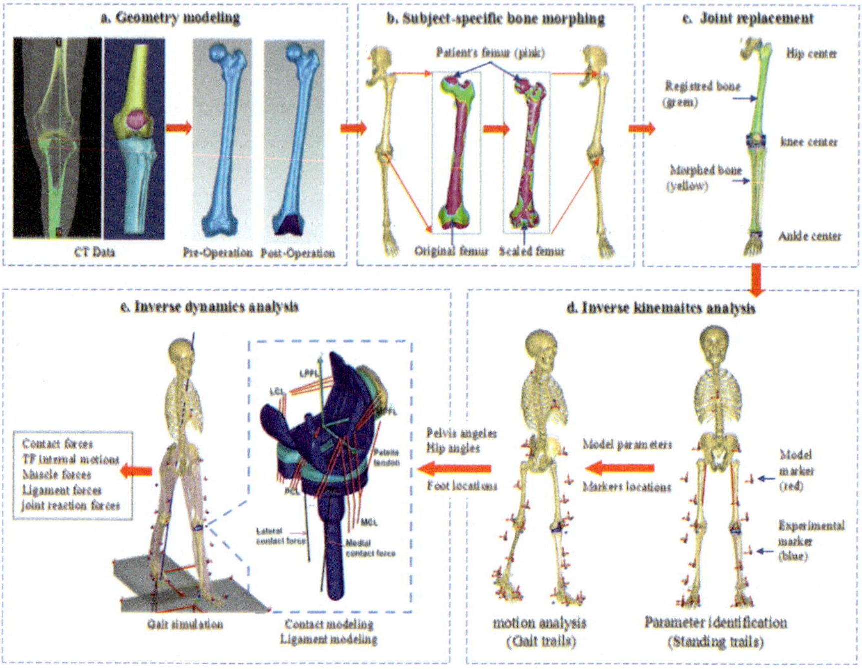

Fig. 7.8 Overview of a patient-specific MSK MBD modeling framework of TKA using FDK in AnyBody. (Reprinted from Ref. [18], Copyright 2016, with permission from Elsevier)

It is difficult to collect all bone CT images of the patient. An inverse kinematic analysis is performed to scale the other remaining bone segments according to patient's height and weight. The marker locations of the MSK model are determined by using patient's standing reference trail in inverse kinematics analysis. In AnyBody, an optimization routine is used to perform the segments' scaling, which uses a set of Karush–Kuhn–Tucker optimal conditions [22] to minimize the differences between the model markers and the experimentally recorded marker trajectories. The local marker coordinates, segment lengths, and other model parameters are simultaneously optimized via the Length–Mass–Fat scaling law.

The patient-specific TKA model is established by registering the postoperative bone geometries to the morphed preoperative bone, aligning the knee implants location, and fitting the joint center and axes of the hip, knee, and ankle. Detailed ligament models are established to maintain knee joint stability. Ligaments surrounding the TF and PF joints include the medial collateral ligament (MCL) and lateral collateral ligament (LCL), posteromedial capsule (PML), posterior cruciate ligament (PCL), and medial and lateral PF ligaments (MPFL and LPFL). Ligaments were set to wrap around the surfaces of the bones and implants to simulate the geometric path. The ligament force was calculated according to a nonlinear piecewise force–displacement relationship [23].

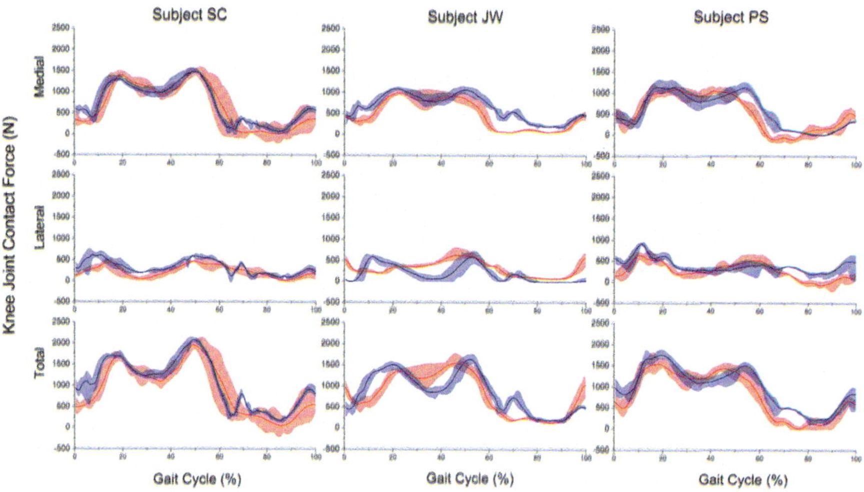

Fig. 7.9 Comparison of the experimental data and computation results on the in vivo medial, lateral, and total contact forces. The solid line represents mean value and the shaded area represents the range for all gait trails. (Reprinted from Ref. [18], Copyright 2016, with permission from Elsevier)

A new knee model was defined using FDK method. Since the patellar ligament was assumed to be rigid, six TF DOFs and five PF DOFs were released in the MSK model. Three deformable contact pairs were defined between the tibial insert and the femoral component and between the patellar button and the femoral component. The tibial insert was divided into medial and lateral compartments for computation of the medial and lateral contact forces. The geometries of the contact surfaces of the implants were represented with the triangles of STL files. A linear force-penetration volume law with a contact pressure module, known as *PressureModule* in N/m^3, was used to compute the knee contact forces. Because the linear force-penetration volume law is quite similar to the elastic foundation theory, the contact pressure module could be determined according to the equations derived by Fregly et al. [24]

The marker trajectories of the gait trails were input to simulate the patient's activities and the pelvis and hip angles as well as foot spatial locations during the gait cycles were obtained after inverse kinematics analysis. Subsequently, an inverse dynamics analysis was performed with ground reaction forces (GRFs) input together with the joint angels and foot locations obtained from the inverse kinematics analysis, and the knee contact forces, TF internal motion, muscle and ligament forces, and joint reactions were output.

As shown in Fig. 7.9, the patient-specific MSK MBD modeling framework was evaluated by comparing the predicted TF total, medial, and lateral contact forces with the measured results over a range of patients. The prediction accuracy was quantified in terms of root mean square error (RMSE) and the Sprague and Geers metrics of magnitude (M), phase (P), and combined error (C) [25]. In general, the predictions were in good agreement with the experimental results for the medial contact forces (174 N < RMSE <224 N, −0.04 < M < 0.20, 0.06 < P < 0.09, 0.08 < C < 0.22) and the lateral contact forces (131 N < RMSE <240 N, −0.25 < M < 0.34, 0.08 < P < 0.22, 0.13 < C < 0.36) [18]. The results reported by Chen et al. [18] suggested that the developed patient-specific MSK MBD modeling framework of TKA using FDK as a powerful tool could be used to investigate the functional outcomes of knee implants.

It is pitiful that the public datasets [20] do not include the fluoroscopy data of gait trials. Only a few fluoroscopy data of the unloaded leg-swing trails can be used to indirectly evaluate the knee kinematics of TKA. The secondary knee kinematics of TKA during the unloaded leg-swing trial had been quantified by Marra et al. [19] with a reasonable accuracy by using a patient-specific MSK model of TKA via the FDK approach. During the leg-swing trial, the predicted PF and patella-tibial (PT) flexion curves were almost linear, in comparison to the experimental knee flexion angle. The predicted flexion motion of the PF and PT overlapped well with $R^2 = 0.99$ and $R^2 = 0.7$, respectively, and the predicted the distal–proximal TF tip and the posterior–anterior TF tip shift overlapped well with $R^2 = 0.96$ [19]. Furthermore, Chen et al. [15] indirectly assessed the predicted in vivo knee kinematics by comparing with the experimental data reported by DesJardins JD et al. [26] (Fig. 7.10). A similar Zimmer NKII knee prosthesis was used in the study reported by DesJardins JD et al. [26]. In general, the predicted knee flexion was consistent with the reported [26] average value of the patient's experimental data. Compared with the reported [26] internal–external (IE) rotation, where the femoral component is not considered with an average 7° external rotational alignment with respect to the tibial component here, the prediction obtained a similar magnitude and general trend. Comparing the reported [26] tibial anterior–posterior (AP) motion, although the prediction had a similar general trend, a larger magnitude was predicted than the reported data. Although a predicted error still exists, the present model could offer a potential way to obtain some useful information of in vivo knee kinematics.

The AP and IE motions on TKA systems produce cross-pathway sliding that usually influences the articular surface wear of UHMWPE components. A center of pressure (COP) method [15] was adopted to successfully quantify the in vivo contact position of the nonconforming TKA. The COP was calculated based on the contact area as an average of position vectors of the penetrating vertices, which was weighed by the force vectors at each of the penetrating vertices [15]. A remarkable difference was found between the medial and lateral COP tracks [15], which may be related to the asymmetric design of tibial insert. As shown in Fig. 7.11, an approximately linear reciprocating movement was observed on the lateral COP track, while an approximately circular movement was observed on the medial COP track. The changes in the multidirectional sliding motions are related to the UHMWPE wear mechanism of tibial insert. According to the relationship between the knee kinematics and the UHMWPE wear, more multidirectional and complex sliding usually produce more wear [27]. The medial and lateral contact tracks (Fig. 7.11) indicate that the TF contact motion is multidirectional and complex. Because the wear rate is

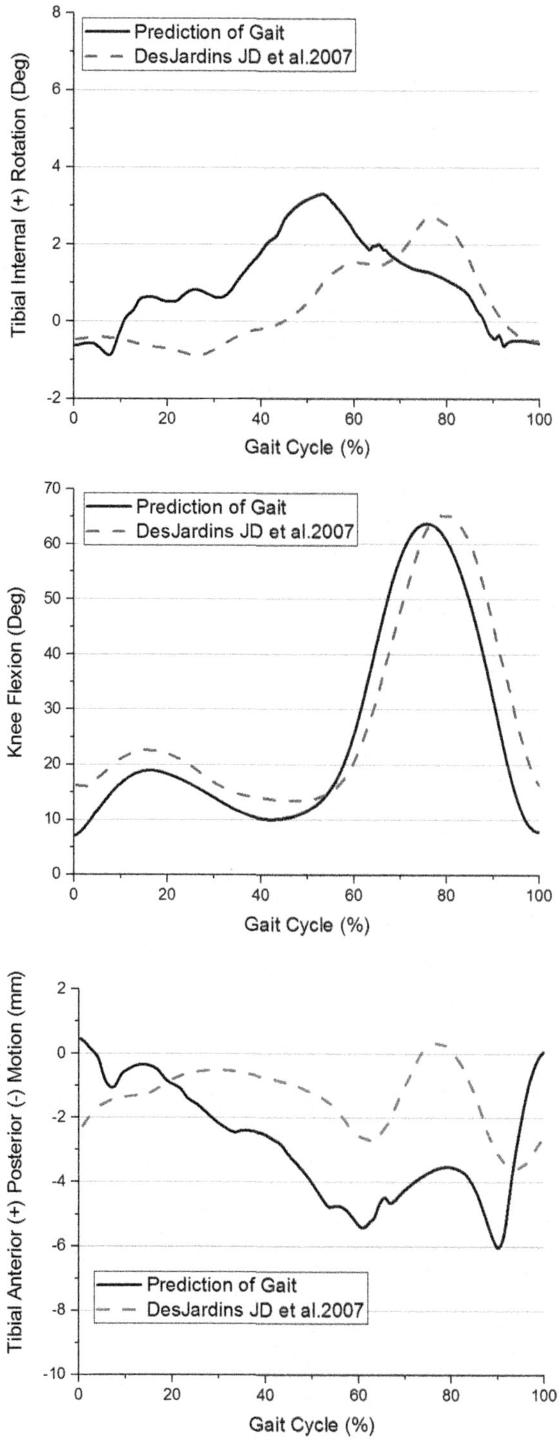

Fig. 7.10 Comparison of the experimental data reported by DesJardins JD et al. [26] and computationally estimated in vivo knee flexion, tibial IE rotation, and tibial AP motion [15]. (Adapted from Chen et al. 2016)

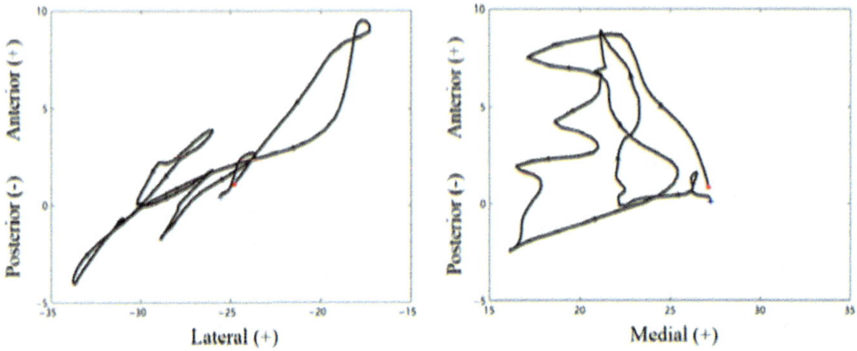

Fig. 7.11 The contact tracks of the medial (right) and lateral (left) COP of the Zimmer NKII TKA insert during the straight gait trial. (Reprinted from Ref. [15], Copyright 2016, with permission from Southwest Jiatong University)

dependent upon the wear path geometry [28], the wear ratio of the medial and lateral bearing compartment may be influenced by the kinematics related to the prosthesis design.

The position changes of the COP during the straight gait cycles are shown in Fig. 7.12. During the different phase of gait cycles, the position changes of the COP were mainly decomposed into anterior or posterior translation and internal or external rotation. The predicted position changes of the COP [15] indicated that motion of the medial femoral condyle in the transverse plane was larger than that of the lateral femoral condyle during the stance phase of the straight gait cycle. The IE rotational pivot points of the femoral component relative to the tibial insert were located predominantly on the lateral side of the TF joint during the stance phase of the straight gait cycle for this TKA design. These findings are consistent with the results reported by Kozanek et al. [11]. Furthermore, based on the fluoroscopic experiments of several patients implanted with the Zimmer NKII right knee prosthesis, DesJardins et al. [26] found that the IE rotation of femoral component had a center-to-lateral condyle pivot point during the gait cycle. Although the medial IE rotational pivot of the femoral condyle was reported in some studies [29, 30], these studies were performed under the lunge or deep knee bend. Kozanek et al. [11] demonstrated that the knee kinematics was activity-dependent and the motion patterns of one activity could not be generalized to interpret the motion patterns of other activities. In addition, the medial–lateral translation should be considered in wear studies for knee implant designs, which aggravated the complex multidirectional sliding motion of UHMWPE bearing surface.

An accurate knowledge of in vivo knee kinetics and kinematics is important for the wear and function assessments of UHMWPE components, which can be further utilized to improve current lifetime of knee prostheses. The developed patient-specific MSK MBD model [18, 19] provided a strong platform for predicting the in vivo knee kinetics and kinematics of TKA.

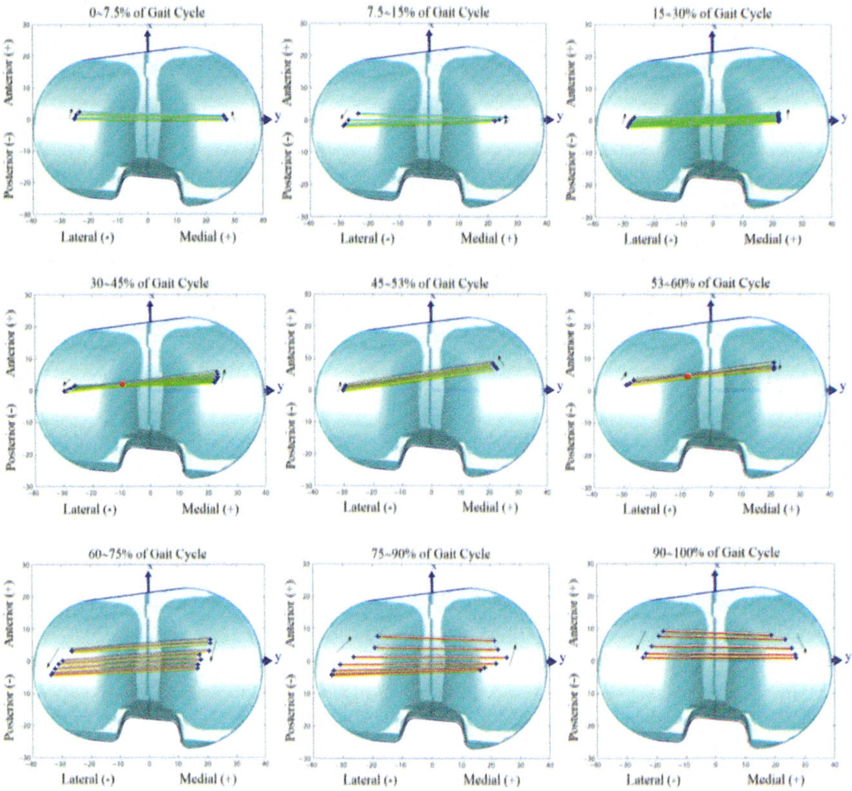

Fig. 7.12 The changes in the COP position during the straight gait trial. (Reprinted from Ref. [15], Copyright 2016, with permission from Southwest Jiatong University)

7.2.2 *Stress and Strain of the UHMWPE Insert*

To our knowledge, researchers have always associated the stress and strain of UHMWPE components with the surface damage and wear failure in TKA. And the stress and strain level acting on UHMWPE components are also adopted to develop and assess the new UHMWPE components design. It is important to understand the stress and strain of the tibial insert during knee loading, and accurately determine the contact pressure and contact area between the articulating bearings surface of artificial knee joint, for accurately assessing the potential wear rate of the UHMWPE components.

7.2.2.1 Experimental Measurement

Fuji pressure-sensitive film is widely used to measure the contact pressure and area in TF joint of a knee prosthesis. In general, the Fuji pressure-sensitive film is usually

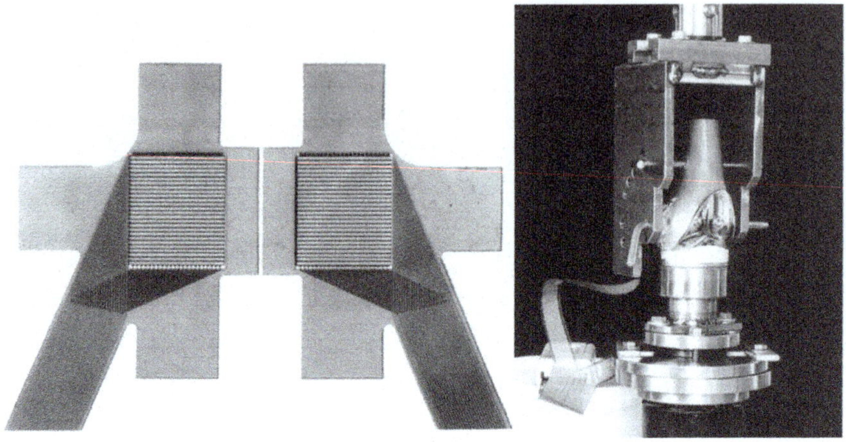

Fig. 7.13 K-scan sensor arrays and the testing apparatus. (Reprinted from Ref. [31], Copyright 1999, with permission from Elsevier)

composed of a microcapsule layer and a color-developing layer. However, Liau et al. [32] found that the contact area was overestimated by 1.2–2.4% when an ultra-super low-grade Fuji film was used to measure the contact area in TF joint, while the contact pressure was underestimated by 8–14% when a medium-grade Fuji film was used to measure the contact pressure in TF joint. In addition, the pressure-sensitive film is primary used to quantify contact stress; the other stress components are not applied.

The K-Scan sensor system is also an important contact pressure and area measurement system (Fig. 7.13). It includes a plastic laminated and thin-film electronic pressure sensor. The thin-film electronic pressure sensor is connected to the hardware and software of a PC with a coupler. K-scan pressure-sensing system is an easier and more reliable technique to provide real-time measurements of TF contact area in TKA than the Fuji pressure-sensitive film [31]. Harris et al. [31] found that the contact area measured by the Fuji pressure-sensitive film was lower than that measured by K-scan system. Moreover, K-scan system is easily used to measure and record contact pressures and areas under dynamic flexion, varus–valgus and torsion, where Fuji pressure-sensitive film is not suitable for such measurements.

7.2.2.2 Finite Element Analysis

Today, finite element analysis (FEA) has become a primary method to determine the stress and strain of the UHMWPE components. FE models take into account the complex three-dimensional geometry of TKA components, materials, and appropriate knee joint loading and kinematics. Contemporarily, there are two finite element analysis (FEA) methods, including explicit FEA method and implicit FEA method. The explicit FEA model has been developed to predict the joint kinematics of TKA in close agreement with the experimental data [33]. The implicit FEA has been considered in predicting material loss caused by wear based on the contact pressure, the contact area, and the sliding distance [34].

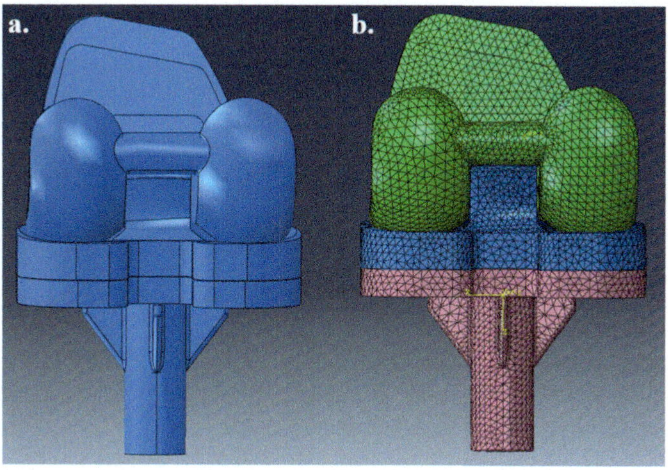

Fig. 7.14 The solid model (**a**) and FE model (**b**) of TKA

FEA is usually performed using Abaqus (Simulia Inc., Providence, RI) and ANSYS (ANSYS Inc., Canonsburg, PA). As shown in Fig. 7.14, an FE model of TKA was developed to simulate the experimental testing under displacement control in Abaqus. In general, the femoral component is modeled as rigid body for saving the computational time. A realistic material model for UHMWPE is incorporated into FEA simulation of TKA. A simple linear elastic material model for UHMWPE also can be adopted depending on the scope and question posed by a finite element analyst. However, for simulation of cyclic loading, the UHMWPE insert is usually modeled as a nonlinear material with a modulus of elasticity of 463 MPa and a Poisson's ratio of 0.46. A stress–strain relationship is usually used in FEA to represent the nonlinear elastic compressive properties of polyethylene, which are derived from modulus–stress data by Cripton et al. [35] An approximate global size is determined and used for the femoral and tibial components according to the mesh sensitivity study. A penalty contact with anisotropic coefficient of friction of 0.04 is adopted to describe the contact between the femoral component and tibial insert.

The experimental data, according to ISO 14243, or the data obtained from the MSK MBD simulation during walking are adopted as the boundary condition for FEA. The boundary conditions (knee joint force and motions) are usually applied to calculate the stresses acting on the UHMWPE insert. In general, the joint force and femoral flexion/extension and internal/external rotations are applied at the femoral control node. The femoral control node is defined at the flexion/extension axis of the femoral component through the center and offset by 5 mm in the medial direction. Adduction–abduction motion is unconstrained, and all other degrees of freedom at this node were constrained. For the UHMWPE insert reference node, only anterior–posterior displacement is applied and other degrees of freedom are constrained. The original position of the implants was determined by bringing the tibial and the femoral contact surfaces into contact at their lowest points.

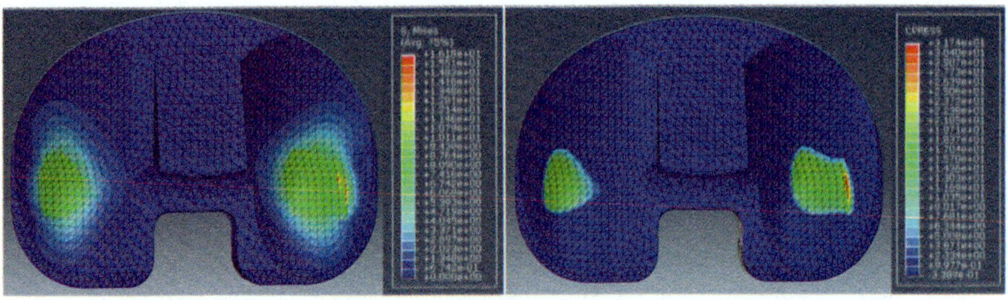

Fig. 7.15 The von Mises stress (left) and contact pressure (right)

The contact pressures and von Mises (effective) stresses of the UHMWPE insert are quantified after FEA simulation (Fig. 7.15). The yield strength of polyethylene ranges from 13 to 32 MPa [36]. A larger magnitude of the von Mises stresses may produce localized yielding and permanent deformation of the UHMWPE insert. The contact pressures over the articulating surface of UHMWPE inserts ranges from 5 to 25 MPa during a gait cycle [8]. Stair climbing can generate a higher contact pressure of 32 MPa and the lunge activity can generate the highest stress of 56 MPa [36]. The peak contact pressure is associated with potential for localized damage of UHMWPE insert, while average contact pressures are used to estimate potential for wear. Contact area may be increased after UHMWPE plastic deformation under high stresses, and these may partially counteract substantial increases in contact pressures of UHMWPE insert [36]. The equivalent plastic strain can be used to determine the potential for damage [36].

For FEA simulation, the predicted stress and strain of UHMWPE insert is not only dependent on the implant geometry, material properties, and boundary conditions but also related to surgical malalignment of components, geometric nonconformities, sclerotin, etc. We will describe those issues later in the chapter.

7.3 Tribology of Artificial Knee Joint

Although the current artificial knee joint can survive in vivo 10–15 years, the wear and aseptic loosening resulted from wear particles still restrict the clinical lifetime of TKAs. The friction, wear, and lubrication of artificial knee joint are extensively investigated through in vitro experimental and computational means. Mild wear is a function of contact pressure, kinematics, and material properties [37]. To understand the wear mechanism, the relationship between contact pressure, kinematics, and wear is investigated via the knee simulator experiment and computational simulation. Section 7.3.1 briefly describes the wear mechanism of the UHMWPE insert. Section 7.3.2 introduces the wear testing of the UHMWPE insert by knee simulator experiment and the gravimetric method and coordinate measuring method of volumetric wear. Section 7.3.2 mainly presented the wear prediction of the UHMWPE insert, including the effects of wear law and boundary condition on wear prediction, as well as the backside wear prediction of the UHMWPE insert (Fig. 7.16).

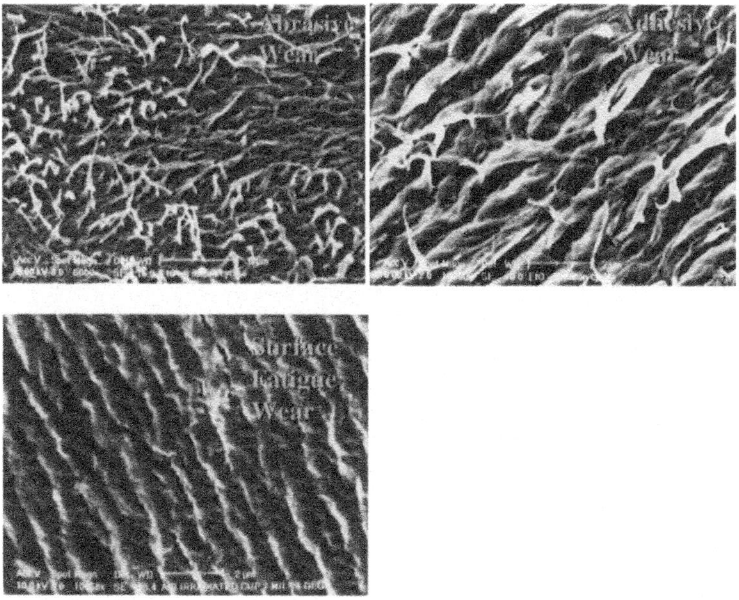

Fig. 7.16 SEM micrographs of the abrasive wear, adhesive wear, and surface fatigue wear. (Reprinted from Ref. [38], Copyright 1998, with permission from Elsevier)

7.3.1 Wear Mechanism of the UHMWPE Insert

The wear mechanism of UHMWPE components includes the abrasive wear, the adhesive wear, and the fatigue wear [38]. During articulation, the wear mechanism of UHMWPE components is affected by loading, sliding speed, sliding distance, lubricant, and surface roughness of grinding materials, etc. and may be related to the thermal degradation and thermal oxidation. The cross-linking and thermal treatment improved the wear properties of UHMWPE components and reduced the wear debris size [2]. For a proper degree of cross-linking, it should be a compromise between mechanical properties, wear properties, and wear debris size [2]. Moderately cross-linked UHMWPE has been used to provide reasonable mechanical and wear properties for TKA prostheses [2].

Cross-shear is an important cause to wear. A significant increase in wear rate was found in the experimental tests when a unidirectional motion was altered to a multidirectional motion [38]. Under unidirectional motion, the UHMWPE chains orientated in the sliding direction become weak in the perpendicular direction, which leads to a strain hardening of the component bearing surface in the direction of sliding [39]. Wear resistance was increased in the sliding direction [39]. Compared with a unidirectional motion, wear increases considerably in a multidirectional motion [40]. The creep and plastic deformation of the bearing surface of UHMWPE components have a significant effect on contact area and wear [40].

Fig. 7.17 The ProSim knee simulator

7.3.1.1 Knee Simulator Experiment

Knee simulator is usually used to test the wear resistance and the function of the knee implant [41–43]. The wear testing method and the ISO standard of knee simulator experiment are presented in this chapter. Figure 7.17 shows a six-station knee simulator designed by ProSim (Manchester, UK). There are two groups, with each group occupying three stations. As shown in Fig. 7.18, all TKA components are mounted at the knee simulator and then loaded with the axial load, flexion–extension, internal–external rotation, and the anterior–posterior translation as the input conditions. The internal–external rotation and anterior–posterior translation of knee simulator can be controlled by using either force or displacement inputs. The load axis and the axis of rotation were aligned when the femoral component was located at the knee simulator.

Current wear testing experiments for artificial knee joints are mainly based on ISO14243 [44–46], including ISO 14243-1:2009, ISO 14243-2:2009, and ISO 14243-3:2014. The relationship between these three parts is shown in Fig. 7.19. The first and third parts are two different control methods: force control and displacement control. The main difference between force control and displacement control is the control mode of AP and IE movement. The movement is achieved by inputting force and torque for force control, while the movement is achieved by inputting displacement and rotation angle for displacement control. ISO14243-2 mainly describes a gravimetric measurement method of volumetric wear for analyzing the wear rate of UHMWPE insert. In addition, for movement and control system of the experimental equipment, both ISO14243-1 and ISO14243-3 put forward the specific requirements. The experiment equipment should meet the requirements of ISO14243.

Figure 7.20 shows the flow diagram of in vitro wear testing of artificial knee joint. Fixtures for mounting the femoral component and tibial tray are designed and UHMWPE insert is soaked in preparatory phase. Different fixture designs are

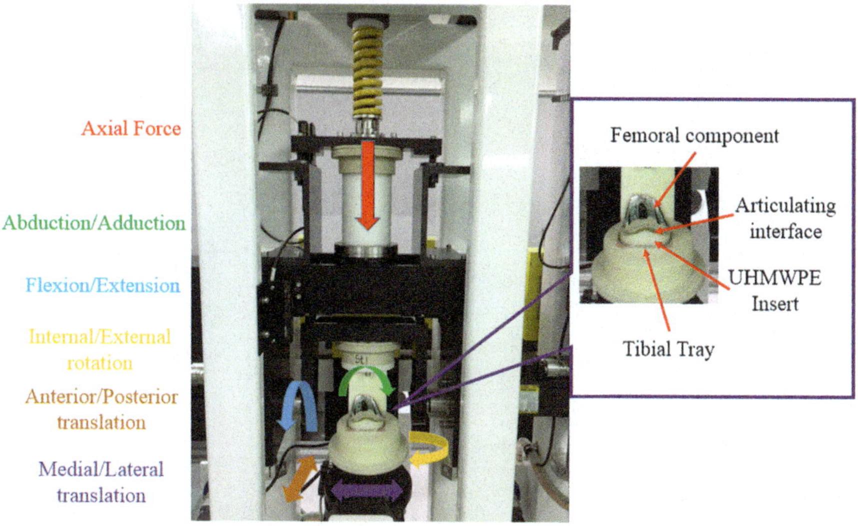

Fig. 7.18 Anatomical mounting of the TKA

Fig. 7.19 The relationship between ISO 14243-1, ISO 14243-2, and ISO 14243-3

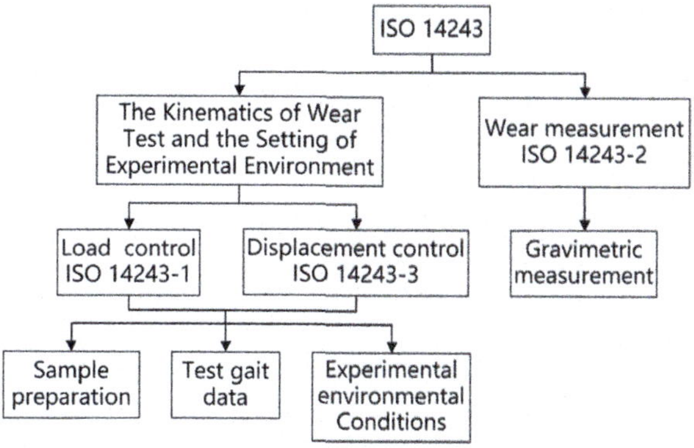

prepared for different knee prosthesis designs according to ISO14243, and the rotation center of the femoral condyle and the initial contact position of femoral condyle on bearing surface of UHMWPE insert are identified. On the other hand, if the gravimetric measurement is used to analyze the wear of the prosthesis, it is necessary to pre-soak UHMWPE insert according to ISO14243-2. Wear testing is carried out until the soaking is stable. For each type of knee prosthesis, a minimum of three test specimens and two control specimens or one loaded control specimen shall be considered in wear testing according to the recommendation of ISO14243. The purpose of the control specimens is to eliminate the effect of complex factors (like water absorption) on the quality of the insert during the experiment.

Besides the pre-market commercial quality testing, in vitro wear testing is also a critical means in the studies of product design, materials, and processing methods

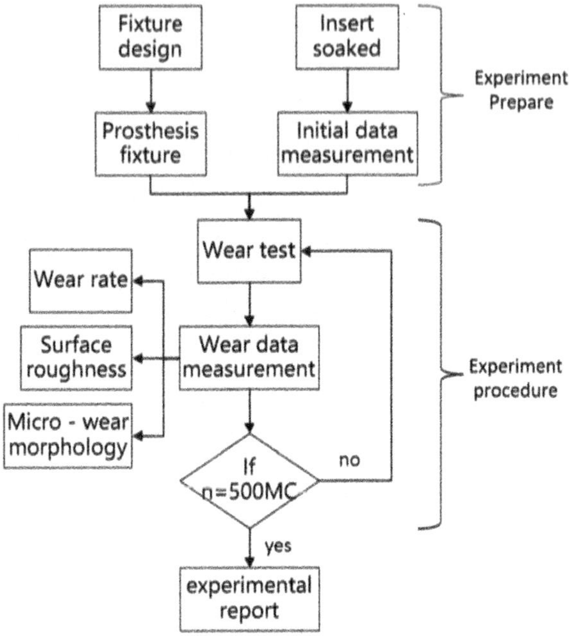

Fig. 7.20 The flow diagram of in vitro wear testing of artificial knee joint

[27, 47–51]. However, if the in vitro testing environment of artificial knee joints does not simulate the bad actual situation in human bodies, it is difficult to observe the prostheses damage such as delamination and fracture. Therefore, artificial knee joints should be tested under realistic loading condition. In order to further simulate the in vivo working environment of knee prosthesis, more harsh conditions of movement are loaded in the experiment and bone cement particles are added in the test medium to simulate the harsh third-party wear. The lubrication system is used to maintain the contact surfaces immersed in the fluid test medium. To test the long-time performance of UHMWPE insert, at least 5×10^6 cycles are carried out.

How to evaluate the volumetric wear during the wear testing? Sections 7.3.1.2 and 7.3.1.3 introduce the gravimetric method and the coordinate measurement method to evaluate the volumetric wear of the UHMWPE inserts.

7.3.1.2 Evaluation of Volumetric Wear by Gravimetric Method

Gravimetric measurement is currently considered to be the golden rule for wear measurements. ISO14243-2 describes the specific operational procedures. Before gravimetric measurement, UHMWPE insert is thoroughly cleaned using an ultrasonic cleaner and fully dried according to standard requirements. The test specimen and control specimen are stored in a dust-free space with constant temperature and humidity. The measuring accuracy is at least 0.1 mg. The weight measurement is performed until the deviation between two readings is within 0.1 mg. The change in weight of each wear specimen is corrected to minimize the measurement error resulted from fluid absorption. The weight gain of the loaded soak control is subtracted at each measurement interval. The wear rates of the inserts are calculated based on linear regression (weight change as a function of cycles) as specified in ISO 14243-2. The zero time point shall not be used in this calculation.

Gravimetric measurement is widely used due to its advantages of high repeatability, easy operation, and low cost. However, the results of gravimetric measurement can only provide information on wear. Wear patterns and linear wear depth data are not known, which hindered further analysis of the product performance. In addition, there will be various degrees of metal particles embedded in the case, which may influence the accuracy of weighing results.

7.3.1.3 Evaluation of Volumetric Wear by Coordinate Measuring Machine

In contrast to the gravimetric method, volumetric measurement is also used in wear testing studies with coordinate measuring machine (CMM). Especially, the volumetric measurement as an international standard (ISO14242-2) is widely used in the in vitro wear evaluation of artificial hip joints. The advantage of volumetric measurement is that the wear morphology on UHMWPE insert can be directly observed and measured, rather than the gravimetric measurement. However, volumetric measurement is not used to evaluate the wear of UHMWPE tibial bearing surface, primarily because of the difficulty in obtaining the volumetric data of tibial bearing surface of UHMWPE insert as limited by the complicated shape factors and the poor reliability. So, there are no international standards for the volumetric measurement method for wear testing of artificial knee joint. Although the volumetric measurement method for artificial knee joint gains the same accuracy as the gravimetric measurement, this method has not yet been widely used because of the complex measurement process and high measurement costs.

The volumetric measurement mainly includes the non-contact and the contact measurement method. It is difficult to accurately measure the volumetric changes using non-contact method because of the high X-ray transmission of UHMWPE. The contact measurement method needs to consider the complexity surface shape of UHMWPE insert, and the coordinate measurement is performed as one of the most suitable methods with high precision. To use the coordinate measurement method (Fig. 7.21), the coordinate system should be established first. It is difficult to find adequate geometric features to establish the coordinate system on the upper surface of UHMWPE insert and guarantee excellent reproducibility. If the coordinate system is built on the fixture, the jog between the prosthesis and the fixture will result in a large measurement error. According Bills et al. [52], an ideal result can be obtained when the coordinate system is established on the insert. But not all of the prosthetic designs allow sufficient places on insert for the establishment of the coordinate system. In order to establish a coordinate system directly on the insert, a secondary processing may be carried out. According to Muratoglu et al. [53], the coordinate

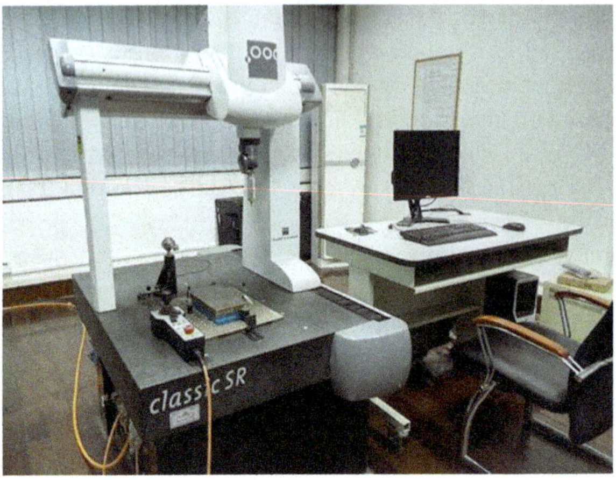

Fig. 7.21 A coordinate measurement machine

system is coincident for the measurements of non-wear status and wear status. However, due to the inevitable measurement errors, the coordinate measurement results cannot be strictly consistent for the unworn area. This makes a high requirement for the post-processing of measurement data.

The volumetric measurements include the coordinate measurement and post-processing of wear volume. The coordinate system should be built on the insert as much as possible to improve the measurement repeatability and reduce the workload of data post-processing. The UV scan command in the PC-DMIS CAD software is used to automatically generate the theoretical values of the touch path and touch vector based on the model of UHMWPE insert. In order to improve the measurement accuracy, an iterative measurement is carried out. That is, the measured values are input to the software as the theoretical value of the required data and then the secondary measurement data are compared with the calculations until satisfaction.

The coordinate data of the upper surface of UHMWPE insert after wear testing are obtained and imported into MATLAB software for wear analysis. First, the point cloud data are processed by triangulation, and the results are shown in Fig. 7.22. Then each triangular facet is connected with its projection triangle in a fixed $Z = m$ plane; the volume of this oblique truncated prism is calculated. The volume of all oblique truncated prisms is the volume of the set of point cloud data. In Figure 7.23, triangles ABC are obtained by triangulation, and the volume V of oblique truncated prism $A_mB_mC_m - ABC$ is the target value. The volume difference between the two data sets before and after wear is the volume wear.

The volume wear rates of the testing specimens are calculated based on linear regression (volume change as a function of wear cycle number). The coordinate information of each touch point can be obtained from this method. Based on these data, the linear wear depth of each part of the UHMWPE articular surface can be easily obtained and the wear pattern of the UHMWPE articular surface can be drawn.

There is creep deformation for UHMWPE insert in the wear testing process, and such creep deformation has a great impact on the wear volumetric measurement. It is

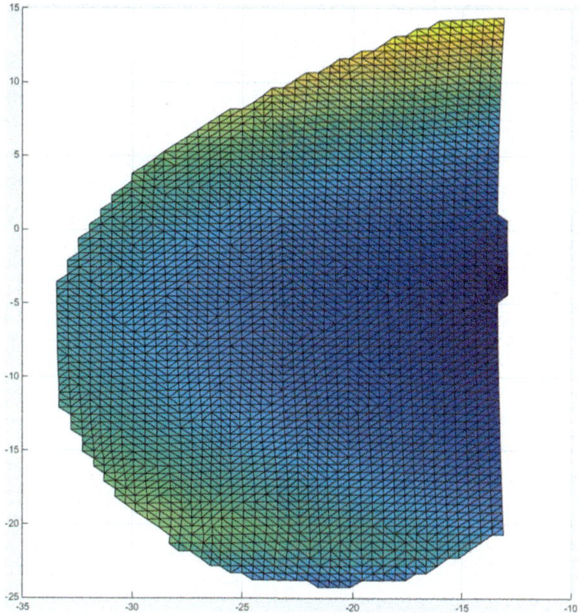

Fig. 7.22 The result of triangulation

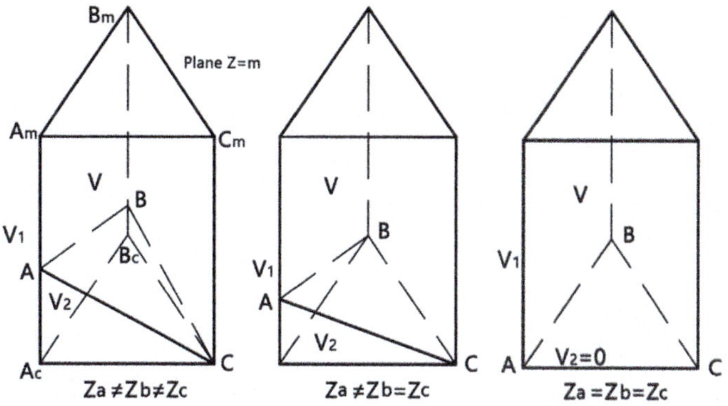

Fig. 7.23 Volume calculation model

necessary to eliminate the effect from creep deformation by comparing with control specimen. The polyethylene liner creep is mainly found for the first one million cycle of wear testing, and the creep tends to be stable after that. So, the linear regression can only be used for the measurement data from one million cycle to five million cycle or later, and the initial value in the Y axis is the creep volume [53].

7.3.2 *Wear Prediction of the UHMWPE Insert*

Knee simulators have been used extensively to investigate the wear mechanisms of polyethylene insert and pre-clinically evaluate new implant designs and bearing materials [54], but wear testing has been very expensive and time-consuming [55]. It is rare to perform wear testing for multiparameter studies. Computational wear prediction is an attractive alternative solution to the in vitro wear simulator tests. Sections 7.3.2.1 and 7.3.2.2, respectively, describe the effects of the differences in wear law and boundary condition on wear prediction. For the past few years, the backside wear of the UHMWPE insert attracts more and more attention. Section 7.3.2.3 describes the backside wear prediction of the UHMWPE insert during gait cycle.

7.3.2.1 Effect of Wear Law on Wear Prediction

The flowchart of computational wear simulation of UHMWPE insert is shown in Figure 7.24. FE contact mechanics model of artificial knee joint is built using the TKA geometries as the description of Sect. 7.2.1.2. Then the knee loading and kinematics are input to the contact mechanics model for FEA. The output contact pressure and sliding track from contact model are used to calculate the wear depth of insert node with the theoretical wear formula. The geometry of bearing surface of UHMWPE insert is updated for the new contact mechanics analysis. This wear iteration is performed until the end of simulation cycles.

As shown in Fig. 7.24, the computational wear prediction depends critically on the theoretical wear formula. The theoretical wear formula is established based on the understanding of the complex wear mechanisms of in vivo UHMWPE insert,

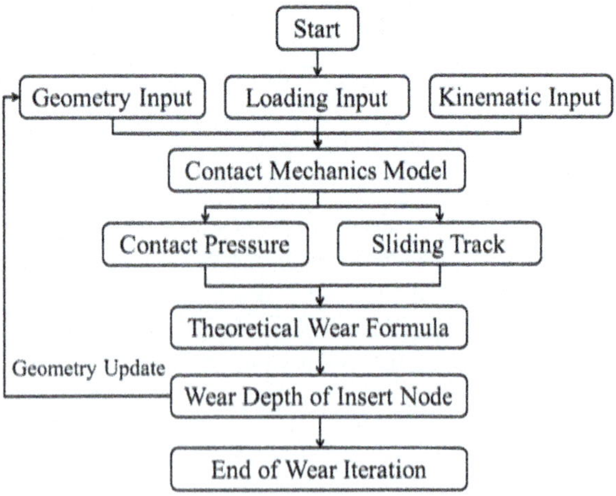

Fig. 7.24 The flowchart of computational wear simulation of UHMWPE insert

which include adhesive, abrasive, three-body, and fatigue wear [56]. Therefore, establishing a precise theoretical wear formula is the main issue of wear prediction.

Archard's wear law as the simplified theoretical wear formula is used to calculate the wear of TKA [37, 57, 58], which is based on the idea that the wear volume (V) is proportional to the load (F) and the sliding distance (S). K is a constant. The formula is described as below:

$$V = KFS \tag{7.1}$$

However, Archard's wear law is not suitable for the description of complex wear mechanisms of UHMWPE insert. Pin-on-disk (POD) tests of UHMWPE have shown that the wear rate of a multidirectional motion is significantly increased compared to a unidirectional motion [58], and it is very important for the UHMWPE wear. Kang et al. [59, 60] used the cross-shear (CS) ratio to quantify the effect of multidirectional motion on wear. In the improved wear formula, the wear factor (K) is not a constant, but the function of the average contact pressure (σ_{AVE}) and the cross-shear ratio, a, b, and c are the coefficients. The wear factor is derived as

$$K(CS, \sigma_{AVE}) = e^{(a + b\ln(CS) + c\sigma_{AVE})} \tag{7.2}$$

Although the contact pressure on bearing surface of UHMWPE insert is related with wear, some POD tests suggest that wear is independent of the contact pressure. A new wear law is proposed based on the idea that wear volume (V) is proportional to the contact area (A) and sliding distance (S) [40]. The wear coefficients (C) were determined according to a multidirectional POD experiment by Kang et al. [59, 60]. The wear formula is defined as

$$V = CAS \tag{7.3}$$

It is traditionally believed that wear is linearly proportional to sliding distance. However, POD test results by Dressler et al. [61] showed that the wear rate was highly dependent on the direction change of sliding motion but dropped to near zero following continued linear sliding. Based on the above finding, new time-dependent wear formula is established that uses an exponential function to express the relationship between wear and sliding distance. The new time-dependent wear formula is expressed as [56]:

$$V = k_M \cdot f(CS) \cdot f(P) \cdot f(S) \cdot A \tag{7.4}$$

where k_M is the final wear depth per cycle with a nominally "long" distance between cycles, $f(CS)$ is a new method to quantify the effect of cross-shear to wear, $f(P)$ is the effect of contact pressure on wear, and $f(S)$ is the effect of sliding distance on wear.

According to the previous study [62], the time-dependent wear formula is described as Eq. (7.5):

$$V = R_w\left[\left(1 - e^{(-C_w/p+B_w)/M_{mc}}\right) - \left(1 - e^{(-B_w/M_{mc})}\right)\right] \tag{7.5}$$

where C_w is the work of a shear force exerted in a single direction over a given time interval for a specific area, M_{mc} describes the mobility of the polymer chain, B_w represents the theoretical work at the time interval of the initiation CS, and R_w is the material resistance to wear.

Wang et al. [63] discovered that the volume wear per unit sliding distance (mm^3/mm) was directly proportional to the curvature of sliding circle (1/r). Curvature is the mathematical equivalence to "cross-shear". Therefore, curvature can be introduced into wear formula to quantify the contribution from multidirectional motion and sliding distance. The new wear formulas are defined as follows:

$$V = KF^{2/3}A^{1/3}S \tag{7.6}$$

and

$$V = CAS \tag{7.7}$$

where K and C are functions of curvature.

For one specific knee implant, the differences of the above wear formulas are compared. Eqs. (7.1), (7.2), (7.3), (7.4), (7.5), (7.6), and (7.7) are denoted as M1 to M7, respectively. The volumetric wear rates of all formulas are shown in Fig. 7.25. Different wear formulas result in different volumetric wear prediction. In order to get more realistic wear prediction, the wear formula should be improved based on in vivo wear mechanism of the UHMWPE insert.

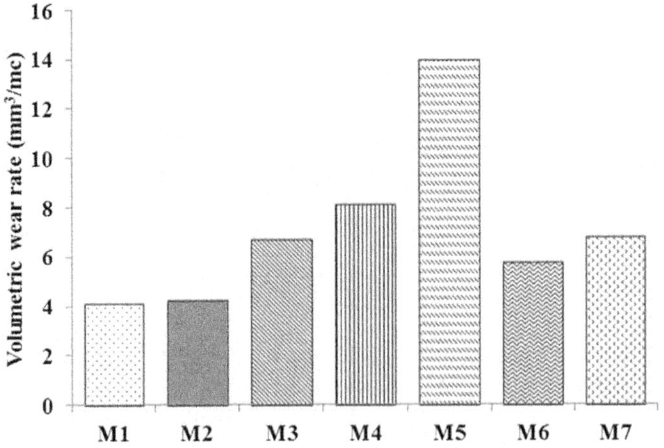

Fig. 7.25 The predicted volumetric wear rates of seven formulas for one specific knee implant after five million cycles

Comparing these seven formulas, Archard's wear law could not account for the complex variations of multidirectional movement and contact pressure. The formulas with cross-shear ratio can consider the effect of multidirectional motion and contact pressure on wear but are unable to explain the POD experiment results reported by Dressler et al. [61]. Strickland et al. [56] described the relationship between sliding distance and wear using exponential function, which provided a more suitable expression. Strickland et al. [56] believed that *CS* was a cycle-averaged measure of the variation in sliding direction, which was fundamentally a causal, time-dependent formula assumed the molecular orientation continually changed as sliding trace. The phenomenon of molecular orientation on the worn surfaces of the UHMWPE insert had been tested using knee simulators and identified by Wang et al. [63]. The chain orientation was driven by plastic strain accumulation at the wear surface by repeated cyclic asperity contact, which was a long-term multicycle result. A cycle-averaged representation is more reasonable to quantify multidirectional motion, but the time-dependent wear model only considers the effect of previous sliding track during one single cycle. The curvature is another cycle-averaged variate to quantify multidirectional motion and also supported by the POD testing results [61]. The sliding distance has a nonlinear relationship with wear rate.

7.3.2.2 Effects of Boundary Condition on Wear Prediction

Besides the wear formulas, the boundary condition is a key parameter for FEA and wear prediction. The majority of previous computational wear studies have adopted the load/motions specified in ISO 14243 (2009) as the input condition. The three mostly recognized movements of knee joint, i.e., flexion–extension, anterior–posterior translation, and internal–external rotation, are specified. As described in Sect. 7.2.2.2, these motions and axial load are applied on TKA components. The patient-specific load/kinematics plays an important role in the in vivo wear behavior. In vivo knee contact forces and joint motions have been measured using instrumented knee prostheses and fluoroscopic imaging system or predicted by MSK MBD modeling of TKA with a reasonable accuracy. It is rare to predict the wear of UHMWPE insert under the patient-specific load/kinematics condition. Moreover, it is also rare to predict the wear of UHMWPE insert by considering more realistic conditions, such as the medial and lateral TF forces and medial–lateral translation.

To compare the wear differences under different boundary conditions, three situations were specified as ISO (Model 1), separate load at the medial and lateral sites (Model 2), and medial–lateral translation (Model 3), respectively. The volumetric wear rates are shown in Fig. 7.26. Although the predicted total volumetric wear rates are similar at three different boundary conditions, the wear asymmetry of the medial and lateral compartments is different. The results may be used to explain the in vivo wear damages on the medial or lateral compartment in retrieved tibial inserts [64, 65]. Thus, the wear prediction is greatly dependent on reasonable boundary conditions.

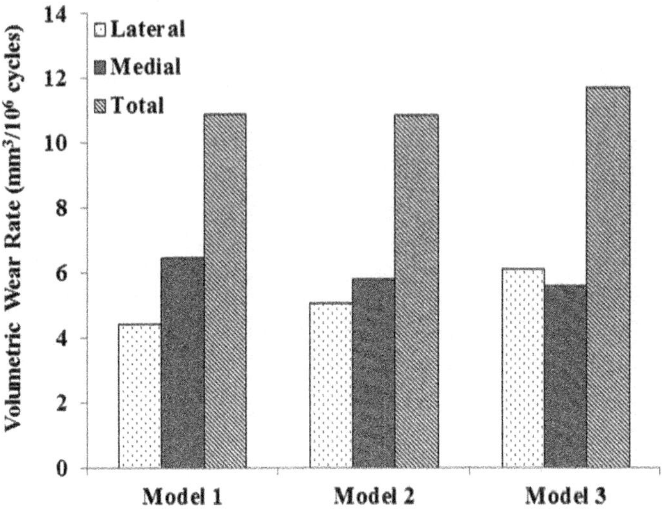

Fig. 7.26 The motion of TKA (**a**), boundary condition specified as ISO (**b**), separate load application at the medial and lateral sites (**c**), and adding medial–lateral translation (**d**)

7.3.2.3 Backside Wear Prediction of the UHMWPE Insert

Wear occurs on the articular surface and the backside surface of the UHMWPE insert for both fixed and mobile bearing knee implants. Backside wear has been shown to account for as much as 30% of total insert wear volume for some TKAs [66], which is thus considered as another important source of wear particles that cause osteolysis and loosening of artificial knee joint. UHMWPE wear particles are generated as a result of the relative movement between contacting components. The relative movement type of backside surface of fixed and mobile bearing knee implants are different. Thus, the wear mechanisms are different. The loading and motions of the femoral component relative to the insert causes the deformation of the insert and relative movement between the backside surface of the insert and the tibial tray [67]. Wear particles produced at this interface due to micro-motion are suggested to be smaller than those originated from the articulating surface [68]. Many factors influence the backside wear of fixed insert, including the contact pressure distribution, tibial tray surface finish, and tibial tray locking mechanism design, kinematics, and loading. The tibial tray surface finish and locking mechanism are most important for backside wear. Burnishing and pitting are the dominant wear damage modes on the backside surface of the UHMWPE insert. Surface deformation and scratching are also noted but few were found in retrieval inserts [69]. The mobile bearing artificial knee joint creates two surface articulations, including the articular surface of polyethylene insert against the femoral component and the backside surface against the tibial tray components. The dual articulation is

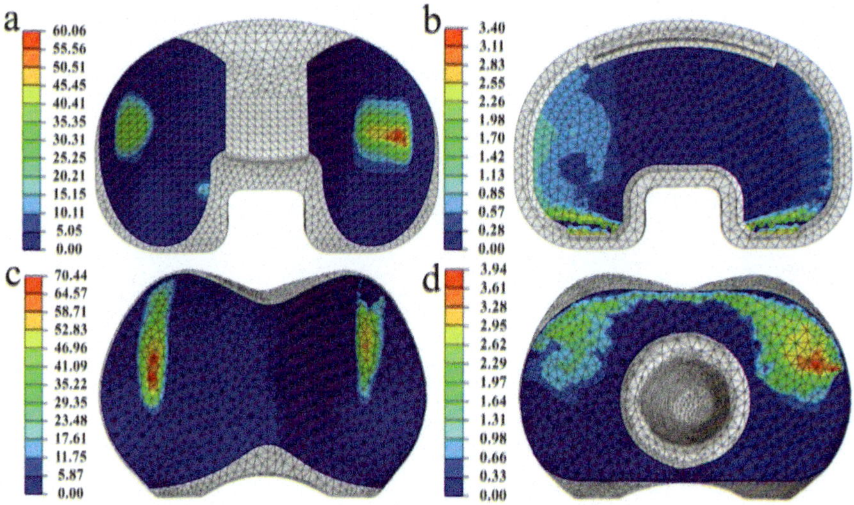

Fig. 7.27 The predicted articular and backside accumulated sliding distance (mm) for both fixed and mobile bearing inserts

designed to reduce the polyethylene wear by splitting the multidirectional motion into unidirectional motion [70]. Mobile bearing knee implants allow the rotational movement between backside surface of insert and tibial tray. It is determined that the most frequent and severe burnishing occurs close to the central peg under the femoral condyles, where most of the load is carried and most of the pitting and scratching occurred near the edges of the bearing [71]. The average volumetric wear rate of the backside surface for some retrieval insert of mobile bearing knee implants was up to 120 mm^3/year [71]. The results indicate that the volumetric wear rate is higher during the initial period for mobile bearing knee implants [71].

Backside wear prediction of the UHMWPE insert requires the information of the surface contact conditions and kinematics, which can be determined through explicit finite element simulation. A fixed and a rotating platform mobile bearing artificial knee joint are simulated for five million cycles. The contribution of backside wear to the total wear is considered. The ISO 14243-3 2009 axial load and kinematics were used to run the explicit finite element simulation. The predicted articular and backside accumulated sliding distance for both fixed and mobile bearing inserts are shown in Fig. 7.27.

The computationally predicted average wear rates using the cross-shear ratio contact pressure independent formula for the fixed and bearing insert (Fig. 7.28). The computationally predicted total average wear rate was 6.8 mm^3/million cycles for fixed bearing insert, and the backside wear rate was 0.4 mm^3/million cycles. While for mobile bearing insert, the total wear rate was 7.3 mm^3/million cycles, and the backside wear rate was 2.1 mm^3/million cycles.

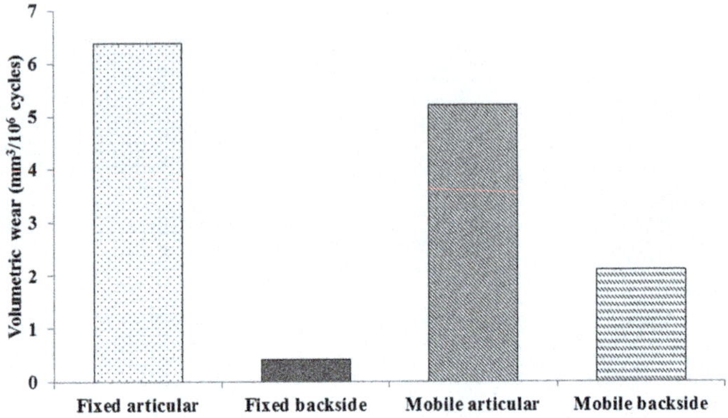

Fig. 7.28 The predicted volumetric wear for both fixed and mobile bearing inserts

The predicted backside wear of fixed bearing insert was far less than articular surface, which is different from the assessment of retrieved tibial insert [72]. The difference in wear mechanisms of articular and backside surfaces may render the difference in prediction and experimental results. Two-body rolling–sliding wear is the dominant wear mode for the articular surface of TKAs [73], while fretting wear is the dominant wear mode for the backside surface [74]. Furthermore, the wear mechanism of the articular surface is dominantly abrasive, while the wear mechanism of the backside surface is dominantly adhesive for the articulating insert against a polished tibial tray. These different wear mechanisms result in significantly different tribological conditions, which in turn need different wear formula to simulate the wear process.

The mobile bearing TKA design is assumed to have a theoretical advantage over the fixed bearing TKA design. However, this theoretical assumption is not confirmed in the wear prediction. There are no obvious advantages in the mobile bearing TKAs according to the retrieved insert study and in vitro wear testing [75, 76]. Figure 7.28 shows that around 30% of the total wear is from the contribution of the backside wear for mobile bearing TKA design. A similar contribution of the backside wear has been reported for retrieved inserts [77].

7.4 The Interaction Between the Biomechanics and Wear of Artificial Knee Joint

To date, a majority of biomechanics and wear studies of artificial knee joint are conducted separately. The knowledge of in vivo performance of UHMWPE insert is obtained from summarizing piecemeal information of biomechanics or wear of artificial knee joint. In vivo loading condition of the UHMWPE insert is complicated; the wear damage of the UHMWPE insert is from the combined action of biomechanics and tribology. However, the interaction between the biomechanics

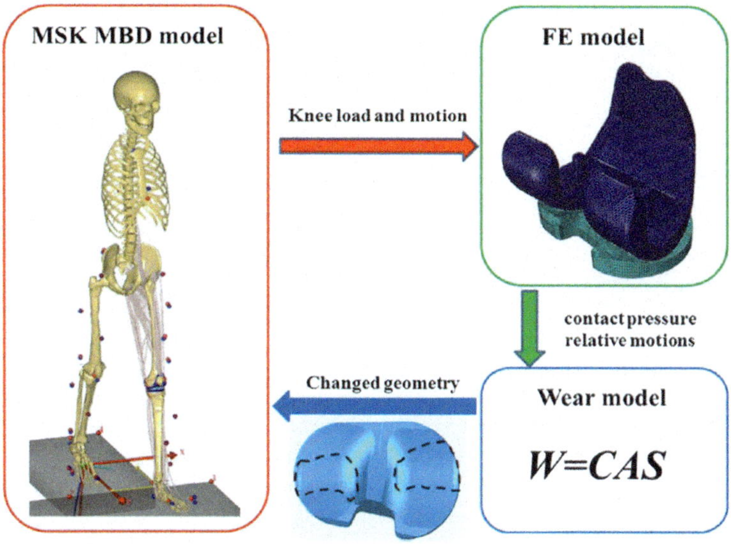

Fig. 7.29 A novel patient-specific computational framework coupling MSK MBD model, FE model, and wear model of TKA. (Reprinted from Ref. [78], Copyright 2017, with permission from Elsevier)

and wear of the UHMWPE insert has been rarely investigated directly. This section mainly describes a patient-specific wear prediction framework for an artificial knee joint with coupled MSK MBD and FEA.

Patient-specific joint dynamic loading and kinematics play an important role in vivo wear behavior of TKA. However, the ISO 14243 (2009) is used as a boundary condition in the majority of previous wear studies. Figure 7.29 shows a novel wear prediction framework by coupling a patient-specific lower extremity MSK MBD model with the FE contact mechanics and wear model of TKA [78]. Patient-specific knee joint forces and motions are input into the FE model as the boundary conditions, and the contact pressure, contact area, and contact sliding distances of the UHMWPE insert during gait cycle are calculated at each node and each time increment. Wear and creep for each node of the contact surfaces are computed using the extracted results from FEA. The worn insert geometry is subsequently imported into the MSK MBD model to obtain new patient-specific boundary conditions for next FEA.

In vivo TF contact forces and motion are influenced by articular surface wear of the UHMWPE insert, and in turn, the above variations in the knee dynamics affect the volumetric wear [78]. Compared with non-coupled model, where the boundary conditions are constant for all cycles, an increasing average contact area and average cross-shear ratio, volumetric wear and wear area, and linear wear depth were observed following the increasing service life in Figs. 7.30, 7.31, and 7.32. Taking the realistic in vivo knee joint dynamics loading conditions and the interaction

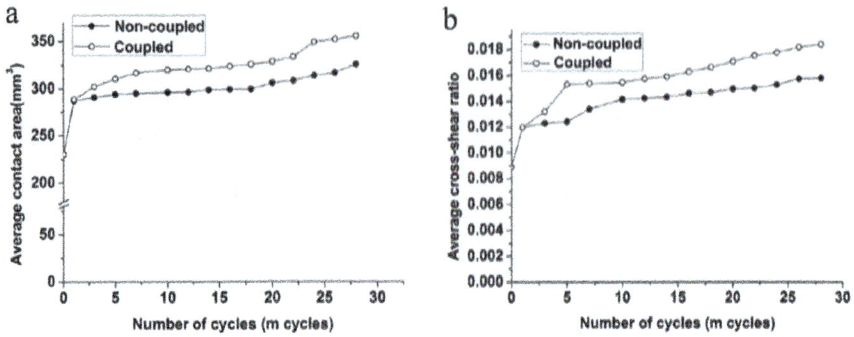

Fig. 7.30 Average contact area (**a**) and average cross-shear ratio (**b**) from wear prediction for coupled and non-coupled models at different wear stages. (Reprinted from Ref. [78], Copyright 2017, with permission from Elsevier)

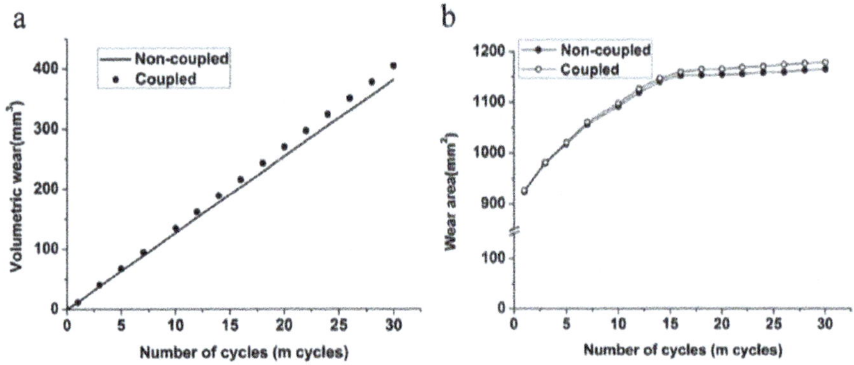

Fig. 7.31 The volumetric wear (**a**) and wear area (**b**) of coupled and non-coupled models at different wear stages. (Reprinted from Ref. [78], Copyright 2017, with permission from Elsevier)

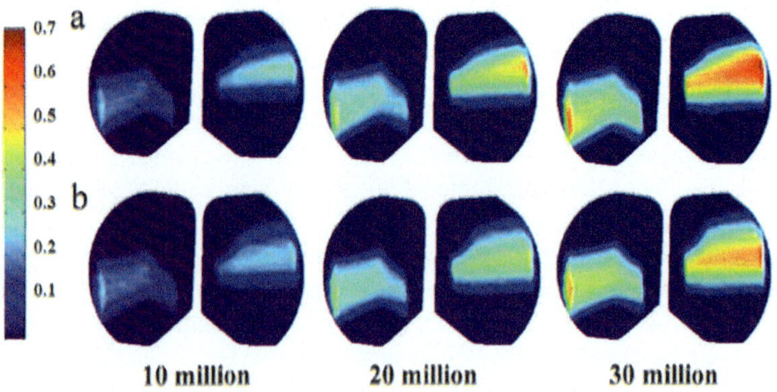

Fig. 7.32 Linear wear (mm) distribution of the insert surface for coupled model (**a**) and non-coupled model (**b**) at different wear stages. (Reprinted from Ref. [78], Copyright 2017, with permission from Elsevier)

between knee joint dynamics, contact mechanics and wear into consideration, the patient-specific coupled wear prediction framework can provide a more realistic functional outcomes of TKA.

7.5 Important Factors on Biomechanics and Wear of Artificial Knee Joint

Clinical reports demonstrate that the differences of wear failure within the TKA patient population are related to the substantial variation in biomechanics and tribology of artificial knee joint [79]. Part of this variation is due to differences in prosthetic design, surgical alignment, and patient-specific joint loading and anatomy [79]. Wear of the UHMWPE insert in TKA resulted from multiple factors. It is important to figure out the factors that are most important to TKA outcome and how these factors accommodate patient-specific variation. These findings will largely help in improving the lifetime of the UHMWPE inserts. Sections 7.5.1, 7.5.2, and 7.5.3, respectively, review the effects of prosthetic design, material, surgical alignment, and patient factors on the biomechanics and wear of artificial knee joints.

7.5.1 Effect of Prosthetic Design and Material Factors

The design factors are the primary contributors to contact mechanics of TKA. Lots of researches have performed to analyze and optimize the design parameters for TKA. The sagittal and coronal plane radii and conformity are the important parameters for implant constraint, flexion kinematics, contact locations, and wear [80, 81]. The articular geometry of artificial knee joints has a strong impact on the loading and kinematics of TKA. The articular geometry with lower contact pressure will impose more constraints on joint kinematics [82].

Figure 7.33 shows the main design parameters of femoral component and tibial insert, including femoral frontal, distal, posterior radii and width of the condylar space and tibial anterior, posterior, distal, and frontal radii, etc. The relative

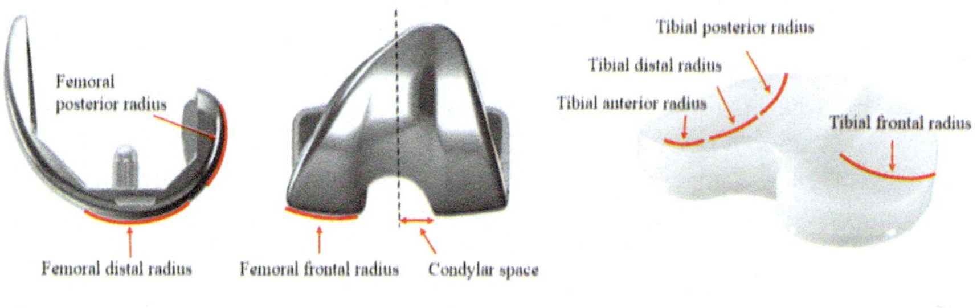

Fig. 7.33 Geometric design variables of knee components

contributions of individual parameter variables to the performance metrics were determined by Marzieh et al. [82] in terms of sensitivity indices (SI). The distal curvature of the femoral and tibial in the sagittal plane influenced both contact pressure. However, the posterior curvature of the femoral component and tibial insert has a higher impact on the internal–external range of motion (SI = 0.72; SI = 0.58) but a smaller impact on the contact pressure (SI = 0.31; SI = 0.23). It is noteworthy that the frontal radius of the femoral component impact both contact pressure (SI = 0.38) and internal–external range of motion (SI = 0.35).

The contact mechanics and kinematics of TKA are simultaneously influenced by the key design parameters of femoral component and tibial insert. Marzieh et al. [82] found that the kinematics of TKA were more sensitive to the femoral posterior radius than to the contact pressure (Fig. 7.34). The femoral posterior radius can be modified to obtain a reduction in the conformity of TKA, which will help enhance the kinematical function of artificial knee joint. Meanwhile, the adverse effect of femoral posterior radius on the contact pressure needs to be tolerated. For the tibial frontal radius, the contact pressure is more sensitive than the kinematics of TKA. An increasing tibia frontal radius may reduce the contact pressure with the minimum adverse effect on the kinematics of TKA. The findings [82] can be used to design the new patient-specific knee prosthesis. When a higher knee kinematical function is demanded by young patients, the femoral posterior, tibial posterior and tibial anterior radii should be designed with less constraint. Because the femoral posterior, tibia posterior and tibial anterior radii have more influence on the internal-external rotation than the contact pressure. When an artificial knee joint is designed for an elderly patient with less physical activity, a more constraint geometry design should be considered based on the change in the tibial frontal radius with the minimum adverse influence on the kinematics.

For posterior cruciate substituting knee implants, the post-cam contact mechanism of tibial insert and femoral component is designed to replace the posterior cruciate ligament for guiding the anterior–posterior motion of artificial knee joint. However, severe wear, early revision due to wear, and fracture of the tibial post have been frequently discussed in clinic. These wear and fracture failures of the UHMWPE post are highly relative to the contact mechanics of post-cam engagement. Fitzpatrick et al. [83] quantified the contact mechanics of post-cam engagement of eight UHMWPE insert designs during the simulated dynamic activity. A considerable range was observed for knee flexion angle and post-cam velocity at engagement (23°–89° and 0.05–0.22 mm/°, respectively) among current eight TKA designs [83]. In vivo performance of the UHMWPE post is greatly dependent on the variation of post-cam designs. According to this analysis reported by Fitzpatrick et al. [83], the design of posterior cruciate substituting knee implants should adopt smooth post-cam engagement in order to reduce edge loading of the post, because a less edge loading can improve the wear resistance of the post. In addition, condylar geometry of the articular bearing surface plays a significant role in minimizing engagement velocity, forces, and stresses in the post [83].

There are many styles of patellar designs commercially available for TKA, including a domed patella (DePuy, Smith & Nephew, Stryker, Biomet, Exactec); an off-set dome patella (Smith and Nephew, Stryker); an uncemented metal-backed

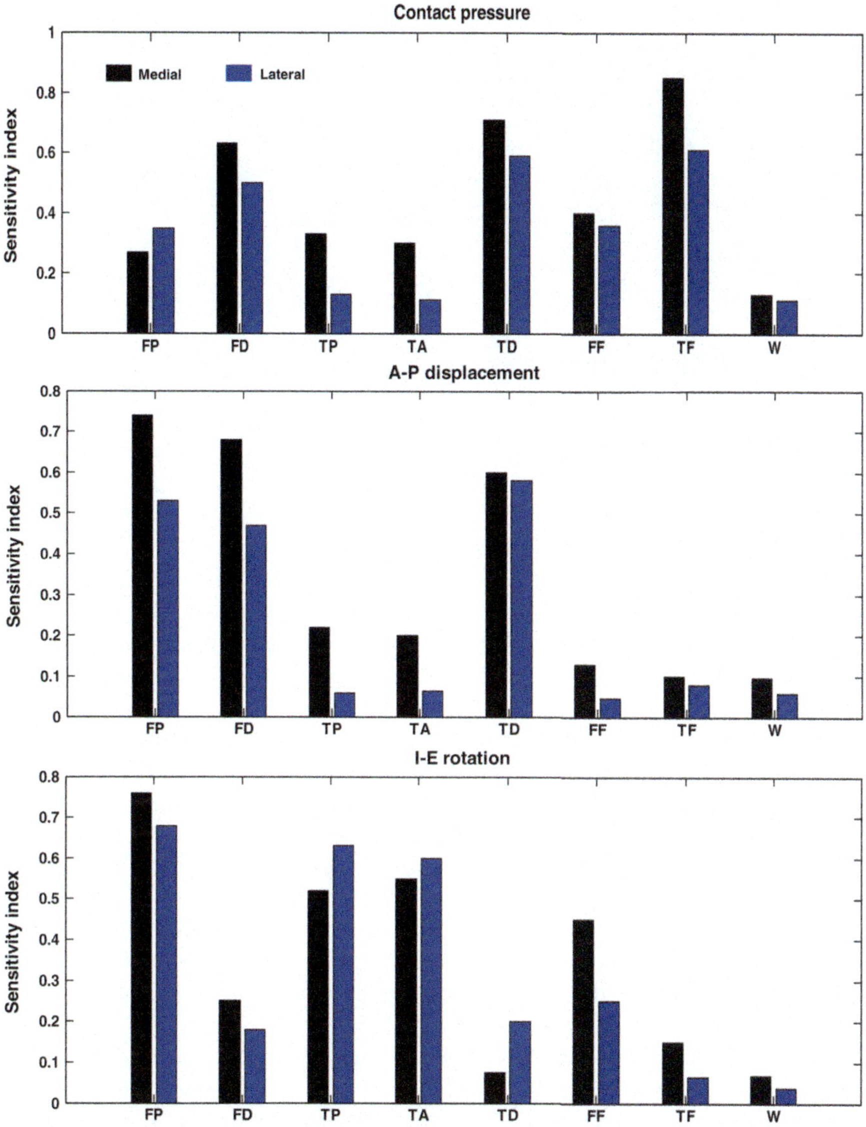

Fig. 7.34 Sensitivity of TKA function due to individual geometric variables. *FP* Femoral posterior, *FD* Femoral distal, *TP* Tibial posterior, *TA* Tibial anterior, *TD* Tibial distal, *FF* Femoral frontal, *TF* Tibial frontal, *W* Condylarspace. (Reprinted from Ref. [82], Copyright 2015, with permission from Elsevier)

dome patella (Biopro); a modified dome (sombrero-style) patella, with either a centralized or medialized apex (DePuy, Stryker, Zimmer); a cylindrical patella (Finsbury); an anatomical fixed bearing patella (DePuy); and an uncemented anatomic rotating platform patella (Endotec). Especially, the dome designs of the

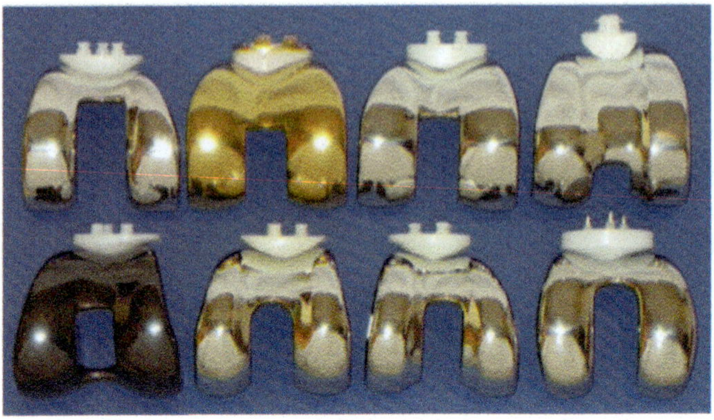

Fig. 7.35 Various femoral arthroplasty components with their respective, designated patellar implant. (Reprinted from Ref. [85], Copyright 2012, with permission from Springer)

patellar UHMWPE component are widely developed. Fitzpatrick et al. [84] investigated the changes in the PF mechanics for three patellar designs (dome patella, modified dome patella, and anatomic patella) under the alignment parameters. The predicted range of motion of the internal–external rotation indicates that the dome patellar design produces more internal–external rotations than the modified dome patellar designs and anatomic patellar designs [84]. For the dome patellar design, the contact mechanics and internal stresses had the least amount of variation with respect to alignment and loading variability [84]. However, while wide bounds of variability were predicted, the anatomic patellar design had a lower contact pressure and a greater contact area than the dome patellar design and modified dome patellar design [84]. Compared with the dome patellar design, the modified dome patellar design had significantly reduced the peak medial–lateral contact force from 141 N to 29 N [84]. The least amount of variation in patellar flexion–extension of 39° was observed for the anatomic patellar design during the cycle, and less variation of 47° was observed for the modified dome patellar design compared with a flexion–extension range of 53° for the dome patellar design [84].

The conformity of the bearing surfaces is always one focus in artificial knee joint design. The insert conformity may affect the wear of TKAs. The conforming designs may reduce the contact pressure on the articulating surface and the structural wear. However, a less conforming design may reduce fatigue wear when the provided contact pressure does not exceed the fatigue limit of the UHMWPE material. Abdelgaied et al. [86] investigated the effect of different insert conformities with the femoral component on the wear in TKAs (Fig. 7.35). The less conforming geometries produced lower wear, which can be explained by the change in cross-shear and contact area. The higher the joint conformity, the higher the contact area is. The contact area and cross-shear are the dimensionless wear coefficient. The relationship between the contact area and the volumetric wear indicated that the lower the contact area, the lower is the wear under the same kinematic conditions.

On the other hand, a higher articular surface conformity may render a higher cross-shear, which may lead to higher wear. The contact radius is increased by increasing the articular surface conformity between the femoral component and the insert in TKA, which results in the higher cross-shears. The predicted average wear rates for the curved insert with highest conformity level were more than three times that for the custom flat insert with the lowest conformity level [86]. When the contact pressure is kept at the strength limit, the lower conformity level results in the reduction in the contact area and eventually reduces wear [86]. A balance between the fatigue strength limit and the wear requirements can be achieved by optimizing the articular surface conformity level [86].

Material is also considered in the design of artificial knee joints. Cross-linking has been introduced to improve the wear and creep resistance of conventional UHMWPE [87]. The wear debris size is reduced along with the improvement of UHMWPE wear resistance through cross-linking. A smaller wear debris size may increase the risk of osteolysis. It is necessary to find a compromise between wear properties, mechanical properties, and wear debris size so that a proper degree of cross-linking should be selected for UHMWPE materials. Abdelgaied et al. [86] investigated the effect of the insert bearing material on the wear rate. Moderately cross-linked UHMWPE and conventional UHMWPE were compared for the DePuy TKA Sigma implant. The moderately cross-linked UHMWPE bearing led to wear rates lower than the corresponding conventional UHMWPE bearings. The predicted wear for all insert designs were reduced up to 50% with moderately cross-linked UHMWPE with respect to that made of conventional UHMWPE [86].

In addition, the material of femoral component also influences the wear property of the insert component. Ueno et al. [91] compared the wear properties of tibial insert in TKA using different femoral component materials and found that, compared with Co–Cr alloy, both alumina ceramics and zirconia are expectable materials for femoral component to remarkably reduce the generation of polyethylene wear debris in TKA.

7.5.2 *Effect of Surgical Alignment Factors*

The malalignment of components in TKA is currently a major cause for postoperative implant loosening or wear. According to the retrospective results of 820 consecutive revision TKAs, Dalury et al. [92] found that the component malalignment is the seventh reason for TKA revision. The component malalignment also affects joint loading, component loosening, and wear, which usually causes revision. Therefore, the effect of component malalignment on TKA revision is more than the retrospective analyses. As shown in Fig. 7.36, the component malalignment may result in localized mechanical overload in the medial or lateral condyle of TKA with increased severe wear of the medial or lateral compartment of the UHMWPE insert, induction of bone cement cracks, prosthetic bone bed damage, component loosening, and ultimately revision surgery [93, 94]. There are approximately 10–30% of

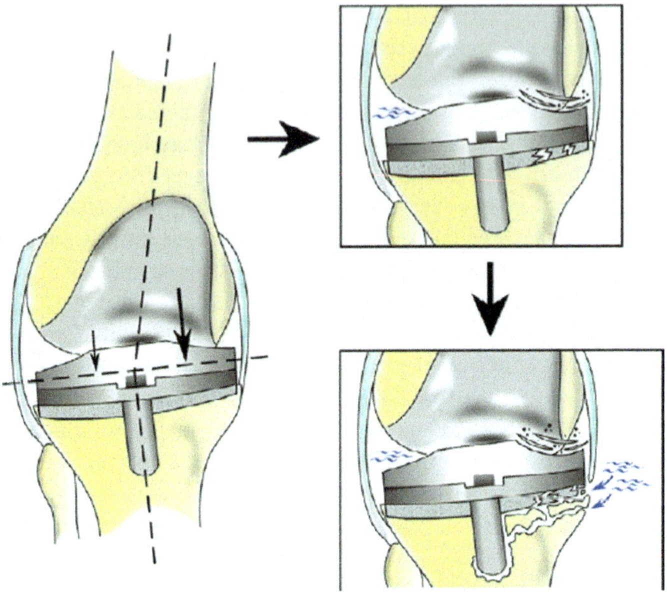

Fig. 7.36 Schematic diagram of the effects of TKA malalignment. Component malalignment may lead to unilateral mechanical overload, aggravate polyethylene wear, causc bone cement crack, bone damage, and the prosthetic loosening. (Reprinted from Ref. [4], Copyright 2013, with permission from Elsevier)

patients with TKA who suffer from a measurable component malalignment [95]. The proportion of the patients with more than ±3° coronal malalignment are approximately 28% even for experienced surgeons [96]. Such deviations in the coronal alignment of the tibial component may lead to early failure and the acceleration of UHMWPE wear [97]. Meanwhile, the risk of medial bone collapse and component loosening failure may be increased [98].

Chen et al. [99] performed a parametric study of the effect of the component malrotation in TKA on the knee loading using a validated MSK MBD model of TKA under a simulated walking gait. For the femoral component, the knee contact forces are more sensitive to variations in the varus–valgus rotation and the internal–external rotation, while for the tibial components, the knee contact forces are more sensitive to variations in the varus–valgus rotation [99]. These varus–valgus and internal–external malrotations of the tibial or femoral component with a 5° variation change the peak medial, lateral, and total contact forces by 17.8–53.1%, 35.0–88.4%, and 5.2–18.7%, respectively [99].

The component malalignments in TKA are the primary causes to the variations in the muscle or ligament forces, where the component malalignments influence the muscle or ligament moment arms [99]. This influence, in turn, eventually results in the changes in the knee load distribution. Chen et al. [99] found that the peak muscle forces of *vastus medialis*, *vastus lateralis*, *medial gastrocnemius*, and *peroneus longus* are obviously increased at a 5° varus alignment of the femoral component [99]. The peak muscle forces of *lateral gastrocnemius*, *tibialis anterior*, and *soleus* are increased, while the peak muscle forces of *medial gastrocnemius* and *peroneus longus* are decreased to zero at a 5° valgus alignment of the femoral component [99]. The changes of the muscle forces eventually alter the TF medial–lateral loading distribution. The external malrotation of the femoral component results in a higher LCL force, whereas the internal malrotation of the femoral component increases the quadriceps and the MCL forces. The tightened quadriceps and MCL are related to the clinically reported unsatisfactory pain. The changes in knee contact forces are resulted from the unbalanced soft tissue loading, especially as the knee moves from flexion into extension. Moreover, the variations in the contact position resulted from component mal-rotation further influence the knee contact mechanics.

Surgeons should cautiously avoid the malrotation of components over 3° variations [99]. The varus–valgus and internal–external malrotation of the femoral component with 3° variations in angulation may markedly change the TF medial–lateral force distribution and total TF contact force. The varus–valgus malrotation of the tibial insert also contributes to the changes in knee contact forces. The investigations by Chen et al. [99] are helpful to understand the relationship between surgical parameters and knee joint biomechanics in TKA and eventually help improve patient satisfaction. These findings are consistent with the clinical observations that the medial bone collapse may result from a varus malrotation greater than 3° and that the unsatisfactory pain levels may result from an internal malrotation greater than 3° of the femoral component [99]. Moreover, as shown in Figure 7.37, the change in the TF meidal contact force was more than 10% for the varus-valgus mal-rotation of the femoral component with as small as a 2° variations in angulation. The varus-valgus mal-rotation of the tibial component with as small as a 2° variations in angulation caused a similar change in the TF meidal contact force.

Besides the component alignment, TKA outcomes may also be influenced by the soft tissues around the knee joint. For patients after TKA, Silva et al. [100] found an

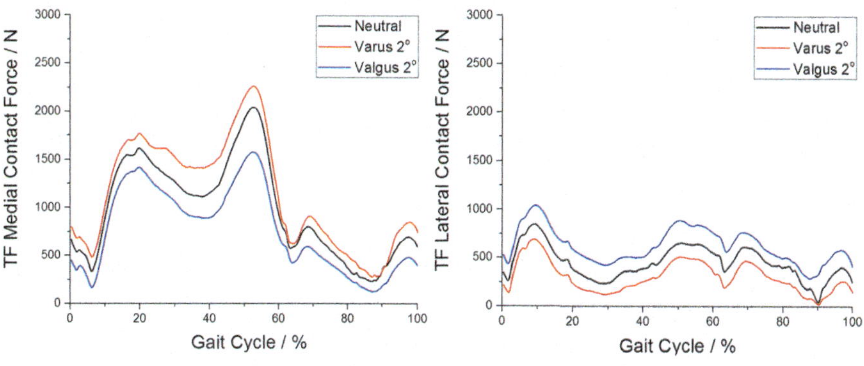

Fig. 7.37 Effects of the femoral varus-valgus mal-rotation of 2° in TKA on the TF medial and lateral contact forces

average of 31% strength reduction for isometric knee flexion and extension and up to 40% reduction for full extension. The variation in muscle strength after TKA can influence the knee contact mechanics. When a patient-specific MSK MBD model of TKA is developed, the effect of the postoperative strength reduction of the soft tissues around the knee joint should be considered in MSK modeling. Marra et al. [19] decreased the knee strengths by 35% to account for the TKA postoperative weakening, which improved the knee contact force prediction. Chen et al. [18] adopted a 35% reduction of the knee flexion/extension muscle strength and found that the predicted RMSE was decreased by 11% and the magnitude error was also decreased from 0.11 to 0.04 for the prediction of medial contact force. Although the RMSE was not improved for the prediction of the lateral contact force, the magnitude error was decreased from 0.32 to −0.04. For the prediction of the total TF contact force, the RMSE was decreased by 31% and the magnitude error was decreased from 0.17 to 0.01. Therefore, the effects of TKA surgery on soft tissues are not negligible for developing and researching artificial knee joint.

7.5.3 Effect of Patient Factors

The patient's gait patterns are characterized by the movement patterns of human limbs, which are potentially affecting the joint biomechanics. Marzieh et al. [101] found that the knee adduction moment is highly sensitive to the changes in pelvis rotation, hip abduction, and knee flexion; the knee flexion moment is highly sensitive to the changes in pelvis rotation, hip flexion, and knee flexion, and the knee rotation moment is highly sensitive to changes in hip abduction, hip flexion, and knee flexion. Therefore, the changes in joint kinematics resulted from different gait patterns may influence the joint loading and eventually the UHMWPE wear. The "cause-and-effect" relationships between joint kinematics and the joint moments play an important role in evaluating the performance of the UHMWPE insert in TKA.

Fregly et al. [102] suggested that there were significant differences in the knee contact mechanics between patients with different gait patterns. Fregly et al. [102] found that a medial thrust gait could produce 16% reduction in medial contact force over stance phase than the patient's normal gait pattern. When a hiking pole was used during walking, the medial contact force could be reduced by 27% in over stance phase compared to the patient's normal gait pattern, and significant 11% and 21% reductions were produced in the lateral and total contact force [102]. The results of the walking pole gait indicated that the patients with TKA can reduce medial, lateral, and total contact force using a walking pole. The changes in knee loading from patient's gait pattern may ultimately lead to different durability of the UHMWPE insert.

Gait modification as a postoperative rehabilitation strategy of TKA is frequently discussed in clinic, which aims to change knee joint loading by altering the patient's walking patterns with minor changes in gait kinematics [103]. The knee joint loading may be changed to minimize UHMWPE wear and prolong the durability of artificial knee joint [103]. Marzieh et al. [104] used a neural network–genetic algorithm approach to predict a gait pattern that would minimize the contact pressure on the bearing surfaces of the UHMWPE insert. Compared to the available gait rehabilitations, the optimized gait pattern could decrease the knee contact pressure by up to 25% [104]. The optimized gait pattern can be used to guide the patient's postoperative rehabilitation.

Excessive body weight has been linked to degenerative disease of the knee [105]. A large proportion of patients after TKA are still obese or morbidly obese [105]. Foran et al. found obesity has a negative impact on the results of TKA and obese patients had lower postoperative Knee Society scores. Obesity may lead to a greater wear due to a larger loading on articular surface of the UHMWPE insert. The bearing surface wear of the tibial insert is directly related to the contact mechanics and kinematics of knee joint, while joint forces and motion are influenced by the patient's body mass index. Knee contact forces will be increased with a 2.5-fold increment for every increase in unit BW during the same walking gait pattern [106]. Harding et al. [107] also found that a higher body mass index was associated with tibiofemoral compressive and shear contact forces as well as muscle forces. When the knee prosthesis is implanted with poor positioning and inadequate balance of soft tissues, the adverse effect of obesity on knee biomechanics will be exaggerated. The survivorship of artificial knee joint will be further diminished.

In addition, activity levels of walking of patients with artificial knee joints are extremely variable. Lavernia et al. [108] found that the activity level of walking was statistically and significantly correlated with deformation and creep in the UHMWPE inserts of TKAs. Creep is a cold flow process of UHMWPE materials that accompanies the wear at early-stage post-implantation [108]. The relationship between the amount of wear observed clinically and the activity level should not be ignored. A young patient with a higher activity level usually requires a longer expected lifetime of artificial knee joint, which remains a significant challenge for contemporary artificial knee joints.

7.6 Summary

To date, UHMWPE is still the best popular bearing material of TKA. Although wear and damage of the UHMWPE components limit the long-term performances of artificial knee joint, there are currently no clinically acceptable alternatives to UHMWPE material.

It is advantageous to improve the wear resistance of UHMWPE components for reducing the risk of osteolysis and aseptic loosening. Understanding in vivo biomechanics and tribology of artificial knee joint is the premise to improving wear and mechanical performance of the UHMWPE components. Combined patient-specific musculoskeletal multibody dynamics and FEA provide a reliable virtual platform for investigating and evaluating the wear and mechanical performance of the UHMWPE components. The in vivo interaction and compromise between the mechanics, kinematics, and wear should be considered as a basis for obtaining higher performances of the UHMWPE components.

The biomechanics and wear of artificial knee joints are influenced by the prosthetic design, surgical alignment, and patient's character factors. In the future, in addition to the improvement of the UHMWPE material performance, patient-specific prosthetic design, surgery, and rehabilitation guidance will be the research directions for improving the in vivo wear performance of the UHMWPE components in TKA.

References

1. Carr BC, Goswami T (2009) Knee implants – review of models and biomechanics. Mater Des 30(2):398–413
2. Abdelgaied AAY (2012) Computational wear modelling of knee joint replacements. The University of Leeds
3. Robertsson O, Sundberg M, Lidgren L, W-Dahl A (2016) Annual report 2016 – the Swedish knee arthroplasty register
4. Gallo J, Goodman SB, Konttinen YT, Wimmer MA, Holinka M (2013) Osteolysis around total knee arthroplasty: a review of pathogenetic mechanisms. Acta Biomater 9(9):8046–8058
5. Kurtz S, Ong K, Lau E, Mowat F, Halpern M (2007) Projections of primary and revision hip and knee arthroplasty in the United States from 2005 to 2030. J Bone Joint Surg 89(4):780–785
6. GBCP E, Bryant JT, Hansson CM, Rudan J, Kennedy LA, Cooke TDV (1995) A quantitative technique for reporting surface degradation patterns of UHMWPE components of retrieved total knee replacements. J Appl Biomater 6(1):9–18
7. Bonaspetti G, Fabio SD, Fenu R, Costa L, Pazzaglia UE (2005) Polyethylene wear in primary knee arthroplasty. J Orthop Traumatol 6(1):15–20
8. Chen Z, Zhang X, Ardestani MM, Wang L, Liu Y, Lian Q, He J, Li D, Jin Z (2014) Prediction of in vivo joint mechanics of an artificial knee implant using rigid multi-body dynamics with elastic contacts. Proc Inst Mech Eng H J Eng Med 228(6):564–575
9. Heinlein B, Kutzner I, Graichen F, Bender A, Rohlmann A, Halder AM, Beier A, Bergmann G (2009) ESB clinical biomechanics award 2008: complete data of total knee replacement loading for level walking and stair climbing measured in vivo with a follow-up of 6–10 months. Clin Biomech (Bristol, Avon) 24(4):315
10. D'Lima DD, Patil S, Steklov N, Chien S, Colwell CW Jr (2007) In vivo knee moments and shear after total knee arthroplasty. J Biomech 40:S11–S17
11. Kozanek M, Hosseini A, Liu F, Van de Velde SK, Gill TJ, Rubash HE, Li G (2009) Tibiofemoral kinematics and condylar motion during the stance phase of gait. J Biomech 42 (12):1877–1884
12. M AS PB, SK S SY, S M CA (2016) Changes in knee kinematics following total knee arthroplasty. Proc Inst Mech Eng H J Eng Med 230:265–278
13. Varadarajan KM, Moynihan AL, D'Lima D, Colwell CW, Li G (2008) In vivo contact kinematics and contact forces of the knee after total knee arthroplasty during dynamic weight-bearing activities. J Biomech 41(10):2159–2168
14. Ji SJ, Zhou YX, Jiang X, Cheng ZY, Wang GZ, Ding H, Yang ML, Zhu ZL (2015) Effect of joint line elevation after posterior-stabilized and cruciate-retaining Total knee arthroplasty on clinical function and kinematics. Chin Med J 128(21):2866–2872

15. Chen Z, Jin Z (2016) Prediction of in-vivo kinematics and contact track of total knee arthroplasty during walking. Biosurface Biotribol 2:86–94
16. Delp SL, Anderson FC, Arnold AS, Loan P, Habib A, John CT, Guendelman E, Thelen DG (2007) OpenSim: open-source software to create and analyze dynamic simulations of movement. IEEE Trans Biomed Eng 54(11):1940–1950
17. Andersen MS, Rasmussen J (2011) Total knee replacement musculoskeletal model using a novel simulation method for non-conforming joints. In: Paper presented at the International Society of Biomechanics conference, Brussels
18. Chen Z, Zhang Z, Wang L, Li D, Zhang Y, Jin Z (2016) Evaluation of a subject-specific musculoskeletal modelling framework for load prediction in total knee arthroplasty. Med Eng Phys 38(8):708–716
19. Marra MA, Vanheule V, Fluit R, Koopman BHFJM, Rasmussen J, Verdonschot N, Andersen MS (2015) A subject-specific musculoskeletal modeling framework to predict in vivo mechanics of total knee arthroplasty. J Biomech Eng 137(2):020904
20. Fregly BJ, Besier TF, Lloyd DG, Delp SL, Banks SA, Pandy MG, D'Lima DD (2012) Grand challenge competition to predict in vivo knee loads. J Orthop Res 30(4):503–513
21. Pellikaan P, van der Krogt MM, Carbone V, Fluit R, Vigneron LM, Van Deun J, Verdonschot N, Koopman HF (2014) Evaluation of a morphing based method to estimate muscle attachment sites of the lower extremity. J Biomech 47(5):1144–1150
22. Andersen MS, Damsgaard M, MacWilliams B, Rasmussen J (2010) A computationally efficient optimisation-based method for parameter identification of kinematically determinate and over-determinate biomechanical systems. Comput Methods Biomech Biomed Engin 13 (2):171–183
23. Blankevoort L, Kuiper J, Huiskes R, Grootenboer H (1991) Articular contact in a three-dimensional model of the knee. J Biomech 24(11):1019–1031
24. Fregly BJ, Bei YH, Sylvester ME (2003) Experimental evaluation of an elastic foundation model to predict contact pressures in knee replacements. J Biomech 36(11):1659–1668
25. Schwer LE (2007) Validation metrics for response histories: perspectives and case studies. Eng Comput 23(4):295–309
26. DesJardins JD, Banks SA, Benson LC, Pace T, LaBerge M (2007) A direct comparison of patient and force-controlled simulator total knee replacement kinematics. J Biomech 40 (15):3458–3466
27. McEwen H, Barnett P, Bell C, Farrar R, Auger D, Stone M, Fisher J (2005) The influence of design, materials and kinematics on the in vitro wear of total knee replacements. J Biomech 38 (2):357–365
28. Turell M, Wang A, Bellare A (2003) Quantification of the effect of cross-path motion on the wear rate of ultra-high molecular weight polyethylene. Wear 255(7–12):1034–1039
29. Walker PS, Arno S, Borukhoy I, Bell CP (2015) Characterising knee motion and laxity in a testing machine for application to total knee evaluation. J Biomech 48(13):3551–3558
30. Bingham JT, Papannagari R, Velde SKVD, Gross C, Gill TJ, Felson DT, Rubash HE, Li G (2008) In vivo cartilage contact deformation in the healthy human tibiofemoral joint. Rheumatology 47(11):1622–1627
31. Harris M, Morberg P, Bruce W, Walsh W (1999) An improved method for measuring tibiofemoral contact areas in total knee arthroplasty: a comparison of K-scan sensor and Fuji film. J Biomech 32(9):951–958
32. Liau J-J, Cheng C-K, Huang C-H, Lo W-H (2002) Effect of Fuji pressure sensitive film on actual contact characteristics of artificial tibiofemoral joint. Clin Biomech 17(9–10):698–704
33. Godest AC, Beaugonin M, Haug E, Taylor M, Gregson PJ (2002) Simulation of a knee joint replacement during a gait cycle using explicit finite element analysis. J Biomech 35(2):267–275
34. Willing R, Kim IY (2009) Three dimensional shape optimization of total knee replacements for reduced wear. Struct Multidiscip Optim 38(4):405–414
35. Cripton PA (1993) Compressive characterization of ultra high molecular weight polyethylene with applications to contact stress analysis of total knee replacements. Master of Science thesis, Queen's University, Kingston, Ontario

36. D'Lima DD, Steklov N, Fregly BJ, Banks SA, Colwell CW Jr (2008) In vivo contact stresses during activities of daily living after knee arthroplasty. J Orthop Res 26(12):1549–1555. https://doi.org/10.1002/jor.20670
37. Fregly BJ, Sawyer WG, Harman MK, Banks SA (2005) Computational wear prediction of a total knee replacement from in vivo kinematics. J Biomech 38(2):305–314. https://doi.org/10.1016/j.jbiomech.2004.02.013
38. Wang A, Essner A, Polineni VK (1998) Lubrication and wear of ultra-high molecular weight polyethylene in total joint replacement. Tribol Int 31(1–3):17–33
39. Wang A (2001) A unified theory of wear for ultra-high molecular weight polyethylene in multi-directional sliding. Wear 248(1–2):38–47
40. Abdelgaied A, Liu F, Brockett C, Jennings L, Fisher J, Jin Z (2011) Computational wear prediction of artificial knee joints based on a new wear law and formulation. J Biomech 44 (6):1108–1116
41. Muratoglu OK, Bragdon CR, Jasty M, O'Connor DO, Von Knoch RS, Harris WH (2004) Knee-simulator testing of conventional and cross-linked polyethylene tibial inserts. J Arthroplast 19(7):887–897
42. Ezzet KA, Hermida JC, Colwell CW Jr, D D'Lima D (2004) Oxidized zirconium femoral components reduce polyethylene wear in a knee wear simulator. Clin Orthop Relat Res 428:120–124
43. Barnett PI, Fisher J, Auger DD, Stone MH, Ingham E (2001) Comparison of wear in a total knee replacement under different kinematic conditions. J Mater Sci Mater Med 12(10–12):1039–1042
44. Kakuta T, Takashima Y, Sano T, Nakamura T, Kobayashi Y, Yamaguchi H, Harada A (2015) Adhesion between semihard polymer materials containing cyclodextrin and adamantane based on host–guest interactions. Macromolecules 48(3):732–738. https://doi.org/10.1021/ma502316d
45. Mihajlovic M, Staropoli M, Appavou M-S, Wyss HM, Pyckhout-Hintzen W, Sijbesma RP (2017) Tough supramolecular hydrogel based on strong hydrophobic interactions in a multiblock segmented copolymer. Macromolecules 50(8):3333–3346. https://doi.org/10.1021/acs.macromol.7b00319
46. Algi MP, Okay O (2014) Highly stretchable self-healing poly(N,N-dimethylacrylamide) hydrogels. Eur Polym J 59:113–121. https://doi.org/10.1016/j.eurpolymj.2014.07.022
47. Zietz C, Reinders J, Schwiesau J, Paulus A, Kretzer JP, Grupp T, Utzschneider S, Bader R (2015) Experimental testing of total knee replacements with UHMW-PE inserts: impact of severe wear test conditions. J Mater Sci Mater Med 26(3):1–13
48. Affatato S, Bracco P, Sudanese A (2012) In vitro wear assessments of fixed and mobile UHMWPE total knee replacement. Mater Des 48(2):44–51
49. Bracco P, Oral E (2011) Vitamin E-stabilized UHMWPE for total joint implants: a review. Clin Orthop Relat Res 469(8):2286–2293
50. Wang A, Yau SS, Essner A, Herrera L, Manley M, Dumbleton J (2008) A highly crosslinked UHMWPE for CR and PS total knee arthroplasties. J Arthroplast 23(4):559–566
51. Young SK, Keller TS, Greer KW, Gorhan MC (2000) Wear testing of UHMWPE tibial components: influence of oxidation. J Tribol 122(1):323–331
52. Bills P, Brown L, Jiang X, Blunt L (2005) A metrology solution for the orthopaedic industry. J Phys Conf Ser 13:316–319
53. Muratoglu OK, Perinchief RS, Bragdon CR, O'Connor DO, Konrad R, Harris WH (2003) Metrology to quantify wear and creep of polyethylene tibial knee inserts. Clin Orthop Relat Res 410(410):155–164
54. Tashiro Y, Uemura M, Matsuda S, Okazaki K, Kawahara S, Hashizume M, Iwamoto Y (2012) Articular cartilage of the posterior condyle can affect rotational alignment in total knee arthroplasty. Knee Surg Sports Traumatol Arthrosc 20(8):1463–1469
55. Knight LA, Pal S, Coleman JC, Bronson F, Haider H, Levine DL, Taylor M, Rullkoetter PJ (2010) Comparison of long-term numerical and experimental total knee replacement wear during simulated gait loading. Springer, New York

56. Strickland M, Dressler M, Taylor M (2012) Predicting implant UHMWPE wear in-silico: a robust, adaptable computational–numerical framework for future theoretical models. Wear 274:100–108
57. Archard JF (1953) Contact and rubbing of flat surfaces. J Appl Phys 24(8):981–988
58. Zhao D, Sakoda H, Sawyer WG, Banks SA, Fregly BJ (2008) Predicting knee replacement damage in a simulator machine using a computational model with a consistent wear factor. J Biomech Eng 130(1):94–97
59. Kang L, Galvin AL, Fisher J, Jin Z (2009) Enhanced computational prediction of polyethylene wear in hip joints by incorporating cross-shear and contact pressure in additional to load and sliding distance: effect of head diameter. J Biomech 42(7):912–918
60. Kang L, Galvin AL, Brown TD, Jin Z, Fisher J (2008) Quantification of the effect of cross-shear on the wear of conventional and highly cross-linked UHMWPE. J Biomech 41(2):340–346
61. Dressler MR, Strickland MA, Taylor M, Render TD, Ernsberger CN (2011) Predicting wear of UHMWPE: decreasing wear rate following a change in direction. In: Transactions of the meeting, pp 2879–2883
62. O'Brien ST, Bohm ER, Petrak MJ, Wyss UP, Brandt J-M (2014) An energy dissipation and cross shear time dependent computational wear model for the analysis of polyethylene wear in total knee replacements. J Biomech 47(5):1127–1133
63. Wang A, Lee R, Herrera L, Korduba L (2013) Modeling and verification of ultra-high molecular weight polyethylene wear in multi-directional sliding. Wear 301(Suppl 1–2):162–167
64. Wasielewski RC, Galante JO, Leighty RM, Natarajan RN, Rosenberg AG (1994) Wear patterns on retrieved polyethylene tibial inserts and their relationship to technical considerations during total knee arthroplasty. Clin Orthop Relat Res 299(299):31–43
65. Currier JH, Bill MA, Mayor MB (2005) Analysis of wear asymmetry in a series of 94 retrieved polyethylene tibial bearings. J Biomech 38(2):367–375
66. O'Brien S, Luo Y, Wu C, Petrak M, Bohm E, Brandt J-M (2012) Prediction of backside micromotion in total knee replacements by finite element simulation. Proc Inst Mech Eng H J Eng Med 226(H3):235–245
67. Gupta SK, Chu A, Ranawat AS, Slamin J, Ranawat CS (2007) Review article: osteolysis after total knee arthroplasty. J Arthroplast 22(6):787–799
68. O'Brien S, Luo Y, Wu C, Petrak M, Bohm E, Brandt JM (2012) Computational development of a polyethylene wear model for the articular and backside surfaces in modular total knee replacements. Tribol Int 59:284–291
69. Jayabalan P, Furman BD, Cottrell JM, Wright TM (2007) Backside wear in modern total knee designs. HSS J 3(1):30–34
70. Mcewen HM, Fisher J, Goldsmith AA, Auger DD, Hardaker C, Stone MH (2001) Wear of fixed bearing and rotating platform mobile bearing knees subjected to high levels of internal and external tibial rotation. J Mater Sci Mater Med 12(10):1049–1052
71. Atwood SA, Kennedy FE, Currier JH, Citters DWV, Collier JP, Collier JP (2006) In vitro study of backside wear mechanisms on mobile knee-bearing components. J Tribol 128 (2):275–281
72. Engh GA, Zimmerman RL, Parks NL, Engh CA (2009) Analysis of wear in retrieved mobile and fixed bearing knee inserts. J Arthroplast 24(6 Suppl):28–32
73. Wimmer MA, Birken L, Sellenschloh K, Schneider E (2013) Damage due to rolling in total knee replacement—the influence of tractive force. Friction 1(2):178–185
74. Brandt JM, Macdonald SJ, Bourne RB, Medley JB (2012) Retrieval analysis of modular total knee replacements: factors influencing backside surface damage. Knee 19(4):306–315
75. Johnson TS, Laurent MP, Yao JQ, Blanchard CR (2003) Comparison of wear of mobile and fixed bearing knees tested in a knee simulator. Wear 255(7–12):1107–1112
76. Haider H, Garvin K (2008) Rotating platform versus fixed-bearing total knees: an in vitro study of wear. Clin Orthop Relat Res 466(11):2677–2685

77. Garcia RM, Kraay MJ, Messerschmitt PJ, Goldberg VM, Rimnac CM (2009) Analysis of retrieved ultra-high-molecular-weight polyethylene tibial components from rotating-platform total knee arthroplasty. J Arthroplast 24(1):131–138
78. Zhang J, Chen Z, Wang L, Li D, Jin Z (2016) A patient-specific wear prediction framework for an artificial knee joint with coupled musculoskeletal multibody-dynamics and finite element analysis. Tribol Int 109:382–389
79. Fitzpatrick CK, Clary CW, Rullkoetter PJ (2012) The role of patient, surgical, and implant design variation in total knee replacement performance. J Biomech 45(12):2092–2102
80. Willing R, Kim IY (2011) Design optimization of a total knee replacement for improved constraint and flexion kinematics. J Biomech 44(6):1014–1020
81. Fregly BJ, Marquezbarrientos C, Banks SA, Desjardins JD (2010) Increased conformity offers diminishing returns for reducing total knee replacement wear. J Biomech Eng 132(2):555–575
82. Ardestani MM, Moazen M, Jin Z (2015) Contribution of geometric design parameters to knee implant performance: conflicting impact of conformity on kinematics and contact mechanics. Knee 22(3):217–224
83. Fitzpatrick CK, Clary CW, Cyr AJ, Maletsky LP, Rullkoetter PJ (2013) Mechanics of post-cam engagement during simulated dynamic activity. J Orthop Res 31(9):1438–1446
84. Fitzpatrick CK, Baldwin MA, Clary CW, Wright A, Laz PJ, Rullkoetter PJ (2012) Identifying alignment parameters affecting implanted patellofemoral mechanics. J Orthop Res 30 (7):1167–1175
85. Schindler OS (2012) The controversy of patellar resurfacing in total knee arthroplasty: ibisne in medio tutissimus? Knee Surg Sports Traumatol Arthrosc 20(7):1227–1244. https://doi.org/10.1007/s00167-012-1985-7
86. Abdelgaied A, Brockett CL, Liu F, Jennings LM, Jin Z, Fisher J (2014) The effect of insert conformity and material on total knee replacement wear. Proc Inst Mech Eng H J Eng Med 228 (1):98–106. https://doi.org/10.1177/0954411913513251
87. Glynjones S, Mclardysmith P, Gill HS, Murray DW (2008) The creep and wear of highly cross-linked polyethylene. J Bone Joint Surg 90(5):556–561
88. Galvin AL, Kang L, Udofia I, Jennings LM, McEwen HMJ, Jin Z, Fisher J (2009) Effect of conformity and contact stress on wear in fixed-bearing total knee prostheses. J Biomech 42 (12):1898–1902
89. Brockett CL, Jennings LM, Fisher J (2011) The wear of fixed and mobile bearing unicompartmental knee replacements. Proc Inst Mech Eng H J Eng Med 225(5):511
90. Brockett CL, Jennings LM, Hardaker C, Fisher J (2012) Wear of moderately cross-linked polyethylene in fixed-bearing total knee replacements. Proc Inst Mech Eng H J Eng Med 226 (7):529–535
91. Ueno M, Ikeuchi K, Nakamura T, Akagi M (2003) Comparison of the wear properties of polyethylene plate in total knee prostheses (TKP) using different femoral component materials. Key Eng Mater 240–242(1):801–804
92. Dalury DF, Pomeroy DL, Gorab RS, Adams MJ (2013) Why are total knee arthroplasties being revised? J Arthroplast 28(8):120–121
93. Liau JJ, Cheng CK, Huang CH, Lo WH (2002) The effect of malalignment on stresses in polyethylene component of total knee prostheses – a finite element analysis. Clin Biomech 17 (2):140–146
94. Werner FW, Ayers DC, Maletsky LP, Rullkoetter PJ (2005) The effect of valgus/varus malalignment on load distribution in total knee replacements. J Biomech 38(2):349–355
95. Zihlmann MS, Stacoff A, Romero J, Quervain IK, Stüssi E (2005) Biomechanical background and clinical observations of rotational malalignment in TKA: literature review and consequences. Clin Biomech 20(7):661–668
96. Ng VY, DeClaire JH, Berend KR, Lombardi AV Jr (2012) Improved accuracy of alignment with patient-specific positioning guides compared with manual instrumentation in TKA. Clin Orthop Relat Res 470(1):99–107
97. Srivastava A, Lee GY, Steklov N, Colwell CW Jr, Ezzet KA, D'Lima DD (2012) Effect of tibial component varus on wear in total knee arthroplasty. Knee 19(5):560–563

98. Berend ME, Ritter MA, Meding JB, Faris PM, Keating EM, Redelman R, Faris GW, Davis KE (2004) The Chetranjan Ranawat award: tibial component failure mechanisms in total knee arthroplasty. Clin Orthop Relat Res 428:26–34
99. Chen Z, Wang L, Liu Y, He J, Lian Q, Li D, Jin Z (2015) Effect of component mal-rotation on knee loading in total knee arthroplasty using multi-body dynamics modeling under a simulated walking gait. J Orthop Res 33(9):1287–1296
100. Silva M, Shepherd EF, Jackson WO, Pratt JA, McClung CD, Schmalzried TP (2003) Knee strength after total knee arthroplasty. J Arthroplast 18(5):605–611
101. Ardestani MM, Moazen M, Jin Z (2015) Sensitivity analysis of human lower extremity joint moments due to changes in joint kinematics. Med Eng Phys 37(2):165–174
102. Fregly BJ, D'Lima DD, Colwell CW (2009) Effective gait patterns for offloading the medial compartment of the knee. J Orthop Res 27(8):1016–1021
103. Ardestani MM, Moazen M, Chen Z, Zhang J, Jin Z (2015) A real-time topography of maximum contact pressure distribution at medial tibiofemoral knee implant during gait: application to knee rehabilitation. Neurocomputing 154(0):174–188
104. Ardestani MM, Moazen M, Jin Z (2014) Gait modification and optimization using neural network–genetic algorithm approach: application to knee rehabilitation. Expert Syst Appl 41 (16):7466–7477
105. Foran JR, Mont MA, Etienne G, Jones LC, Hungerford DS (2004) The outcome of total knee arthroplasty in obese patients. J Bone Joint Surg 86-A(8):1609–1615
106. Knarr BA, Higginson JS, Zeni JA (2015) Change in knee contact force with simulated change in body weight. Comput Methods Biomech Biomed Engin 19(3):1–4
107. Harding GT, Dunbar MJ, Hubley-Kozey CL, Stanish WD, Astephen Wilson JL (2016) Obesity is associated with higher absolute tibiofemoral contact and muscle forces during gait with and without knee osteoarthritis. Clin Biomech 31:79–86
108. Lavernia CJ, Sierra RJ, Hungerford DS, Krackow K (2001) Activity level and wear in total knee arthroplasty. J Arthroplast 16(4):446–453

Chapter 8
Biomechanics and Biotribology of UHMWPE Artificial Hip Joints

Yongchang Gao and Zhong-Min Jin

Abstract Well function of hip joints ensures daily movements such as walking, standing, climbing, or lifting. However, joint diseases such as osteoarthritis, rheumatoid arthritis, and trauma often require the natural bearings to be replaced by artificial ones. John Charnley pioneered the first metal-on-polyethylene artificial hip joints in the 1960s, when he articulated a femoral head against the ultrahigh molecular weight polyethylene (UHMWPE) liner. Although ceramic-on-ceramic and metal-on-metal artificial hip joints have been widely used in clinic, the UHMWPE hip implants are most prevailing with great success. Currently, over one million patients accept total hip replacement around the world every year, and the demand remains increasing with the accelerated aging population. However, unlike natural synovial hip joints with excellent elastohydrodynamic lubrication, artificial hip joints overall experience boundary lubrication or mixed lubrication. Under such lubrication conditions, direct contact between femoral head and acetabular liner is inevitable and finally generates extensive micro-wear debris. Then bioreaction of soft tissues rendered by UHMWPE wear particles occurs, which eventually leads to aseptic loosening of hip implants in the long term. In the past decades, much research enhancing wear resistance of the UHMWPE hip implants has been done by polymer scientists, biomedical engineers, orthopedic surgeons, and manufacturers. This chapter aims to review the latest research on wear performance of UHMWPE artificial hip joints from both biomechanics and biotribology.

Y. Gao
School of Construction Machinery, Chang'an University, Xi'an, Shaanxi, China

Z.-M. Jin (✉)
Tribology Research Institute, School of Mechanical Engineering, Southwest Jiaotong University, Chengdu, Sichuan, China

State Key Laboratory of Manufacturing System Engineering, School of Mechanical Engineering, Xi'an Jiaotong University, Xi'an, Shaanxi, China

Institute of Medical and Biological Engineering, School of Mechanical Engineering, University of Leeds, Leeds, UK
e-mail: zmjin@mail.xjtu.edu.cn

J. Fu et al. (eds.), *UHMWPE Biomaterials for Joint Implants*, Springer Series in Biomaterials Science and Engineering 13,
https://doi.org/10.1007/978-981-13-6924-7_8

Keywords UHMWPE · Hip joint implants · Lubrication · Wear · Biomechanics · Biotribology · Hip joint simulator · Wear debris

8.1 Introduction

8.1.1 UHMWPE Artificial Hip Joints

Natural synovial hip joints are one of the most important joints for human to achieve all kinds of movements. It is expected to function well in the human body for a lifetime while transmitting large dynamic loads and yet accommodating a wide range of movements. However, diseases such as osteoarthritis, rheumatoid arthritis, and trauma often require these natural bearings to be replaced by artificial hip implants [1]. Total joint replacement has proved to be the most successful surgical treatment for hip joint diseases for more than 50 years [2].

At present, more than one million hip joint replacements are carried out every year all over the world [3, 4]. Up to now, three kinds of material combinations are introduced and widely used in clinic for artificial hip joints: metal/ceramic femoral head vs polymer (UHMWPE) cup, ceramic femoral head vs ceramic cup, and metal femoral head vs metal cup [5–7]. However, the majority of these devices utilize a material combination of ultrahigh molecular weight polyethylene (UHMWPE) as the acetabular cup articulating against either a ceramic (alumina) or a metallic (stainless steel, cobalt-based alloy) femoral head (Fig. 8.1). This mostly applied

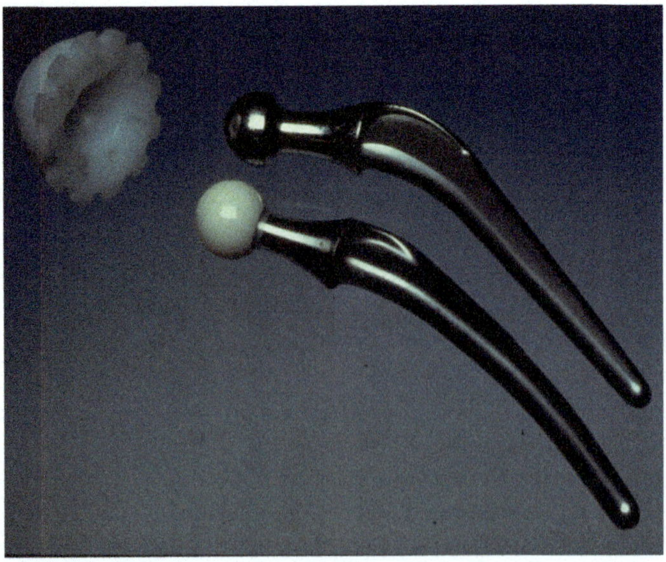

Fig. 8.1 A typical Charnley hip prosthesis consisting of an UHMWPE acetabular cup against either a metallic (stainless steel) or a ceramic (alumina) femoral head

combination, however, encountered with severe wear problems, which will be described in Sect. 8.2. This section concentrates on UHMWPE artificial hip joints.

8.1.2 Complications After Total Hip Replacement

The patient who has adopted total hip replacement (THR) could fully or almost relieve from serious pain and recover daily movement. The hip implants can last 15 years in about 90% of those who received them [5]. However, there are still some complications with total hip replacement because of the huge demands. After a few weeks of total hip replacement, the most common and serious complication are blood clots in the legs and infection of operation side [8], respectively. These short-term complications can be reduced with improvements in sterile technique and medical treatment in surgery.

The most important long-term complication is aseptic loosening, which occurs over time and can cause problems with the function of the hip prosthesis. What's worse, the loosened hip implants need to be revised. Dislocation is another long-term common complication for THR [9, 10], which ranks only second to aseptic loosening. Berry et al. [9] investigated the dislocation rate of THR and observed a gradual increase over time.

8.1.3 Relationship Between Aseptic Loosening and Wear for THR

Hip implants experience daily movements such as walking, running, climbing, lifting, and even standing and sitting. The implant takes large forces and a range of motions [11, 12]. Generally, patients with hip implants take about one million steps per year in average [13], and younger patients move more than this average level. On the one hand, such motions and starting/stopping lead to both collision and relative sliding between the femoral head and the cup. On the other hand, in vivo hip implants mainly undergo mixed lubrication [14], which results in a poor separation of the articulating surfaces by the lubricant. Thus, the wear of hip implants is inevitable to generate wear debris, as has been verified both in vivo and in vitro. The majority of wear debris ranges from 0.1 μm to 1 μm [15, 16]. These UHMWPE microparticles released from the articulating interface enter the periprosthetic tissues and are taken up by macrophages. The macrophages release a range of mediators of inflammation including cytokines in an attempt to eliminate these bio-inert particles. Meanwhile, such mediators have osteolytic effects and cause the gradual loss of bones surrounding the hip implants, which eventually lead to aseptic loosening of the prosthesis.

The aseptic loosening is the result of long-term accumulation of hip implant wear. Thus, most prostheses could still function well in several years. For conventional UHMWPE hip implants, 90% of them could last 15 years after surgery [17]. However, the wear performance of UHMWPE hip implants plays a major role in the long-term success of the medical devices. Much research work in this field is focusing on wear and wear debris-induced osteolysis in currently used artificial hip joints, by using both experimental and theoretical tools to conduct tribological studies on contact mechanics, friction, lubrication, and wear.

8.1.4 Relationship Between Biomechanics and Wear for THR

Wear of hip implants resulted from contact and relative motion between the femoral head and acetabular cup is described in Sect. 8.3. Hence, the contact mechanics [18] and kinematics [19] need to be carefully investigated during the research of hip implant wear. Biomechanics covers the study of the structure and function of biological systems such as humans, animals, plants, organs, fungi, and cells by using the methods of mechanics. Both contact mechanics and kinematics of hip implants belong to biomechanics. Specially, biomechanics of hip implants mainly refer to the mechanical properties of soft tissue and bones and their preoperative and postoperative movements [20]. More details about biomechanics of hip implants are described in Sect. 8.2.1.

8.2 Biomechanics of Artificial Hip Joints

8.2.1 Introduction

The main objectives of THR are to reduce pain, improve range of motion, and restore joint function of patients [21–24]. Recovery of muscle strength and mechanical performance as well as motions are the basis to achieve these aims of THR. The following three paragraphs will give a brief introduction to biomechanics of artificial hip joints before detailed research is presented.

The hip implant supports quite large loads generated by muscle activity pulling the prosthesis, the weight of the limbs and trunk. Therefore, it is of great importance to recover the strength of muscles surrounding the hip implant. The postoperative strength of most muscles shows a gradual increase in the first year of surgery and then keeps stable over time. However, the muscle's strength could not recover to the level of the healthy side. It is difficult to measure the hip contact force related to muscle strength. The most common way to obtain hip contact force is to use musculoskeletal system [25, 26], which is often used to estimate the internal joint loading, joint kinematics, and muscle forces. Different musculoskeletal multibody dynamic models have been introduced to predict both hip contact force and motions by using different models, which will be further introduced in the latter sections.

Contact mechanics is a study of load transfer between two contacting solids. The main parameters determined from contact mechanics analysis include contact area and stresses. Contact stresses are generally related to structural failure and fatigue-related wear mechanisms. A large contact area is required to produce a low contact stress under a given load. However, if the contact area is too large, the contact may be extended to the equatorial region and the edge of the cup, not only leading to stress concentrations, limiting the normal movement of the hip joints, but also blocking the lubricant entry and causing lubricant starvation and depletion. In addition to the tribological studies at the bearing surfaces, contact mechanics can be used to simulate the implantation of the prosthetic components (e.g., press fitting) to examine the stresses in the bone and the deformation of the prosthetic components. Much research on contact stresses and contact area has been done and will be introduced in detail in the latter sections.

Kinematics is the study of motion of bodies without referencing to mass or force. In the hip, the geometry of the ball and socket prevents any translation but allows rotation during all kinds of human movements [27]. The relative rotations in all three planes make large sliding distance, which is directly proportional to wear of the prosthesis. The magnitude of sliding distance may be different under different movements and differ in the same movement because of individual characteristics such as walking speed, stride length, etc. What's more, the range of motion is related to actions of muscles, which also influences the relative sliding. The corresponding studies will be presented in the latter sections.

8.2.2 *Musculoskeletal Multibody Dynamic Simulation*

8.2.2.1 Musculoskeletal Modeling Based on Force-Dependent Kinematics

It is important to understand the performance of muscles after THR because of their influences on both hip contact force and range of hip joint motions as described in Sect. 8.2.1. Currently, the common way to determine hip contact force and range of hip joint motions is musculoskeletal multibody dynamic analysis. Different models of the human musculoskeletal system have been developed over the past decades [28–30] to determine the hip contact force and motions. The majority of musculoskeletal multibody dynamic modeling has been driven by the commercial musculoskeletal modeling system. Hip implants were simplified as a hinge to fulfill relative rotations without considering microseparation state of the hip joint center (HJC) of the femoral head and geometries or material properties [31].

Based on previous musculoskeletal multibody dynamic modeling, Zhang et al. [32] recently introduced a force-dependent kinematics (FDK) approach that considers both the position and the geometry of the femoral head of the hip joint to simultaneously predict hip contact force and hip joint translation. In this study, a typical hip implant was directly implanted into the left lower limb to replace disabled natural hip joint, and material properties were also taken into account in the model.

The contact force predicted by this method changed over gait cycle and reached the maximum value of about three times body weight. The predicted results of hip contact force agreed with experimental data.

8.2.2.2 Effect of Surgical Position on Hip Contact Force

Artificial hip joint is expected to be fixed into the anatomic position to obtain the biggest range of motions for the patient, which is mainly determined by the hip implant center position. For most patients, it is easy to implant the prosthesis into the anatomic position. However, for persons whose hip joint is congenital defect such as developmental dysplasia of the hip, it is rather hard to reposition the hip implant into the anatomic position. Using the FDK method, Zhang et al. investigated the influence of hip implant position on contact forces under both walking and stair climbing.

According to clinic observations, the deviation distance of hip joint center from its anatomic position should not exceed 15 mm. In these studies, the hip implant was set in the maximum deviation distance of 15 mm in six directions. The results of the hip contact force (HCF) under different hip implant centers when the patient walks are shown in Fig. 8.2. The hip contact force in anterior–posterior direction increased if the prosthesis center position deviated 15 mm from the standard position in the lateral, inferior, and posterior directions. In contrast, the hip contact force would decrease under same conditions in the medial, superior, and anterior directions. The variation trend of hip contact force in both lateral–medial direction and superior–inferior direction under different prosthesis positions was in accordance with that in anterior–posterior direction. Furthermore, the hip contact force deviated remarkably increased due to the deviation from the anatomic position (Table 8.1).

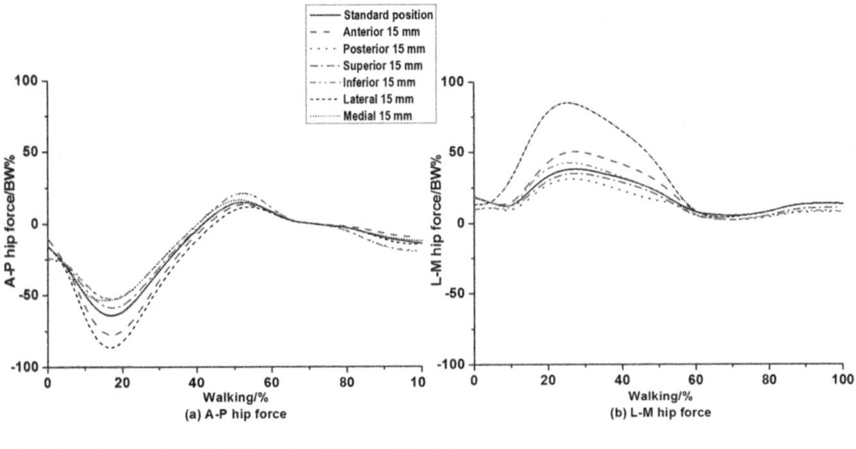

Fig. 8.2 The predicted HCF under different hip joint centers

Table 8.1 Increasing of the maximum hip contact force when the prosthesis deviated 15 mm from the anatomic position during walking

Unit/ %	Anterior 15 mm	Posterior 15 mm	Superior 15 mm	Inferior 15 mm	Lateral 15 mm	Medial 15 mm
A-P	−17.96	20.86	−8.45	16.59	35.62	−17.71
L-M	−18.38	34.12	−7.75	11.79	125.2	−47.69
S-I	−11.62	16.99	−4.35	3.91	17.39	−14.07

Table 8.2 Increasing of the maximum hip contact force when the prosthesis deviated 15 mm from the anatomic position during squatting

Unit/ %	Anterior 15 mm	Posterior 15 mm	Superior 15 mm	Inferior 15 mm	Lateral 15 mm	Medial 15 mm
A-P	0.34	−4.81	3.77	−25.72	−17.06	3.04
L-M	5.47	−8.44	4.43	−5.72	−6.38	9.53
S-I	−9.40	6.23	3.23	−12.00	10.93	−19.30

The hip implant force is also influenced if the prosthesis deviates from the anatomic position under squatting movement. The results are shown in Table 8.2. It can be also concluded that the hip implant center would apparently influence the hip contact force.

Therefore, the hip contact force would be apparently influenced by positioning of the hip implant in surgery for both walking and squatting movements. This should be carefully considered in surgery, and the hip implant is prospected to be fixed into the anatomic position.

8.2.2.3 Effect of Surgical Approach on Hip Contact Force

For different patients, different surgical approaches may be used during the process of THR. Up to now, posterior approach, lateral approach, and anterior approach are widely used in clinic. For different approaches, the damage of muscles is different. Taking the posterior approach as an example, the gluteus maximus is damaged, but the function of the abductor is kept well. Thus, the hip implant could possess a large range of motions. Zhang et al. used the FDK method to study the influence of surgical approaches on hip contact force.

In this study, all muscle's forces that would be damaged totally or partially under different surgical approaches were multiplied by a damage coefficient (Table 8.3). During walking, the effect of surgical approach on the maximum hip contact force is shown in Table 8.4. It reveals that the maximum hip implant forces of anterior approach and superior approach are obviously different from the normal hip contact force. Using these two kinds of surgical approaches would lead to sharp decrease in the lateral–medical direction but slight increase in the superior–interior direction for the maximum hip contact force. However, the lateral approach had little influence on the hip contact force comparing to the other two surgical approaches.

Table 8.3 Damage coefficient of muscles under different surgical approaches

Muscles name	Lateral approach	Posterior approach	Anterior approach
Gluteus medius anterior (6)	0.6	0.6	0.8
Gluteus medius posterior (6)	0.6	0.6	0.8
Gluteus minimus anterior (1)	0.6	0.6	0.5
Gluteus minimus medial (1)	0.6	0.6	0.5
Gluteus minimus posterior (1)	0.6	0.6	0.5
Gluteus maximus (12)	1	0.6	1
The tensor fascia lata muscle (2)	0.6	1	1

Table 8.4 Effect of surgical approach on the maximum hip contact force during walking

	Different surgical approaches			
	Anterior approach		Posterior approach	Lateral approach
Unit/%		Dev. of Max F	Dev. of Max F	Dev. of Max F
A-P	Heel strike	−7.22	−10.46	−9.98
		±1.56	±2.19	±3.18
	Toe off	+24.91	+24.19	−19.88
		±9.27	±8.73	±6.25
S-I (swing)		+4.04	+4.04	−0.64
		±1.30	±0.39	±0.25
L-M (swing)		−38.47	−44.82	−22.53
		±2.84	±0.84	±1.06

Table 8.5 Effect of surgical approach on the maximum hip contact force during squatting

Unit/%	Anterior approach	Posterior approach	Lateral approach
A-P	−0.61	1.84	−0.6
L-M	−6.84	0.11	−2.76
S-I	−2.63	1.01	−1.11

The effects of surgical approach on the maximum hip contact force during squatting are shown in Table 8.5. The maximum hip contact force only slightly changed for all three surgical approaches during squatting in comparison with that during walking.

In summary, different surgical approaches would affect hip contact force during walking, and the lateral approach may have much less influence on the hip contact force. Therefore, it is better to consider this factor when choosing surgical approaches.

8.2.3 Investigation of Contact Mechanics

As mentioned above, contact pressure and contact area are very important for hip implants. It is preferred to test both contact pressure and contact area in vivo or in vitro, which has been done in previous studies [33–35]. However, experiments are expensive and time-consuming, which are not suitable for parameterized design of hip implants. Therefore, the numerical method becomes a most appropriate alternative method to predict contact pressure and contact area. To date, the finite element method is the most common numerical method [36, 37]. The related research will be presented in details in the following sections.

Relative sliding, as well as contact, between the femoral head and the acetabular cup is needed to be determined for hip implants. The analytical method, Euler method, has been introduced to calculate accumulated sliding distance of hip implant under different movements. In the latest study, a dynamic finite element method has been developed to predict sliding distance. These studies will be described in the later sections.

What's more, the dynamic finite element method was used to investigate both contact mechanics and relative sliding of the dual mobility hip implant. This will also be fully introduced in the next sections.

8.2.3.1 Computational Prediction of Contact Pressure

During the process of finite element modeling, simplified model is generally used to represent artificial hip. Typical finite element model consists of pelvic bone, bone cement, UHMWPE cup, and femoral head, as done by Hua et al. [38] (Fig. 8.3). This model is comprised of an UHMWPE cup, a stainless femoral head, a pelvic bone, and a bone cement. The dimensions were accordance with real implants. The bone cement fixed both pelvic bone and UHMWPE cup together, contact pairs was set between the UHMWPE cup and the stainless femoral head. All parts were meshed using proper element type and size: the femoral head was treated as rigid body

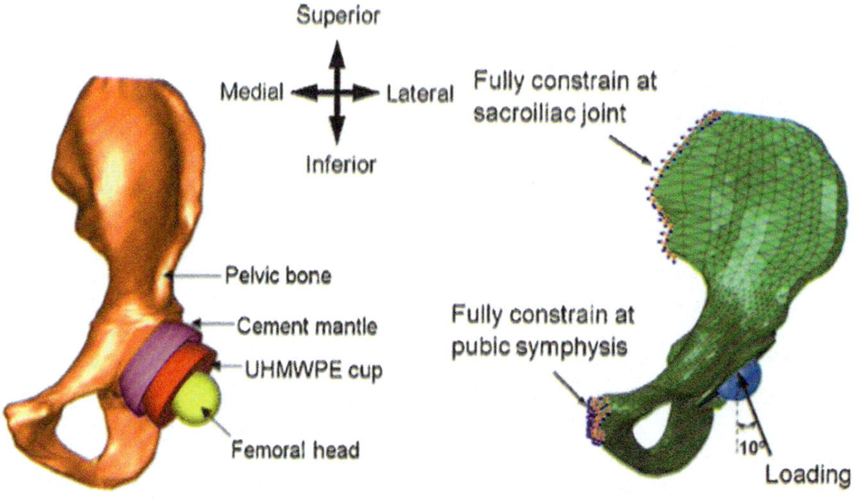

Fig. 8.3 The geometry (left) and boundary conditions (right) of the hip joint model. (Reprinted from Ref. [38], Copyright 2012, with permission from Elsevier)

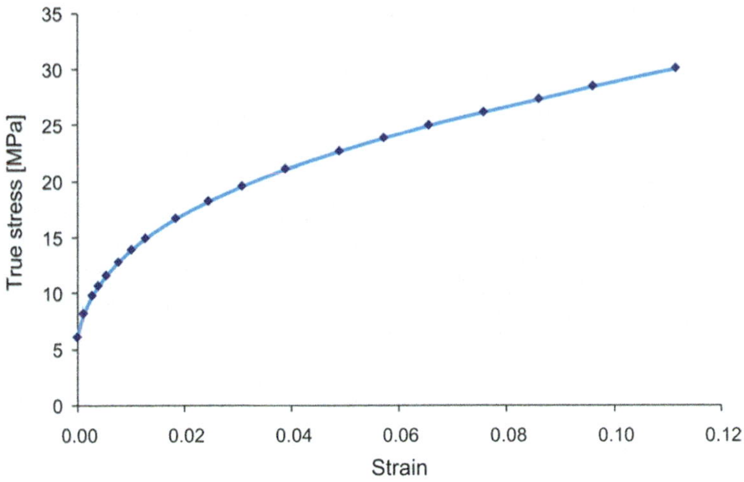

Fig. 8.4 The plastic stress–strain relation for UHMWPE. (Reprinted from Ref. [38], Copyright 2012, with permission from Elsevier)

because its Young's modulus was about two orders of magnitude higher than other parts. According to practical restraints, two regions of pelvic bone were fully constrained as shown by the finite element model. A contact force of 2500 N was applied at the femoral head center and with the direction of 10° medially.

The UHMWPE was modeled as nonlinear elastic–plastic based on the plastic stress–strain constitutive relationship (Fig. 8.4). The plastic stress–strain data were taken from Liu [39] for a similar polyethylene material. The other material properties used in this study are given in Table 8.6. Then performing finite element contact statics analysis, both contact pressure and contact area could be predicted.

Using this static finite element method, contact pressure and contact area of hip implants could be predicted. The following several paragraphs show some applications of this method for conventional hip implants.

The in vivo results show that the orientation of the acetabular cup was important for hip implant. When an artificial hip joint is implanted into a patient, the orientation of UHMWPE cup is carefully determined because both higher inclination and anteversion angles will lead to poor contact mechanics and severe wear [40–42]. The suggested safe inclination angle is below 55°. The first application using the static finite element method was to investigate the contact mechanics of UHMWPE hip implant under adverse edge loading.

Table 8.6 Material properties used in study of Hua et al.

Component	Materials	Young's modulus (GPa)	Poisson's ratio
UHMWPE cup	UHMWPE	0.85	0.4
Bone cement	PMMA	2.5	0.254
Cortical shell	Cortical bone	17	0.3
Cancellous bone	Cancellous bone	0.8	0.2

Reprinted from Ref. [38], Copyright 2012, with permission from Elsevier

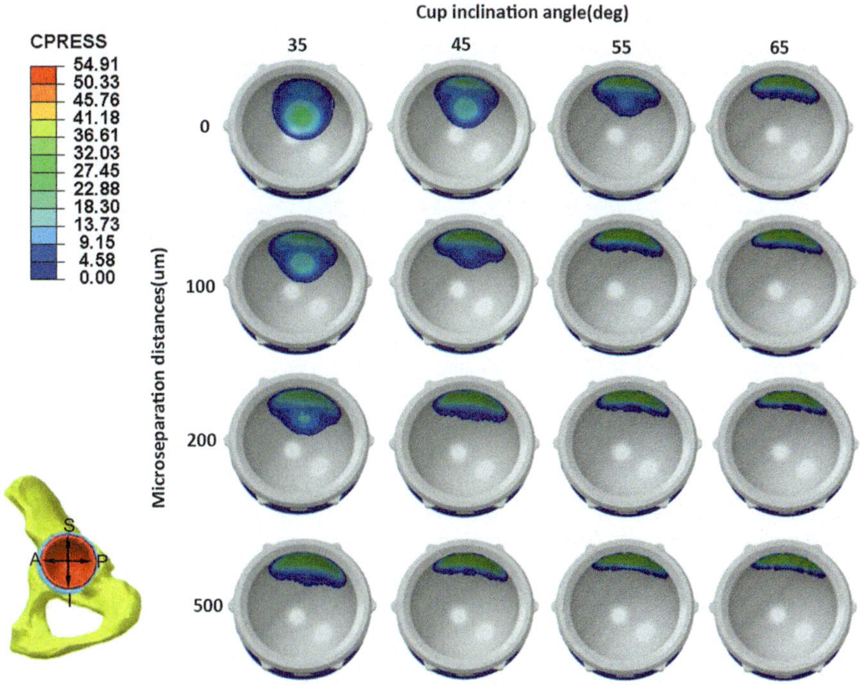

Fig. 8.5 The distribution of contact pressure (MPa) on the frontside articulating surface as a function of cup inclination angles and microseparation distances [36]

Recently, Hua et al. investigated the contact mechanics of modular metal-on-polyethylene THR under adverse edge loading conditions [36]. Different cup inclination angles coupled with different microseparations, contact pressure, and contact zone (contact area) of UHMWPE cup inner surface were shown in Fig. 8.5. The contact pressure increased with inclination angles and microseparation distance, but the contact zone decreased at the same time. Furthermore, edge contact occurred under high inclination angle and big microseparation distance. Edge contact and higher contact pressure are harmful for both stability and wear performance of artificial hip joints. The predicted result agreed with the clinical observation that a high inclination angle is harmful to the conventional artificial hip joints. Similar results were also reported by Wang et al. [37], who investigated the effects of both inclination and anteversion angles on the contact pressure.

In addition, wear volume and depth directly depend on contact mechanics; the static finite element method has been widely used during computational simulation of wear process of artificial hip joints. Kang et al. [43] and Liu et al. [44] have calculated contact pressure of UHMWPE cup as input of wear calculation of artificial hip joints. Contact pressure varied with different gait cycles and overall decreased with increasing gait cycles, but it nearly did not vary after the first

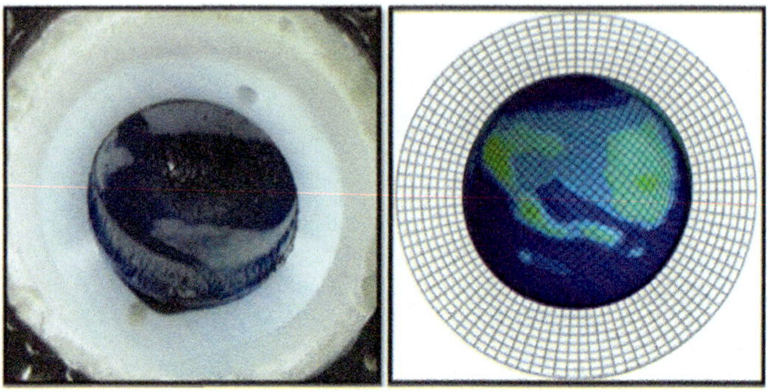

Fig. 8.6 Comparison of contact area between the finite element model and compression test for one sample of custom-coated hip implant after 5 Mc (million cycles) simulator testing (12.6% cycle, loading 3000 N). (Reprinted from Ref. [45], Copyright 2015, with permission from Elsevier)

10 million gait cycles. This agrees with the fact that the contact pressure will decrease with the increasing contact area.

The contact mechanics of UHMWPE cup after different walking gait cycles has been investigated [45]. The worn surfaces of the cups under different stages of hip simulator testing were reconstructed and were applied to the simulation. The contact areas predicted by finite element method and observed in experiments are compared in Fig. 8.6. It is revealed that contact zone tested in experiment and predicted by static finite element method are consistent. Besides, the results of contact pressure at different walking gait cycles are shown in Fig. 8.7. The results reveal that the contact pressure of UHMWPE cup after wear became much higher than that of unworn (the maximum value at 4 Mc was about 2.5 times than that of unworn). The finding of variation of contact pressure by this study is different from the result by Kang et al.

8.2.3.2 Computational Prediction of Sliding Distance

Wear is expected to be proportional to the sliding distance. However, it is hard to determine the sliding distance of implants by experiments. Therefore, it is usually predicted using numerical method, including the Euler method developed by Saikko et al. [46] and Kang et al. [43]. The details of this method are described in the following paragraphs.

When calculating the accumulated sliding distance between the femoral head and the acetabular cup, the hip implant could be simplified to a ball-in-socket model (Fig. 8.8a, b). The transformed simplified spherical coordinates were built at the center of the head, and the spherical coordinates were (R_1, θ, ϕ). A separate moving coordinate system, x', y', z', was placed at the center of the head and assumed to be fixed relative to the head and to coincide with the center of the cup. Three rotation motions of flexion–extension (FE), abduction–adduction (AA), and internal–external rotation (IER) were assumed to move around the moving axes z', x', y',

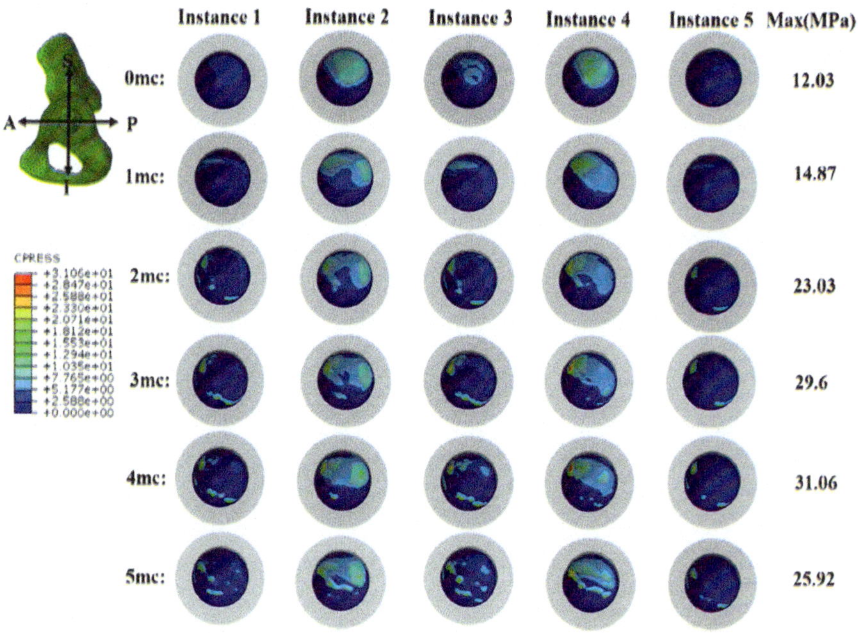

Fig. 8.7 Contour plot for the contact pressure of the articulating surfaces in realistic model at a series of testing points (0–5 Mc). (Reprinted from Ref. [45], Copyright 2015, with permission from Elsevier)

respectively. The corresponding motion data under walking have been reported by Johnston and Smidt [47]. The hip joint rotated following the Euler sequence of FE→AA→IER, which was used in the computation of sliding distance.

Given that θ and ϕ are the spherical coordinates of any point on the head at instant i in the x, y, z coordinate system (Fig. 8.8b), the position vector $\{P_i\}$ for this point is expressed as

$$\{P_i(\theta, \phi)\} = \begin{Bmatrix} R_1 \sin\theta\cos\phi \\ R_1 \sin\theta\sin\phi \\ R_1 \cos\theta \end{Bmatrix} \tag{8.1}$$

where R_1 is the radius of the head.

As mapped onto the x, y, z coordinates system (Fig. 8.8a), $\{P_i(\theta, \phi)\}$ is rewritten as a new position vector $\{Q_i(\theta, \phi)\}$ according to

$$\{Q_i(\theta, \phi)\} = \begin{bmatrix} \cos\kappa & -\sin\kappa & 0 \\ \sin\kappa & \cos\kappa & 0 \\ 0 & 0 & 1 \end{bmatrix} \begin{bmatrix} 1 & 0 & 0 \\ 0 & \cos\lambda & -\sin\lambda \\ 0 & \sin\lambda & \cos\lambda \end{bmatrix} \times \{P_i(\theta, \phi)\} \tag{8.2}$$

where κ and λ are the anteversion and inclination angles of the cup. The anteversion angle generally varies between 15° and 20° and is fixed as 0° in the present study.

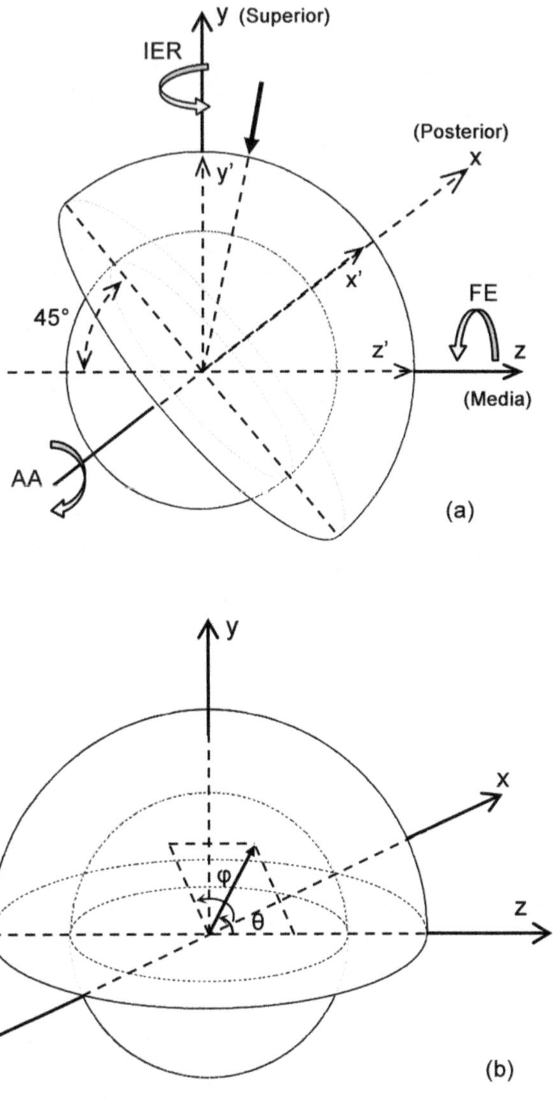

Fig. 8.8 (**a**) Ball-in-socket geometry in the anatomical spherical coordinate axes; (**b**) transformed simplified spherical coordinates

Assuming $\{Q_i(\theta, \phi)\}$ as the preceding position vector, the new position vector $\{Q_{i+1}(\theta, \phi)\}$ after one set of rotations at instant i+1 is expressed as

$$\{Q_{i+1}(\theta, \phi)\} = \left(R_{z'x'y'}\right)_i \{Q_i(\theta, \phi)\} \tag{8.3}$$

where $(R_{z'x'y'})_i$ is the relative rotation matrix for the rotation sequence FE→AA→IER. Assuming α_i, β_i, and γ_i as the rotation angles corresponding to FE, AA, and IER, respectively, $R_{z'x'y'}$ was given by Craig [39] as

$$\left(R_{z'x'y'}\right)_i = \begin{bmatrix} -\sin\alpha_i \sin\beta_i \sin\gamma_i + \cos\alpha_i \cos\gamma_i & -\sin\alpha_i \cos\beta_i & \sin\alpha_i \sin\beta_i \cos\gamma_i + \cos\alpha_i \sin\gamma_i \\ \cos\alpha_i \sin\beta_i \sin\gamma_i + \sin\alpha_i \cos\gamma_i & \cos\alpha_i \cos\beta_i & -\cos\alpha_i \sin\beta_i \cos\gamma_i + \sin\alpha_i \sin\gamma_i \\ -\cos\beta_i \sin\gamma_i & \sin\beta & \cos\beta \cos\gamma_i \end{bmatrix} \tag{8.4}$$

The accumulated sliding distance $\Delta S_j(\theta, \phi, t)$ was calculated as

$$\Delta S_j(\theta, \phi, t) = \left|\left\{Q_{i+1}(\theta, \phi) - Q_i(\theta, \phi)\right\}\right| \tag{8.5}$$

All accumulated sliding distance could be calculated by combining all above formats. This is so-called numerical Euler method.

Kang et al. has computed the maximum accumulated sliding distance of UHMWPE cup under normal walking gait. The predicted sliding distance by Kang el al. agreed well with that by Calonius and Saikko.

Besides, the Euler method is also used to predict the sliding trace of artificial hip joints, as has been done by Saikko et al. [46] (Fig. 8.9). It can be seen that the sliding trace of most points on femoral head was oval. The results predicted by numerical Euler method generally agreed with the experimental results of the zone. The Euler method is appropriate and convenient to predict sliding distance and trace of conventional hip implant because of the single articulation for this prosthesis.

However, for dual mobility hip implant consisting of two articulations, this method is not suitable any more. To solve this problem, a dynamic finite element method was developed by Gao et al. [47]. This method could be performed as follows.

In this study, both conventional and dual mobility hip implants were modeled using the dynamic finite element method, as shown in Figs. 8.10 and 8.11,

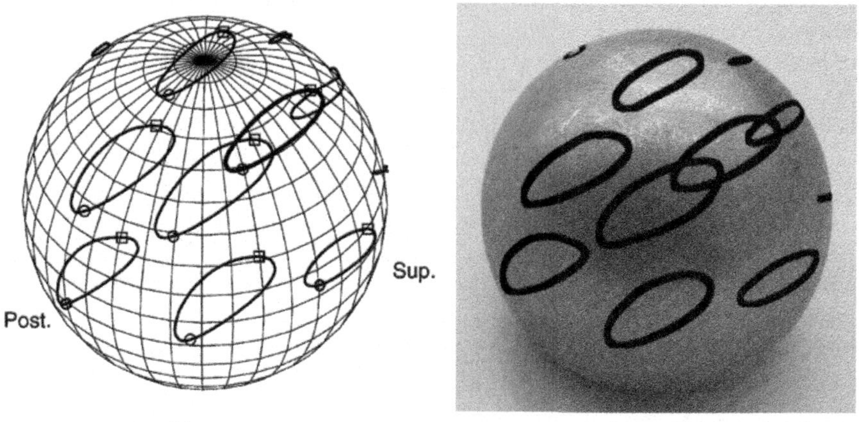

(a) (b)

Fig. 8.9 Comparison of sliding distance of femoral head between (**a**) numerical results and (**b**) experimental data by Saikko et al. (Reprinted from Ref. [46], Copyright 2015, with permission from Elsevier)

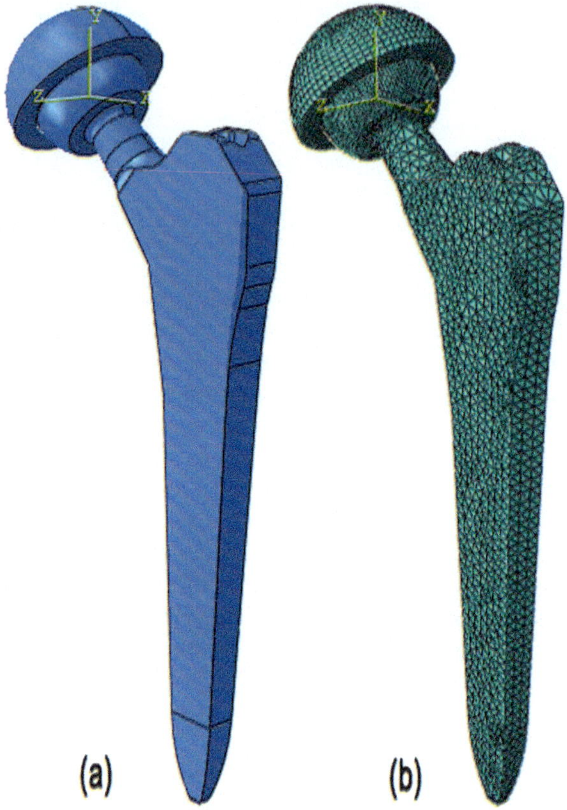

Fig. 8.10 (**a**) CAD and (**b**) FE models of conventional artificial hip joint using dynamic finite element method

respectively. The conventional hip implant was comprised of UHMWPE cup, CoCr alloy femoral head, and Ti alloy stem; the dual mobility hip implant was comprised of metal back shell, UHMWPE liner, CoCr alloy head, and Ti alloy stem. For both hip implants, the diameter of femoral head was 28 mm. The entire stem was used for both hip implants. UHMWPE was also modeled as nonlinear elastic–plastic material as Hua et al. [38] has done, but other materials were linear elastic. The detailed material parameters for dual mobility hip implant are shown in Table 8.7.

For conventional hip implant, there is only one contacting interface between the femoral head and the cup inner surface. In contrast, two contacting interfaces were modeled for dual mobility hip implant, one between the femoral head and the liner inner surface and the other between the liner outer surface and the back shell inner surface. For these two types of hip implants, all components including contacting surface were meshed by eight-node structured hexahedron elements, while stem was meshed using coarse four-node tetrahedron elements. The UHMWPE cup outer surface was fully constrained for conventional hip implant, but that of the back shell was fixed for dual mobility hip implant.

Both three-dimensional forces and three rotation motions were applied at the center of the femoral head including flexion–extension (FE), abduction–adduction

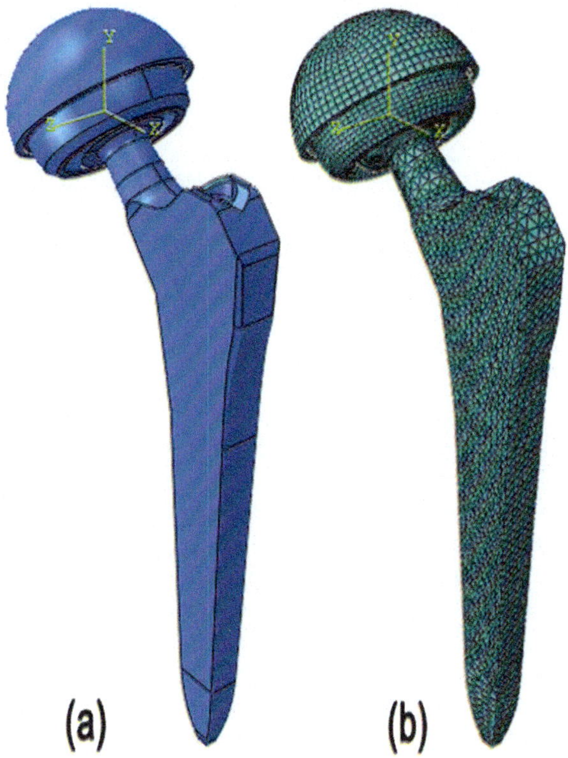

Fig. 8.11 (**a**) CAD and (**b**) FE models of dual mobility hip implant using dynamic finite element method

Table 8.7 Material parameters of the dual mobility hip implant

Materials	Density (g/mm$^{3)}$	Elastic modulus (GPa)	Poisson's ratio	Yield strength (MPa)
CoCr alloy	7.61	217	0.30	\
UHMWPE	0.932	1	0.45	\
Ti alloy	4.52	110	0.3	23.56

(AA), and internal–external rotation (IER). Forces could be directly applied, but rotation motions could not be used directly. Kang et al. used FE, AA, and IER to describe a typical walking gait. At any instant of a walking gait cycle, FE, AA, and IER refer to the rotation angles from the initial position to the current position. However, during a continuous dynamic process, the stem and femoral only could rotate to a new position from last position rather than from initial position. Therefore, a method was introduced to convert the initial FE, AA, and IER data to a new data so as to simulate continuous dynamic rotations. This method can be used to calculate all incremental rotation vectors between any two adjacent instants of a gait cycle.

Then the method to calculate incremental rotation vector between two adjacent instants was developed. The original FE, AA, and IER angles at a time instant were represented through the Euler rotation angles to enable the stem to rotate continuously from the beginning position to a new position. A moving coordinate system

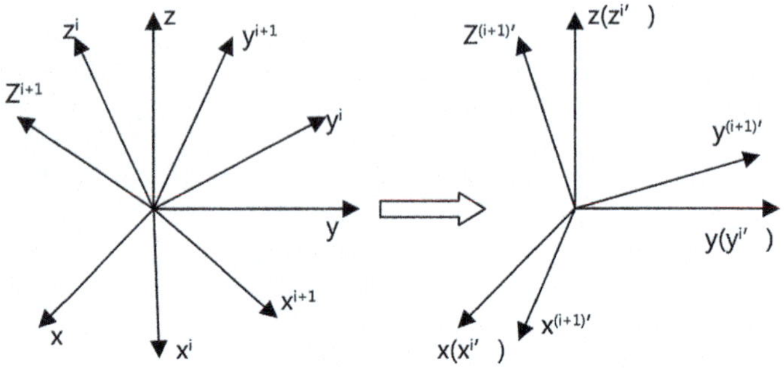

Fig. 8.12 Incremental rotation vectors calculation by inverse Euler rotation

XYZ was fixed too and located at the center of the femoral head. This moving coordinate system was rotated with the head during a gait cycle, and its initial orientation was in accordance with the fixed coordinate system in Sect. 8.2.1. The Euler rotation started from the FE around the X-axis, followed by the AA and the IER along with Y-axis and Z-axis of the moving coordinate system, respectively [43]. Incremental rotation vectors were therefore calculated, according to the static movement wave forms. The movement wave forms in Sect. 8.2.1 were divided into N instants. For arbitrary two adjacent instants i and $i+1$, both Euler rotation matrices R_i and R_{i+1} were calculated according to Saiko and Calonius [46], and then the incremental rotation vector between these two instants was obtained from the known R_i and R_{i+1} [48]. Then inverse Euler rotation R_i^{-1} was applied to both R_i and R_{i+1}, and the incremental rotation vector between R_i and R_{i+1} was converted to a new incremental rotation vector corresponding to the fixed coordinate systems. In this way, all incremental rotation vectors corresponding to the fixed coordinate system (Fig. 8.12) were calculated, which could be used to continuous dynamic process [47].

Then, the sliding distance of any position could be calculated by performing the dynamic finite element method.

The predicted sliding distance of the UHMWPE acetabular cup of the conventional hip implant during one walking gait cycle using this method is shown in Figs. 8.13 and 8.14. Sliding distance increased with the gait cycle and reached the

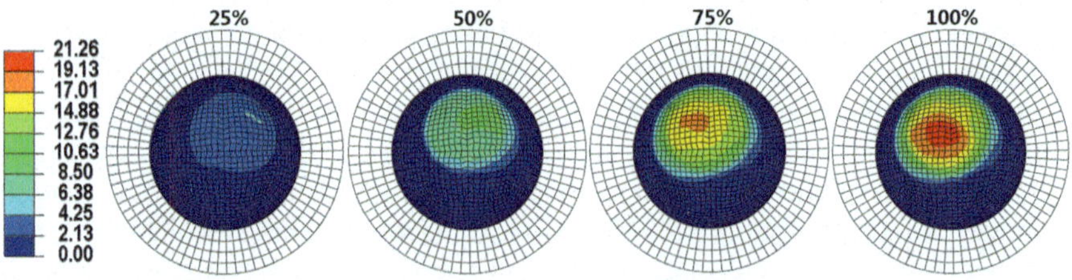

Fig. 8.13 Contours of the cup inner accumulated sliding distance (mm) at different percentages of the gait cycle for the conventional artificial hip joint model predicted by the dynamic finite element method under a fixed element size of 1.5 mm

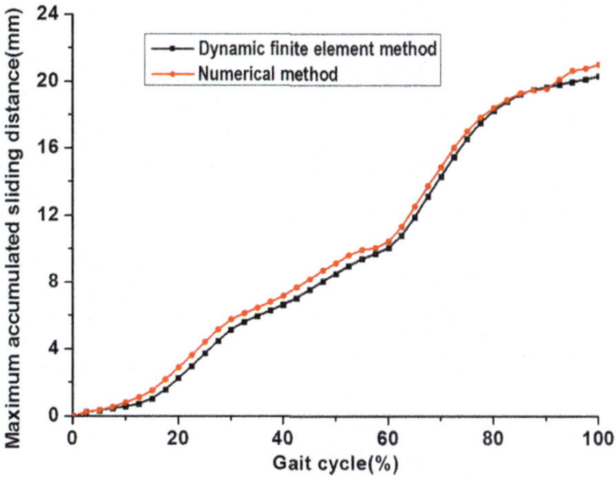

Fig. 8.14 Comparison of the maximum accumulated sliding distance of the cup as a function of the gait cycle for the conventional artificial hip model using different methods

maximum value at the end of gait cycle. The maximum value predicted at each instant was highly consistent with the numerical Euler method. This implies that the dynamic finite element method could be used to calculate sliding distance of conventional artificial hip joints.

8.2.3.3 Contact Mechanics of Dual Mobility Hip Implants

As mentioned in Sect. 8.1.2, dislocation is another common long-term complication for artificial hip joint. However, conventional hip implants could not prevent dislocation effectively. Dual mobility hip implant has been introduced by Bousquet in the 1970s [49] and has proved successful in preventing dislocation [50, 51]. For this kind of hip implant, the intermediate UHMWPE liner is not fixed onto the metal cup any more, and thus it could rotate with femoral head under some extreme condition. It is important to understand the biomechanics performance of this kind of implant. The following paragraphs present some research on its biomechanics.

Using the dynamic finite element method, Gao et al. investigated contact mechanics of dual mobility hip implant under different frictions [52]. The influence of friction on contact mechanics for dual mobility hip implant is shown in Figs. 8.15, 8.16, 8.17, and 8.18. For different friction coefficient ratios of dual mobility hip implant, (1) contact pressure and contact area varied with applied loads. However, contact zone was obviously different with a friction coefficient ratio of 1 comparing to that with a friction of coefficient ratio of 1.6. Only relative sliding occurred for inner articulation if friction coefficient ratio was 1, but outer relative sliding happened when friction coefficient ratio increased to 1.6; (2) for different coefficient ratios, the maximum contact area at liner outer surface was evidently higher than that of inner surface, which finally led to contrasting results for the maximum contact pressure of these two surfaces.

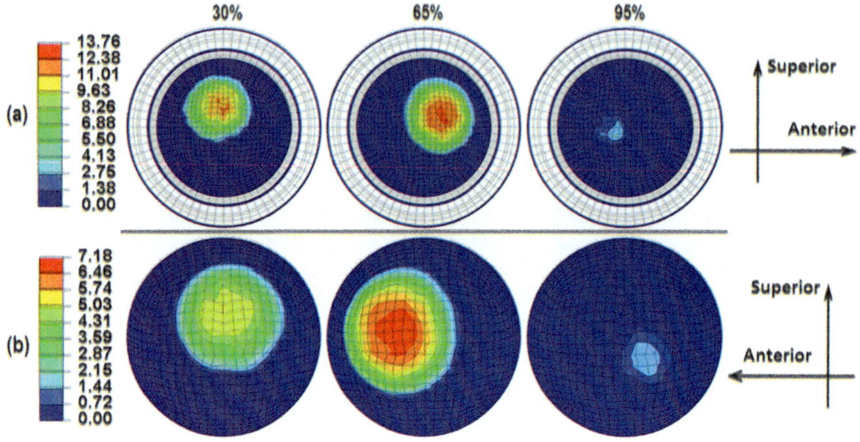

Fig. 8.15 Contours of the liner contact pressure under a friction coefficient ratio of 1 during different walking instants: (**a**) inner contact pressure (MPa), (**b**) outer contact pressure (MPa)

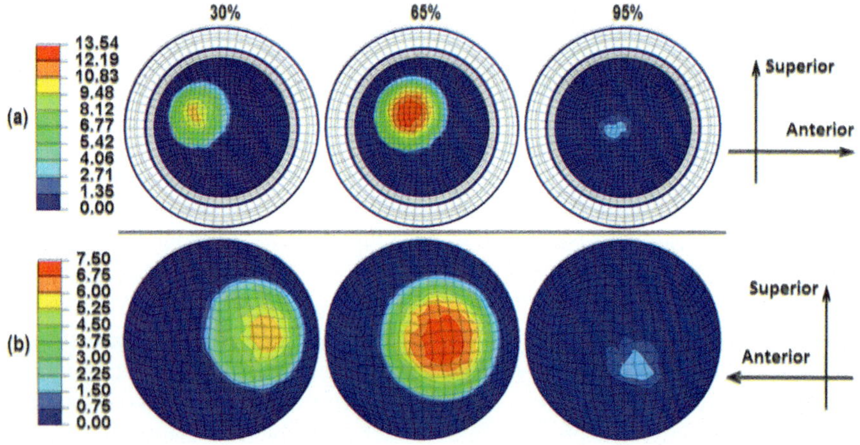

Fig. 8.16 Contours of the liner contact pressure under a friction coefficient ratio of 1.6 during different walking instants: (**a**) inner contact pressure (MPa), (**b**) outer contact pressure (MPa)

The results of accumulated sliding distance with different friction coefficients are shown in Figs. 8.19 and 8.20. It is concluded that the liner kept static when the ratio of friction coefficient was 1 or 1.23. However, outer relative sliding occurred when the ratio of friction coefficients between two articulations (inner vs outer) increased to 1.6 for dual mobility hip implant. Comparing the maximum sliding distance of the liner outer and inner surface under different ratio of friction coefficient, the outer maximum value was much larger than the inner maximum value. What's more, the liner outer surface possesses a larger contact area than the inner surface when the liner rotated under high ratio of friction coefficient. This means that, if the relative sliding happens for the liner outer surface owing to friction coefficient, it will lead to

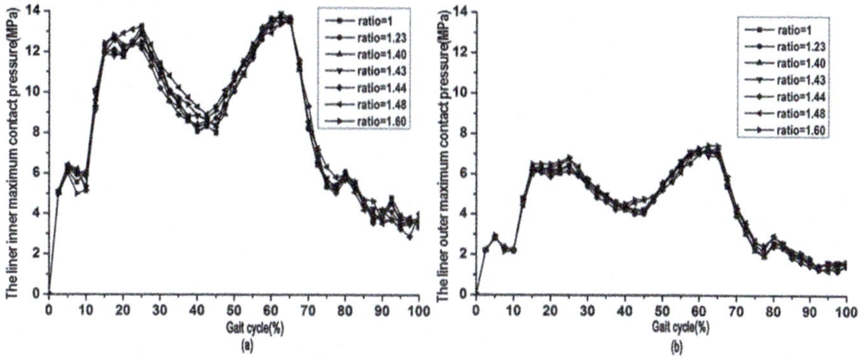

Fig. 8.17 Maximum contact pressure of the liner as a function of the gait cycle under different friction coefficient ratios of the inner to the outer articulation: (**a**) inner maximum contact pressure and (**b**) outer maximum contact pressure

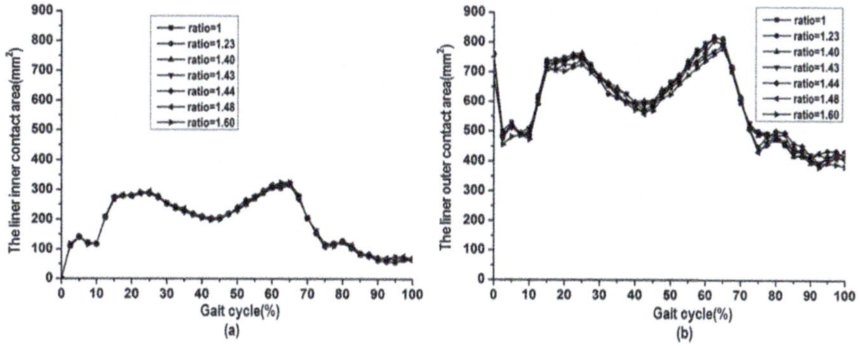

Fig. 8.18 Contact area of the liner as a function of the gait cycle under different friction coefficient ratios of the inner to the outer articulation: (**a**) inner contact area and (**b**) outer contact area

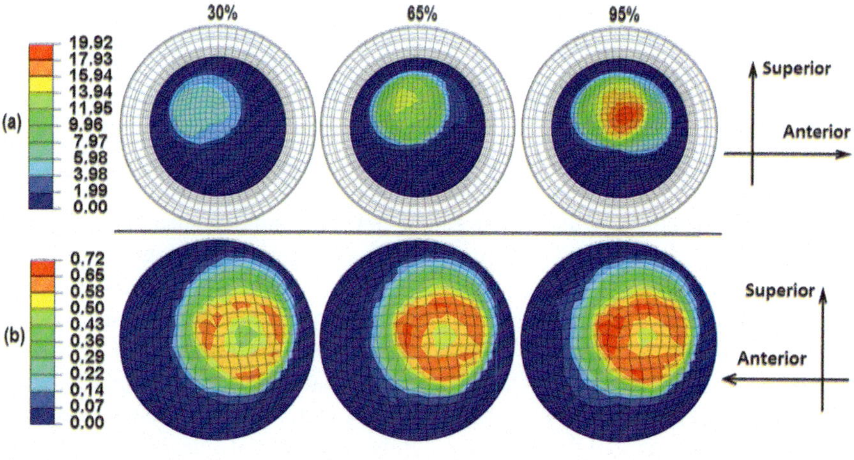

Fig. 8.19 Contours of the liner accumulated sliding distance under a friction coefficient ratio of 1 during different walking instants: (**a**) inner accumulated sliding distance(mm) and (**b**) outer accumulated sliding distance (mm)

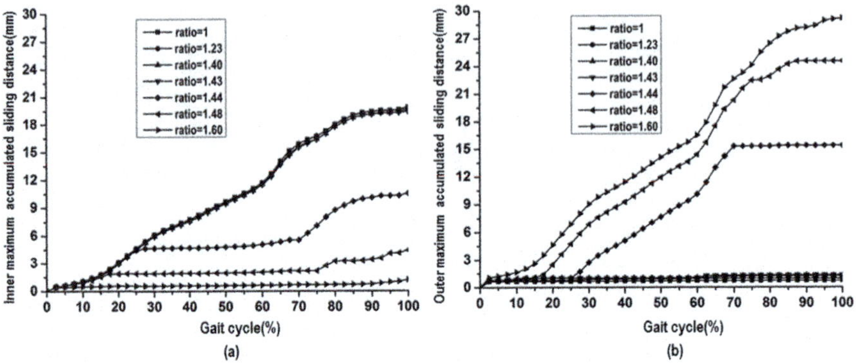

Fig. 8.20 Maximum accumulated sliding distance of the liner as a function of the gait cycle under different friction coefficient ratios of the inner to the outer articulation: (**a**) inner maximum accumulated sliding distance and (**b**) outer maximum accumulated sliding distance

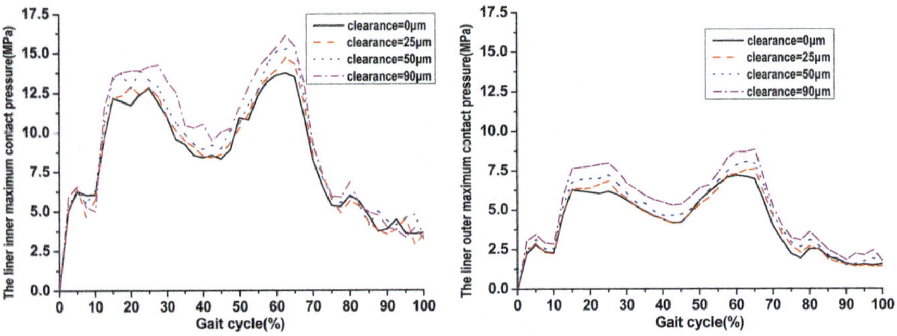

Fig. 8.21 Maximum contact pressure the liner as a function of the gait cycle under different initial clearances of the outer articulation and a fixed friction coefficient ratio of 1.40: (**a**) inner maximum contact pressure and (**b**) outer maximum contact pressure

more wear volume because wear volume is proportional to sliding distance and contact area. The liner outer wear loss of the retrieved dual mobility hip implants from patients is larger than that of the inner side, which could be explained by the rotation of the liner.

Different clearances will influence both the maximum contact pressure and contact area for conventional artificial hip joint according to previous studies. The clearance may also affect the contact mechanics of dual mobility hip implant. In a study by Gao et al., different initial clearances from 0 to 90 μm were set between liner outer surface and back shell inner surface under a constant ratio of friction coefficient of 1.4. The results of the maximum contact pressure and contact area of both surfaces of the liner are shown in Figs. 8.21 and 8.22. The maximum contact pressure of both surface of the liner was slightly increased with clearance increasing. In contrast, the influence of clearance on contact area of liner outer surface was significant. The outer contact area of the liner sharply decreased with the increasing

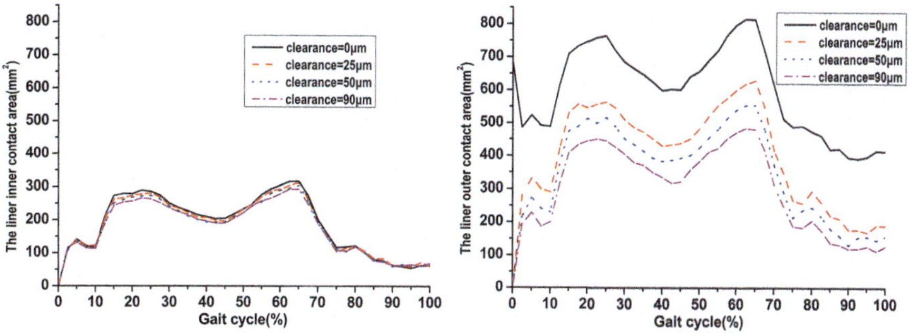

Fig. 8.22 Contact area of the liner as a function of the gait cycle under different initial clearances of the outer articulation and a fixed friction coefficient ratio of 1.40: (**a**) inner contact area and (**b**) outer contact area

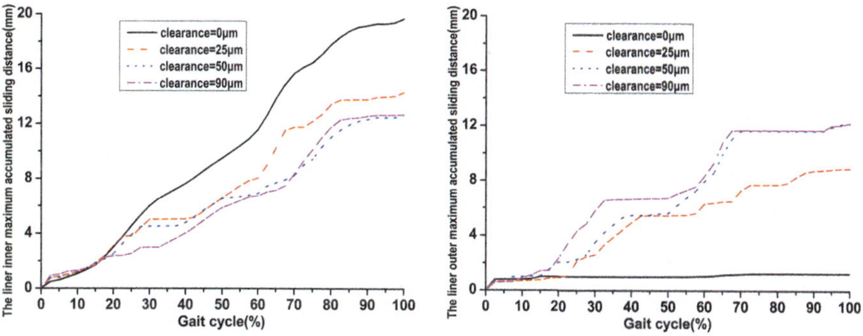

Fig. 8.23 Maximum accumulated sliding distance of the liner as a function of the gait cycle under different initial clearances of the outer articulation and a fixed friction coefficient ratio of 1.40. (**a**) Inner maximum accumulated sliding distance. (**b**) Outer maximum accumulated sliding distance

clearance, but the inner contact area almost remained the same. With different initial interferences between liner outer surface and back shell inner surface, both contact pressure and contact area were different. It may suggest that a tiny clearance could be designed for outer articulation of dual mobility hip implant to decrease the contact area without affecting other performances of contact mechanics.

The maximum sliding distances of the liner inner and outer surfaces under different clearances are shown in Fig. 8.23. It is apparent that the liner kept static when the clearance was zero, but it started to gradually rotate with the head with the increasing clearance. It means that, if the ratio of friction coefficient is slightly lower than the critical value, the clearance of the outer articulation would be helpful for the outer relative movement for the dual mobility hip implant. However, if the friction coefficients were the same at both articulations, increasing outer surface clearance could not lead to the outer movement for dual mobility hip implant.

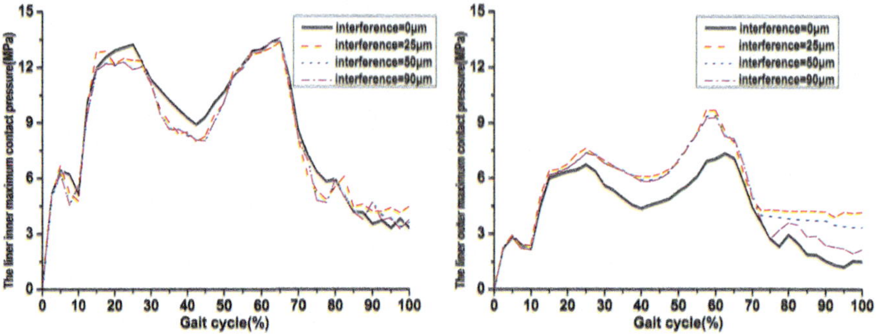

Fig. 8.24 Maximum contact pressure the liner as a function of the gait cycle under different initial interferences of the outer articulation and a fixed friction coefficient ratio of 1.48: (**a**) inner maximum contact pressure and (**b**) outer maximum contact pressure

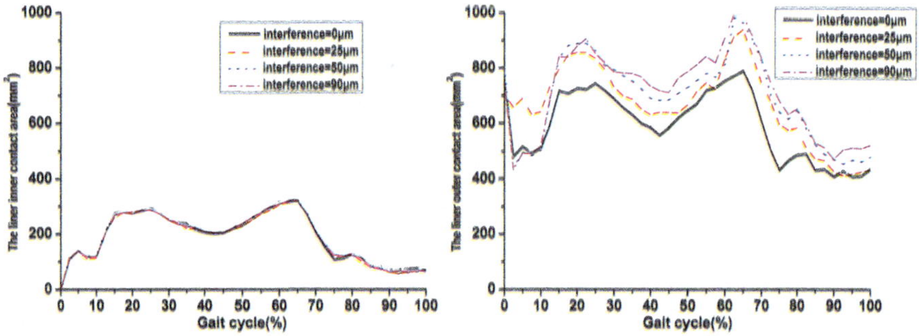

Fig. 8.25 Contact area of the liner as a function of the gait cycle under different initial interferences of the outer articulation and a fixed friction coefficient ratio of 1.48: (**a**) inner contact area and (**b**) outer contact area

Similarly, different interferences from 0 to 90 μm were set between the liner and the back shell to investigate its effect on contact mechanics and kinematics for dual mobility hip implant. The results of contact pressure and contact area under different initial interferences are shown in Figs. 8.24 and 8.25. The results are different from those with different initial clearances for dual mobility hip implant. Initial interferences between the liner outer surface and the back shell inner surface did not affect the liner inner maximum contact pressure and contact area but led to apparent increase in both liner outer maximum contact pressure and contact area. The results of the sliding distance with different initial interferences are shown in Fig. 8.26. It is obvious that the liner rotated with the femoral head at the fixed friction coefficient ratio of 1.48 if the interference was 0. However, even an interference of 25 μm prevented the liner rotating with the head. It means that, when the liner tends to rotate

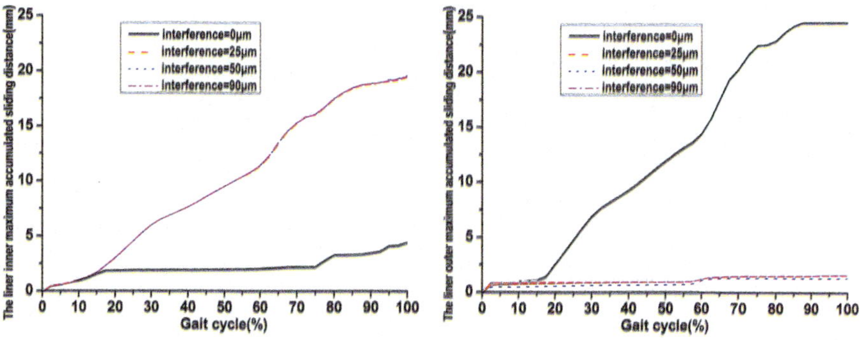

Fig. 8.26 Maximum accumulated sliding distance of the liner as a function of the gait cycle under different initial interferences of the outer articulation and a fixed friction coefficient ratio of 1.48. (**a**) Inner maximum accumulated sliding distance. (**b**) Outer maximum accumulated sliding distance

with the femoral head, it is convenient to prevent its rotation by setting a tiny interference at the outer articulation for dual mobility hip implant.

For dual mobility hip implant, there is another possibility to cause relative sliding of the outer articulation. This may be induced by the contact between stem neck and liner inner rim. Once this happens, the impingement also occurs to UHMWPE liner, and then the liner will rotate with stem until the impingement stops. What's more, the contact between the stem and the UHMWPE liner of the dual mobility hip implant would happen under extreme motions, such as climbing stairs and standing up. Therefore, the dynamic finite element method has been used by Gao et al. to predict the patient movements that led to contact between stem and UHMWPE liner and finally the relative sliding of outer articulation for dual mobility hip implant. The results under chair down/chair up and combining movements are shown in Figs. 8.27, 8.28, 8.29, and 8.30.

It is obvious that contact between the stem neck and the UHMWPE liner inner rim occurred for both cases. The outer accumulating sliding increased with this contact, and the increase of the outer accumulating sliding stopped once the contact ended. Therefore, the dual mobility from different movements of patient after surgery is an unquestionable reason to the dual wear for dual mobility hip implant. However, this contact was rather short during the whole movement, and the outer maximum accumulating sliding distance was much shorter than that of the inner value. This means that the outer wear caused by this case is much lower than that of inner wear. This does not agree with the observations by Geringer et al. [53] and Adam et al. [54] that the outer wear and the inner wear were almost the same. Therefore, more studies are needed to further investigate the true reasons to the wear volume of both articulations for dual mobility hip implant.

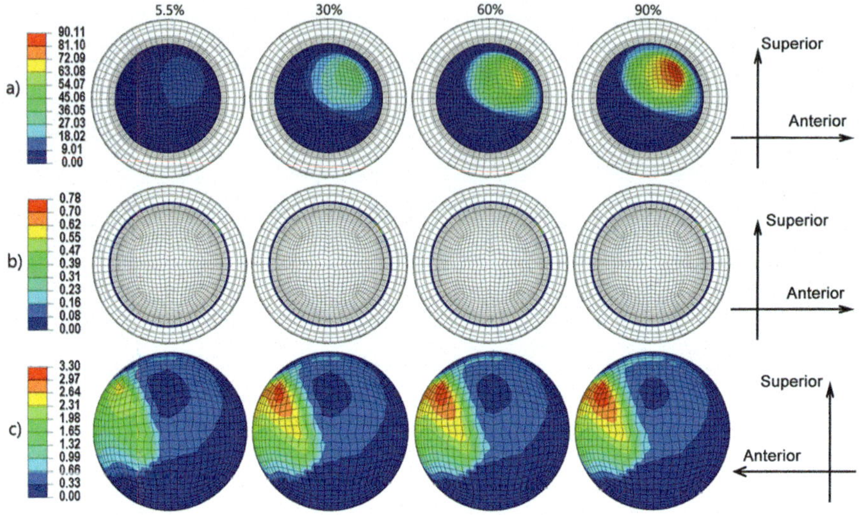

Fig. 8.27 Contours of the liner accumulated sliding distance during different chair down/chair up instants: (**a**) inner accumulated sliding distance (mm), (**b**) liner inner rim accumulated sliding distance (mm), and (**c**) outer accumulated sliding distance (mm)

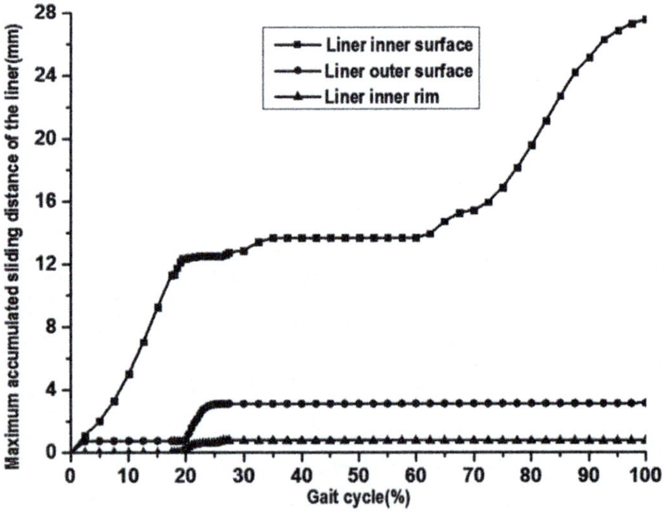

Fig. 8.28 Maximum accumulated sliding distance of the liner as a function of the gait cycle under chair down/chair up movement

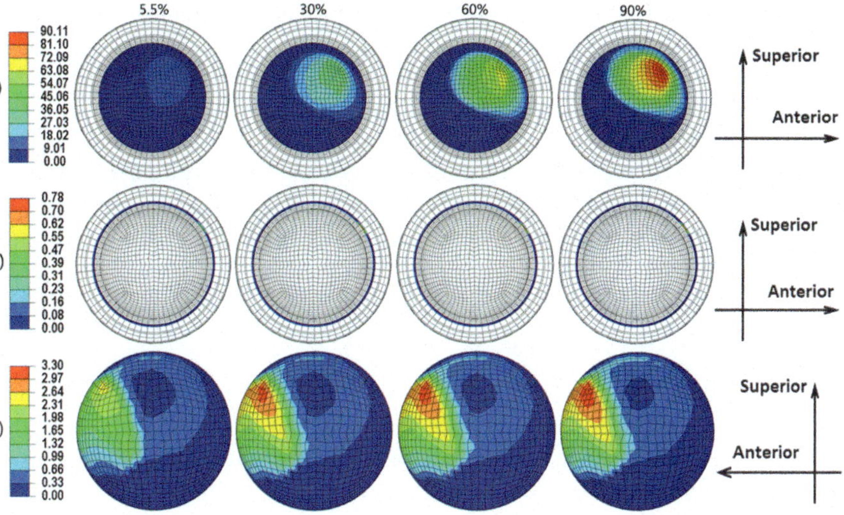

Fig. 8.29 Contours of the liner accumulated sliding distance during combining gait instants: (**a**) inner accumulated sliding distance (mm), (**b**) liner inner rim accumulated sliding distance (mm), and (**c**) outer accumulated sliding distance (mm)

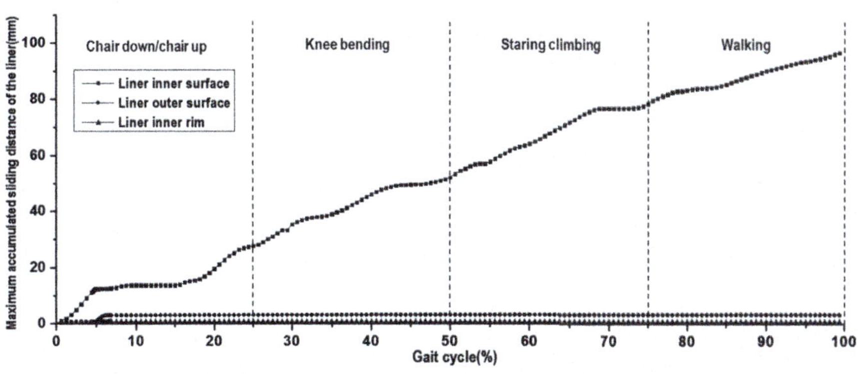

Fig. 8.30 Maximum accumulated sliding distance of the liner as a function of the gait cycle under combining movements

8.3 Biotribology of Artificial Hip Joints

8.3.1 Introduction

Biotribological studies of the bearing surfaces of artificial hip joints include friction, wear, and lubrication, which have been shown to mainly depend on the lubrication regimes involved. There are three lubrication regimes, boundary lubrication, fluid-film lubrication, and mixed lubrication. In the boundary lubrication regime, a significant asperity contact is experienced, and consequently both friction and wear are high. In the fluid-film lubrication regime where the two bearing surfaces are completely separated by a continuous lubricant, minimal friction and wear are expected. The mixed lubrication regime consists of both fluid-film lubricated and boundary contact regions. Friction and lubrication studies are usually performed to understand the wear mechanism involved in artificial hip joints. However, friction forces may be important in determining the stresses experienced at the interface between the implant and the bone cement [55] as well as temperature rise [56]. The following paragraphs will introduce research on friction, lubrication, and wear of artificial hip joints.

8.3.2 Friction

Friction in artificial hip joints is usually measured in a pendulum-like simulator with a dynamic load in the vertical direction and a reciprocating rotation in the horizontal direction. The coefficient of friction is usually expressed as friction factor defined as

$$\mu = \frac{T}{w(d_{\text{head}}/2)} \tag{8.6}$$

where T is the measured friction torque and w the load.

The measured coefficient of friction in a particular hip prosthesis itself can generally reveal the nature of the lubrication regime, since each mechanism is associated with broad ranges of the coefficient of friction. The variation in the coefficient of friction against a Sommerfeld number, $S = (\eta u d_{\text{head}}/w)$[1], can further indicate the lubrication regime. If the measured friction factors remain constant, decrease, or increase as the Sommerfeld number increases, the associated modes of lubrication are boundary, mixed, or fluid film, respectively [2].

The experimental studies of metal-on-UHMWPE hip implants by Unsworth et al. [46] clearly indicated boundary or mixed lubrication under steady load, although a suddenly applied load was found to invoke squeeze-film lubrication action. Similar conclusions were drawn by O'Kelly et al. [47] under dynamic loading conditions. In a study by Scholes and Unsworth [57] in 2000, a simple harmonic oscillatory motion and dynamic loading were applied in a hip function simulator. The results indicated that the friction factors were in the range 0.02–0.06 for 28 mm diameter femoral heads and UHMWPE cups, which are representative of mixed lubrication. Then a study by Banchet et al. [58] in 2007 tested the friction coefficients for different materials vs UHMWPE, and the results are shown in Fig. 8.31.

Figure 8.32 shows the friction factor of metal-on-UHMWPE artificial hip joints by different manufacturers by using Leeds ProSim hip simulator. It can be seen that the friction factor fluctuated during the whole walking process in both F/E and A/A orientations.

[1]Where η is viscosity and u velocity

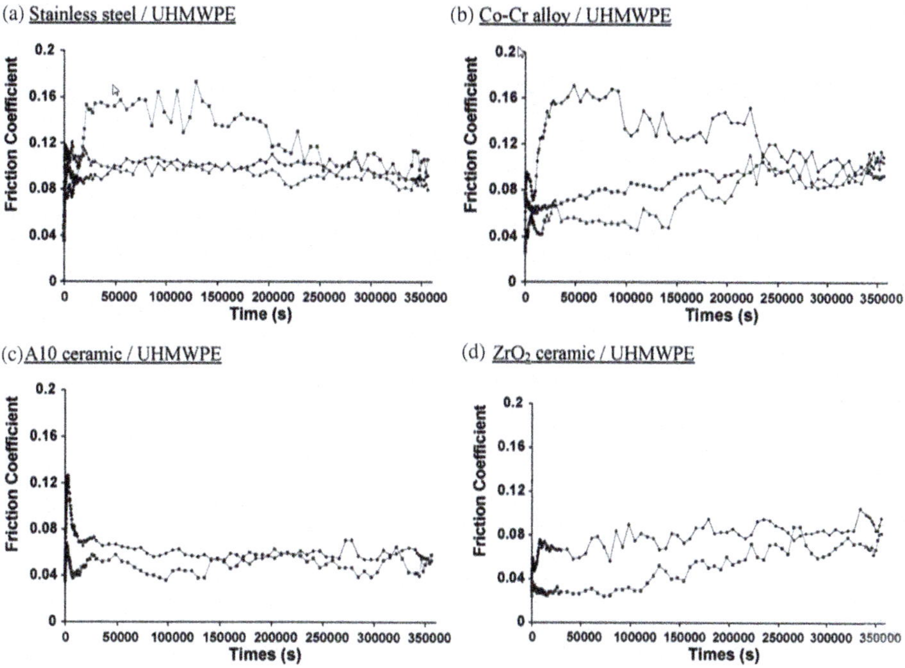

Fig. 8.31 Friction coefficient evolutions vs time. (Reprinted from Ref. [58], Copyright 2007, with permission from Elsevier)

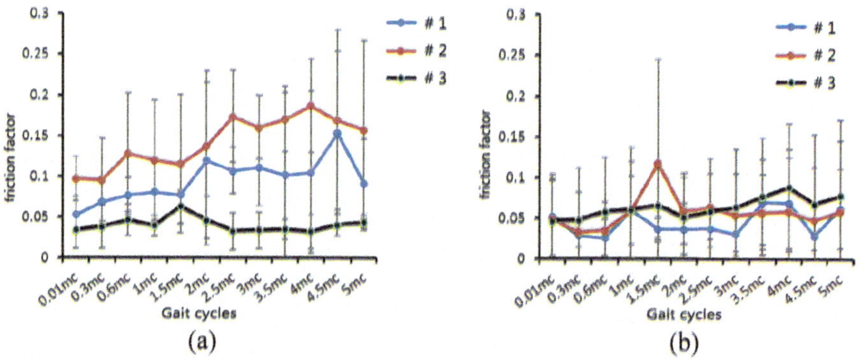

Fig. 8.32 Friction factor (mean ± SD) variation in (**a**) FE and (**b**) AA during walking by Leeds ProSim hip simulator

8.3.3 Lubrication

Lubrication studies of artificial hip joints are generally carried out using both experimental and theoretical approaches. The experimental measurement usually involves the detection of the separation between the two bearing surfaces using a simple resistivity technique. A large resistance may imply a thick lubricant film, while a small resistance is attributed to the direct surface contact. Such a technique is directly applicable to metal-on-metal bearings as well as metal-on-UHMWPE and ceramic-on-ceramic bearings if appropriate coatings are used [59]. The theoretical analysis generally involves the solution to the Reynolds equation, together with the elasticity equation subjected to the dynamic load and speed experienced during walking. The predicted film thickness (h_{min}) is then compared with the average surface roughness (*Ra*) using the following simple criterion:

$$\lambda = \frac{h_{\min}}{\left[{Ra_{\text{head}}}^2 + {Ra_{\text{cup}}}^2\right]^{1/2}} \tag{8.7}$$

The lubrication regime is then classified as fluid film, mixed, or boundary if the predicted λ is greater than 3, between 1 and 3, or less than 1, respectively.

Fluid-film lubrication studies in metal-on-UHMWPE hip implants have been largely based on the elastohydrodynamic mechanism associated with the relatively compliant polyethylene material. Initial attempts to predict the lubricating film thickness were made by using an equivalent ball-on-plane configuration with the equivalent radius, a quasi-static condition, and a semi-infinite solid model, and the Hamrock and Dowson [60] film thickness formulae were applied accordingly [59]. The effects of the ball-in-socket configuration and the finite thickness of the acetabular cup were subsequently examined, initially under quasi-static conditions [61] and later extended to transient walking conditions [14]. It was shown that the average of the predicted transient lubricating film thickness over one normal walking cycle was remarkably close to that estimated under the quasi-static condition based on the average load and speed. The predicted average lubricating film thickness was in the range between 0.1 and 0.2 μm. Therefore, a mixed lubrication regime was predicted, since typical average surface roughness between 0.1 and 1 μm has been reported for UHMWPE bearing surfaces [62]. This also demonstrates the importance of the mixed lubrication analysis of metal-on-UHMWPE hip implants.

8.3.4 Wear

Our current understanding of the mechanisms of wear particle-induced osteolysis has been developed from over 30 years' experience with metal-on-UHMWPE implants. The major factors limiting the longevity of initially well-fixed UHMWPE total joint replacements are osteolysis and aseptic loosening [63]. There are evidences from in vivo and in vitro studies that osteolysis is an UHMWPE particle-related phenomenon.

After total hip arthroplasty, a pseudo-capsule forms around the joint, and this may have a pseudo-synovial lining. A thin fibrous interfacial tissue with few vessels or cells may form at the bone–cement or bone–prosthesis interface [64]. At revision surgery for aseptic loosening, the fibrous membrane is thickened and highly vascularized and contains a heavy infiltrate of UHMWPE-laden macrophages and multinucleated giant cells [65]. There is a correlation between the number of macrophages and the volume of UHMWPE wear debris in the tissues adjacent to areas of aggressive osteolysis [66, 67]. There is a direct relationship between the particle concentration and the duration the implant, and there are billions of particles generated per gram of tissue [68, 69]. Osteolysis likely occurs when the threshold of particles exceeds 1×10^{10} per gram of tissue [67]. Each mg of polyethylene wear has been estimated to generate 1.3×10^{10} particles [68].

The UHMWPE particles isolated from retrieved tissues vary in size and morphology, from large platelet-like particles up to 250 μm in length, fibrils, shreds, and sub-micrometer globule-shaped spheroids 0.1–0.5 μm in diameter [68, 70]. The vast majority of particles are the globular spheroids, and the most frequent distribution is about 0.1–0.5 μm, although the larger particles may account for a high proportion of the total volume of wear debris. Analysis of the mass distribution as a function of size is therefore necessary to discriminate between patient samples [68, 71]. UHMWPE wear particles generated in vitro in hip simulators have a larger proportion of the mass of particles in the 0.01–1 μm-sized range than those isolated from periprosthetic tissues [71, 72]. This may indicate that in vivo, the smaller particles are disseminated more widely away from the implant site. Recently, improvements to particle imaging techniques have revealed nanometer-sized UHMWPE particles generated in hip simulators. These particles have yet to be identified in vivo. These nanometer-size particles account for the greatest number of particles generated but a negligible proportion of the total volume [73].

Studies of the response of macrophages to clinically relevant, endotoxin-free polyethylene particles in vitro have clearly demonstrated that particle-stimulated macrophages elaborate a range of potentially osteolytic mediators and bone resorbing activity [74–78]. The activity to induce bone resorption in particle-stimulated macrophage supernatants has been shown dependent on particle size and concentration with particles in the 0.1–1.0 μm size range at a volumetric concentration of 10–100 m^3 per cell, which are the most biologically reactive [74, 76]. This effect of UHMWPE particle size on osteolysis has been demonstrated in animal studies [79]. These findings have enabled preclinical prediction of the functional biological activity of different polyethylenes by analyzing the wear rate and mass distribution of the particles as a function of particle size [72, 80].

8.3.4.1 Wear Test

Wear of artificial hip joints has been investigated extensively, due to its direct relevance to biological reactions and clinical problems of osteolysis and loosening. Volumetric wear and wear particles can be measured using the following machines:

(a) Pin-on-disc machine
(b) Pin-on-plate machine
(c) Hip joint simulators

A unidirectional sliding motion is usually used in the pin-on-disc machine, and the reciprocating motion is added to the pin-on-plate machine. Both of these machines are used to screen potential bearing materials under well-controlled and often simplified conditions. Generally, it is necessary to introduce additional motion in order to produce a multidirectional motion. Using this simple device, wear rate could be tested for different material combinations. For metal-on-UHMWPE hip implants, both linear wear rate and volumetric wear rate were tested. The volumetric wear rate of metal-on-UHMWPE was about $1.066 \times 10^{-6} mm^3 N^{-1} m^{-1}$ by Maxian [81] and Teoh [82].

The next stage of wear testing is usually carried out in hip joint simulators with a varied degree of complexity of the three-dimensional loading and motion patterns experienced by hip joints while immersing the test joints in a lubricant mimicking the synovial fluid. Wear can be evaluated by either dimensional or gravimetric means. Representative commercial hip joint simulators of HUT-4 and Leeds ProSim are shown in Fig. 8.33. These simulators are consisted of at least six stations that operate simultaneously.

Through wear testing, both weight loss and morphology of wear debris could be investigated. When UHMWPE was combined with different materials, the wear rate varied a lot according to Buford et al. [83], as shown in Fig. 8.34. The 316 L steel has the highest wear depth, and the Co–Cr–Mo has the highest wear resistance. Wear volume of conventional and γ-irradiated UHMWPE has been tested [85] (Fig. 8.35). The wear volume of highly cross-linked UHMWPE was much lower than that of conventional UHMWPE. Essner et al. [84] demonstrated that the majority length of wear debris was at the range of 0.1~1 μm (Fig. 8.36).

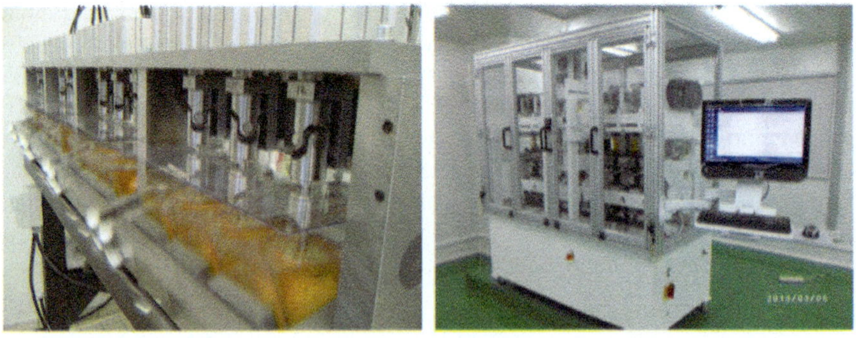

(a) (b)

Fig. 8.33 Hip simulators (**a**) HUT-4 and (**b**) Leeds ProSim

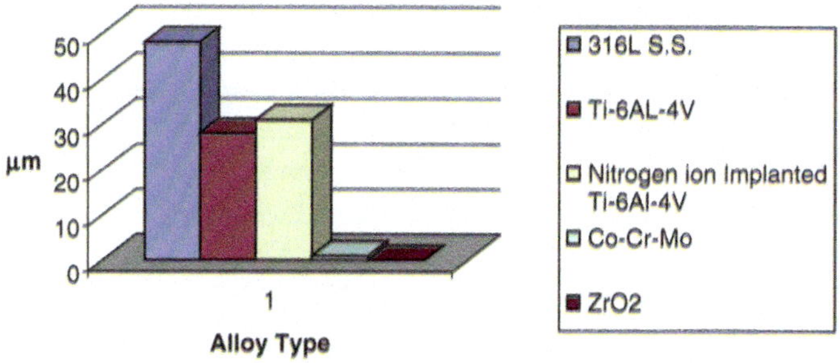

Fig. 8.34 Wear depth in various biomaterials. (Reprinted from Ref. [83], Copyright 2004, with permission from Elsevier)

Fig. 8.35 Hip simulator wear comparison of non-radiated and γ-irradiation in air UHMWPE cups. (Reprint for Ref. [85], Copyright 2005, with permission from Elsevier)

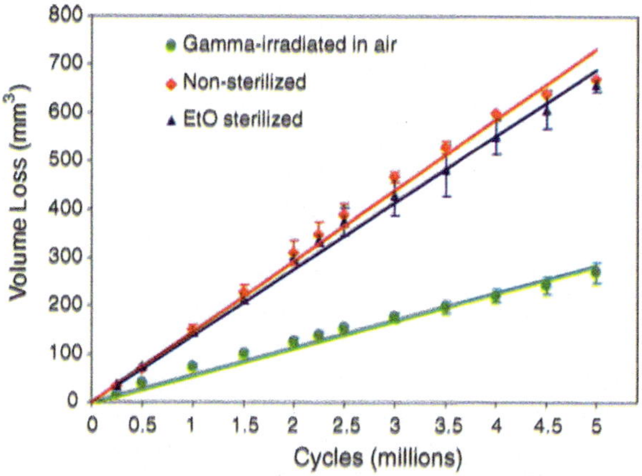

Fig. 8.36 Hip simulator Duration® UHMWPE acetabular liner wear debris. (Reprinted from Ref. [84], Copyright 2005, with permission from Elsevier)

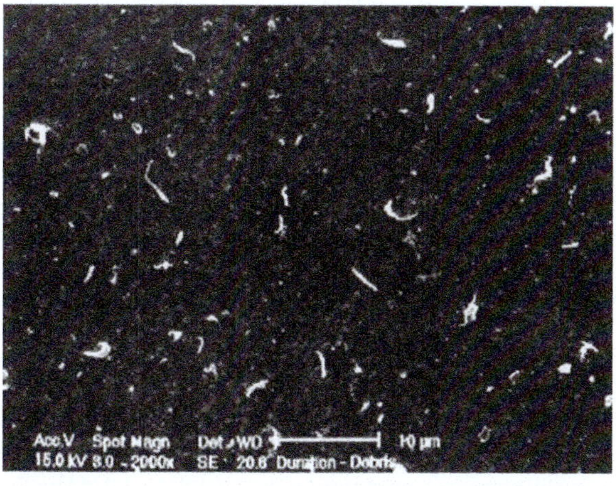

Oral et al. [86] tested wear loss and measured wear trace of artificial hip joints of different UHMWPE materials. The result of wear loss for different UHMWPE materials is shown in Fig. 8.37. The wear loss of UHMWPE sharply decreased if

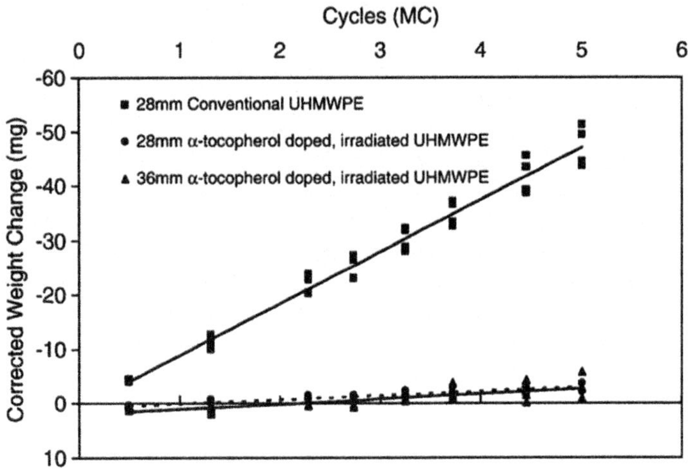

Fig. 8.37 Average total weight change of tested liners ($n = 4$) corrected for fluid absorption shown with corresponding linear regression lines. (Reprinted from Ref. [85], Copyright 2006, with permission from Elsevier)

it was irradiated. Besides, there were still machining marks on irradiated UHMWPE surface, while this was polished for conventional UHMWPE cup surface, as shown in Fig. 8.38.

Hip simulator indeed could test both wear volume and morphology of artificial hip joints, and experimental data are consistent with clinic outcomes. However, wear test is rather time-consuming; a short-term test of 5 million walking gait cycle will last at least 3 months. Besides, it is also somewhat expensive. Furthermore, it is hard to use this method to do parametric studies. Thus, wear numerical computation has been introduced and widely used to predict wear of artificial hip joints.

8.3.4.2 Wear Numerical Computation

The numerical method of wear prediction for artificial hip joints was firstly introduced by Maxian et al. [81] in 1996. In that study, the Archard's wear law was used to calculate wear depth and volume of UHMWPE cup, and corresponding data were about 0.1 mm/year and 15 mm^3/year, respectively. In the first wear model, because wear rate UHMWPE was assumed constant, it could only be used to predict wear of the artificial hip used by Maxian et al. [81]. However, since this numerical method was introduced, parametric analysis of wear for artificial hip joints has become available. Then, the elastic–plastic property of UHMWPE is considered for the wear numerical computation for metal-on-UHMWPE artificial hip joints. Besides, parameters that could affect wear rate of UHMWPE material are taken into account in the wear rate derived, such as contact pressure, surface roughness, and sliding velocity. The procedure of that numerical method could be illustrated by using the following flowchart in Fig. 8.39.

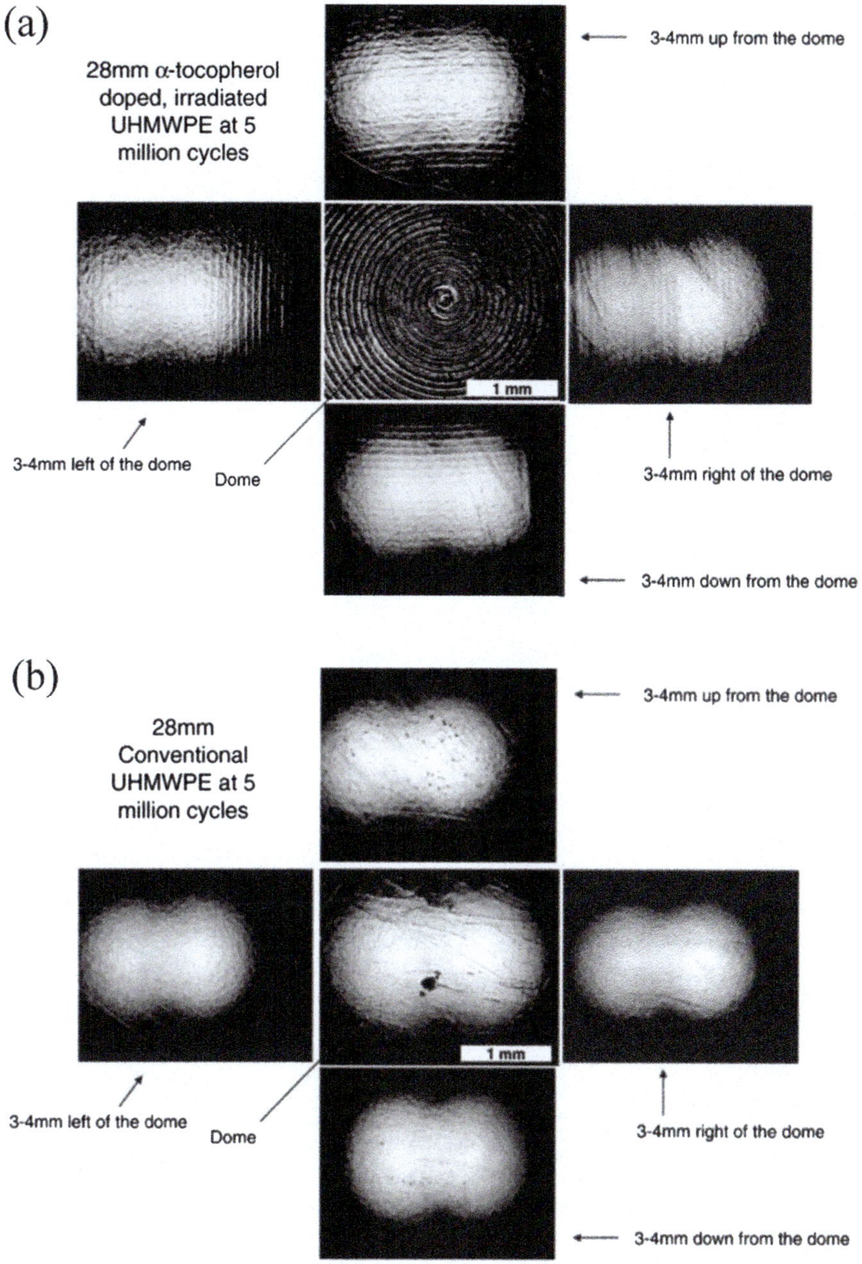

Fig. 8.38 Compilation of photos showing a 28-mm vitamin E doped, (**a**) irradiated, and (**b**) conventional UHMWPE liner after 5Mc of testing on the hip simulator. Machining marks are present in all four quadrants and also at the dome. (Reprinted from Ref. [85], Copyright 2006, with permission from Elsevier)

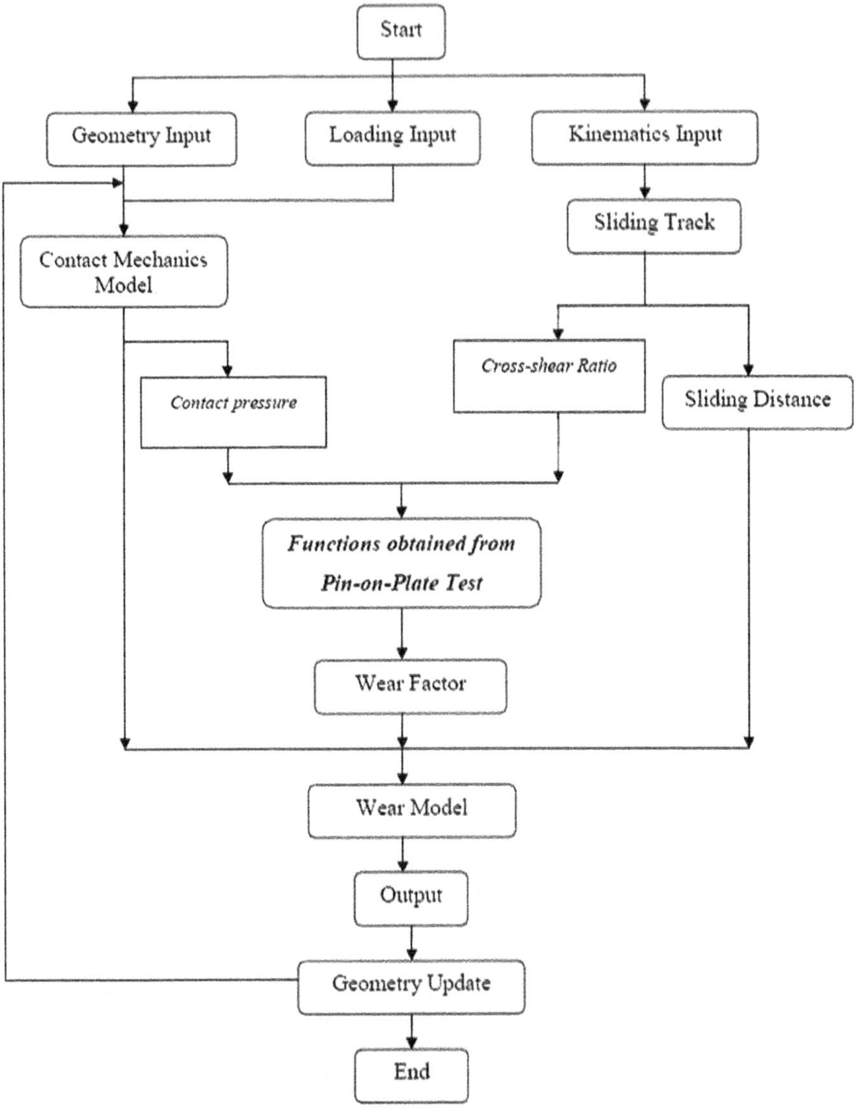

Fig. 8.39 Computational and experimental elements in wear simulation

The effect of multidirectional sliding of artificial hip joints on the wear of UHMWPE cups has been considered. According to experimental observations, the polymeric chains of UHMWPE acquire a principal molecular orientation (PMO) and are thus hardened along this direction, which makes the polymers harder to fail and wear. In contrast, in the direction perpendicular to the PMO, there is a strain softening phenomenon which is predominantly responsible for the detachment of fibrous wear debris from the worn surfaces (Fig. 8.40). Thus, the so-called cross-

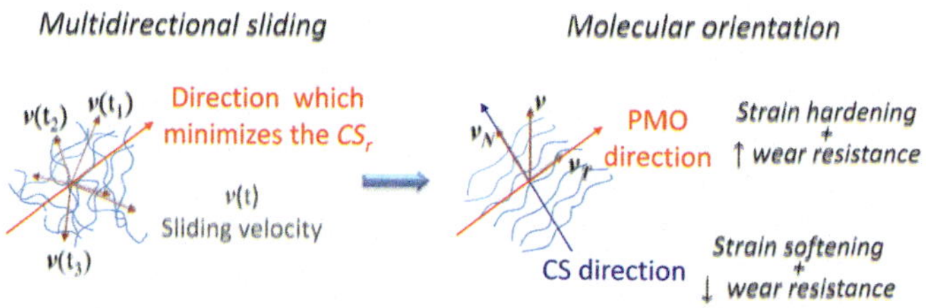

Fig. 8.40 Schematic representation of the cross-shear effect. UHMWPE polymeric chain, initially randomly oriented (**a**) and reoriented in the PMO direction because of multidirectional sliding on a harder counterface (**b**). (Reprinted from ref. [86], Copyright 2008, with permission from Elsevier)

shear ratio, *CS*, is defined to represent the effect of this multi-direction motion on wear rate of UHMWPE. The cross-shear ratio, *CS*, could be determined by frictional work released perpendicular to the PMO dividing the total frictional work.

The latest studies by Kang et al. [86, 87] and Liu et al. [44] have further developed the method to calculate cross-shear ratio, *CS*, and the wear coefficient k_c (or wear factor k_f) that is related to *CS*. When different forces and rotation motions are used, the *CS* distributions could be predicted for UHMWPE cup, as described by Mattei et al. (Fig. 8.41). It can be seen that wear coefficient (or wear factor) would increase with increasing *CS*.

Kang et al. also compared both *CS* and wear volume of conventional and highly cross-linked UHMWPE artificial hip joints [88]. The relationship between wear factor and *CS* for different UHMWPE materials is shown in Fig. 8.42. The wear factor under same *CS* of conventional UHMWPE was much higher than that of highly cross-linked UHMWPE. The wear volume results by Kang for both UHMWPE materials are shown in Fig. 8.43. It is obvious that the wear volume of conventional UHMWPE under same conditions was much higher than that of highly cross-linked UHMWPE.

By using different wear laws, both wear depth and volume could be calculated, and the corresponding data by Mattei et al. [89] are shown in Fig. 8.44. It can be seen both wear depth and wear volume are different by using different wear coefficients (or wear factors) in spite of the nearly same wear zones.

Liu et al. [90] also used a new wear law introduced by themselves to predict wear volume of UHMWPE with creep considered, only to find that the total wear volume of UHMWPE cup sharply increased.

8.4 Conclusions and Future Trends

8.4.1 Conclusions

The UHMWPE artificial hip joint is widely used in clinic, but wear performance limits its long-term usage. This chapter has introduced attempts to enhance the

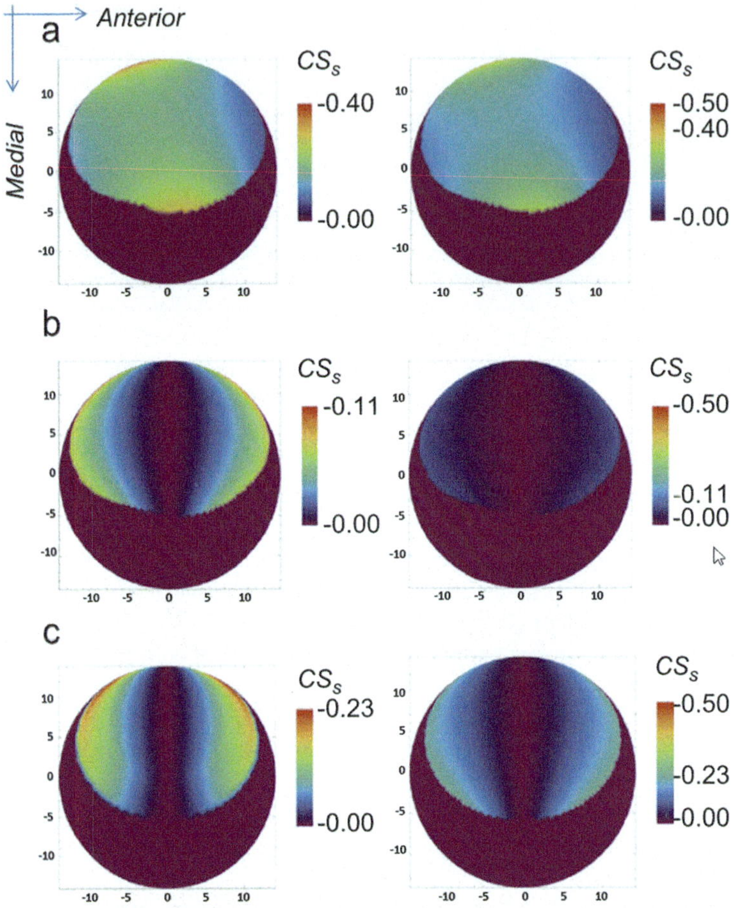

Fig. 8.41 CS_s maps for three different kinematics conditions: complete in vivo gait velocity (**a**), simplified case with ω_y–ω_y (**b**), and Leeds ProSim simulator (**c**). In the left column, each plot is represented within its own minimum and maximum values, while in the right column all the *CS* plots are depicted in the same range (0–0.5). (Reprinted from Ref. [86], Copyright 2013, with permission from Elsevier)

UHMWPE cup wear resistance ability and research on wear mechanism and performance from both biomechanics and biotribology. The main conclusions obtained from these studies can be mainly summarized as:

(a) The musculoskeletal multibody dynamics of the UHMWPE hip implant is necessary, and both hip joint center and surgical approach would influence hip contact force; thus the hip implant position and surgical approach should be considered according to patient during surgery.
(b) The finite element method has been widely used to investigate both contact mechanics and kinematics. The contact pressure and contact area are related to

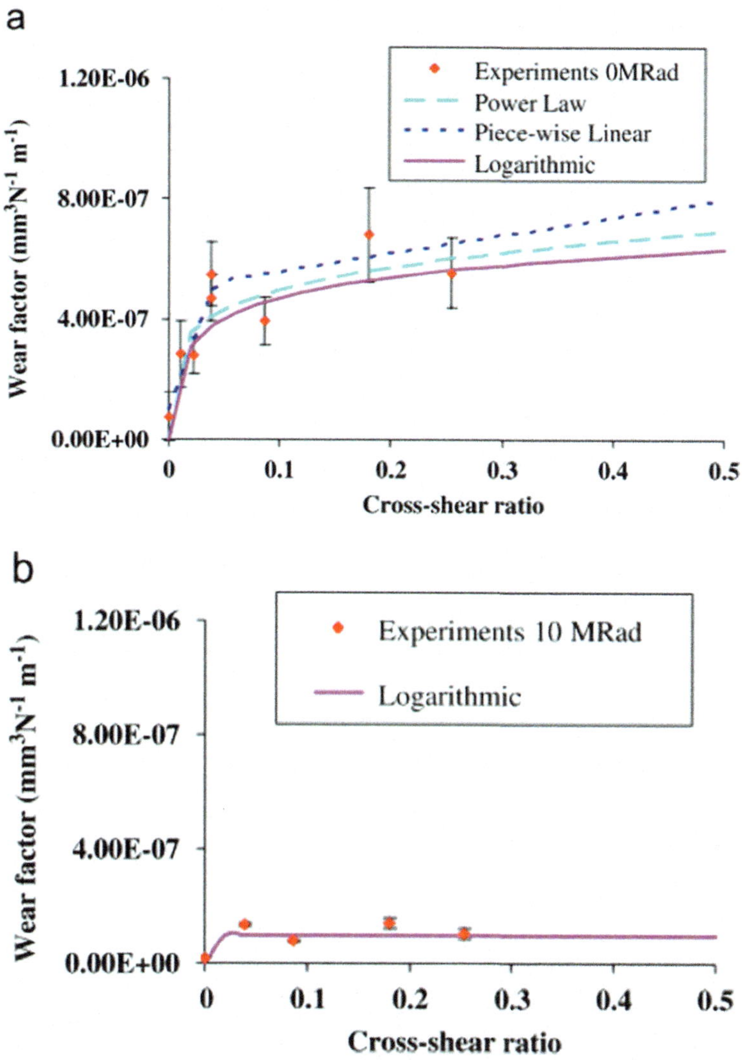

Fig. 8.42 Wear factor versus average cross-shear ratio for (**a**) 0 MRad conventional UHMWPE and (**b**) 10 MRad highly cross-linked UHMWPE (dispersion bar: mean ($n \geq 3$) ± 95% CI) [88]

hip contact force, clearance/interference, friction coefficient, and so on, while the sliding distance would be mainly affected by patient movements.

(c) Biotribology performances including friction, lubrication, and wear of the UHMWPE hip implant have been widely investigated; the UHMWPE hip implant experiences boundary lubrication, and its wear volume is different for the conventional UHMWPE and the high cross-linked UHMWPE material.

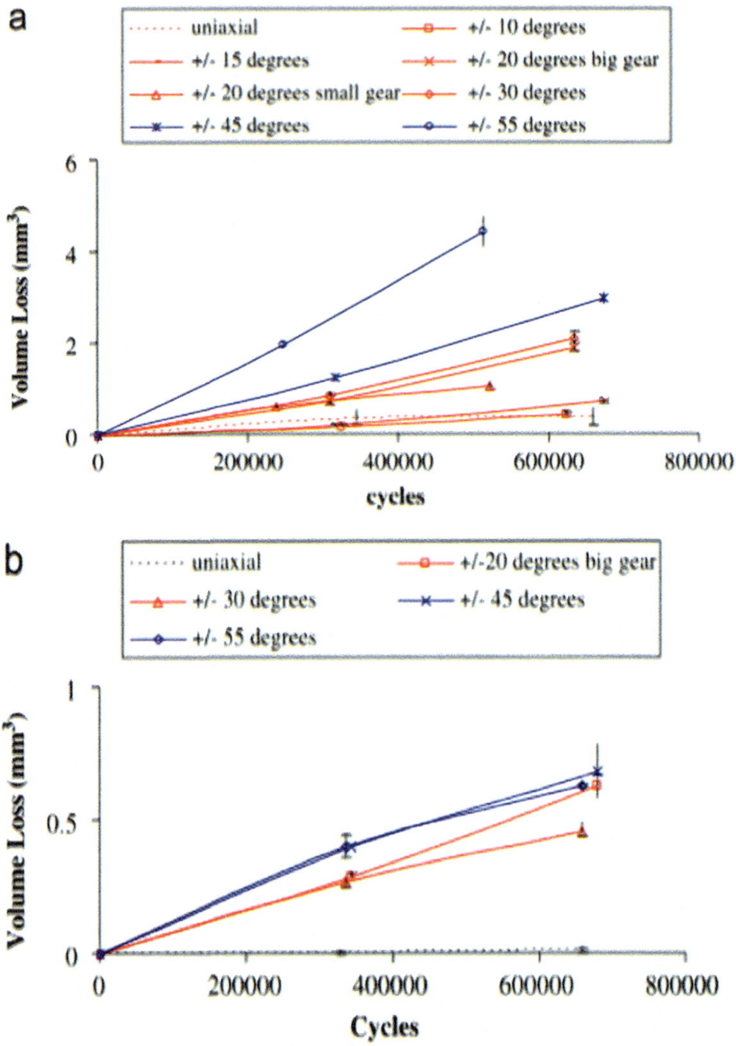

Fig. 8.43 Volume loss at various stages of cycles for (**a**) 0 MRad conventional UHMWPE and (**b**) 10 MRad highly cross-linked UHMWPE (dispersion bar: mean ($n \geq 3$) $\pm$ 95% CI) [88]

8.4.2 *Future Trends*

Currently, much research has been done to understand wear mechanisms and improve its resistant performance as represented in this chapter. However, longer lifetime is expected in the future for the UHMWPE hip implant because of diverse demands from young and active patients. This may be achieved by the following attempts:

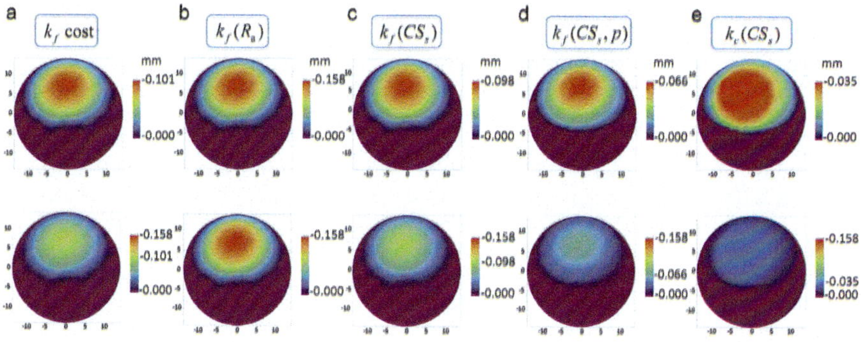

Fig. 8.44 Acetabular cup linear wear maps (in the x_c-y_c plane) and volumetric wear after 1Mc are predicted using different wear laws, under in vivo gait conditions. (Reprinted from Ref. [89], Copyright 2013, with permission from Elsevier)

(a) Improvement of Wear Performance of the UHMWPE Material

As done in the past decades, both wear depth and wear volume of the UHMWPE hip implant have been dramatically decreased by replacing conventional UHMWPE material with highly cross-linked UHMWPE material. In the future, the UHMWPE with even lower wear could be expected.

(b) Individual Preoperative Surgical Planning

The wear performance of an UHMWPE hip implant is related to the prosthesis design, surgical accuracy, and patient characteristics. Therefore, it is better to simulate wear performance of the hip implant before surgery using the current computational methods to predict the optimal parameters to guide the surgery.

References

1. Gluscevic BM, Kraljevic BD, Jovanovic VA, Stosic PB, Milosavljevic DM, Radivojevic RM (2006) Primary total hip arthroplasty in patients with rheumatoid arthritis. Acta Chir Lugosl 53 (4):4
2. Dowson D (2001) New joints for the millennium: wear control in total replacement hip joints. Proc Inst Mech Eng H J Eng Med 215(4):335–358
3. De Martino I, Triantafyllopoulos GK, Sculco PK, Sculco TP (2014) Dual mobility cups in total hip arthroplasty. World J Orthop 5(3):180–187. https://doi.org/10.5312/wjo.v5.i3.180
4. Tarabolsi M, Klassen T, Mantwill F, Gartner F, Siegel F, Schulz AP (2013) Patterned CoCrMo and Al O surfaces for reduced free wear debris in artificial joint arthroplasty. J Biomed Mater Res A 101(12):3447–3456. https://doi.org/10.1002/jbm.a.34644
5. Ge S, Wang S, Huang X (2009) Increasing the wear resistance of UHMWPE acetabular cups by adding natural biocompatible particles. Wear 267(5–8):770–776. https://doi.org/10.1016/j.wear.2009.01.057
6. Dumbleton JH, Manley MT (2005) Metal-on-metal total hip replacement: what does the literature say? J Arthroplast 20(2):174–188

7. Greene JW, Malkani AL, Kolisek FR, Jessup NM, Baker DL (2009) Ceramic-on-ceramic total hip arthroplasty. J Arthroplast 24(6 Suppl):15–18. https://doi.org/10.1016/j.arth.2009.04.029
8. Lindeque B, Hartman Z, Noshchenko A, Cruse M (2014) Infection after primary total hip arthroplasty. Orthopedics 37(4):257–265. https://doi.org/10.3928/01477447-20140401-08
9. Berry DJ, von Knoch M, Schleck CD, Harmsen WS (2004) The cumulative long-term risk of dislocation after primary Charnley total hip arthroplasty. J Bone Joint Surg Am 86-A(1):9–14
10. Kosashvili Y, Drexler M, Backstein D, Safir O, Lakstein D, Safir A, Chakravertty R, Dwyer T, Gross A (2014) Dislocation after the first and multiple revision total hip arthroplasty: comparison between acetabulum-only, femur-only and both component revision hip arthroplasty. Can J Surg/Journal canadien de chirurgie 57(2):E15–E18
11. Carriero A, Zavatsky A, Stebbins J, Theologis T, Lenaerts G, Jonkers I, Shefelbine SJ (2014) Influence of altered gait patterns on the hip joint contact forces. Comput Methods Biomech Biomed Engin 17(4):352–359. https://doi.org/10.1080/10255842.2012.683575
12. Schwachmeyer V, Damm P, Bender A, Dymke J, Graichen F, Bergmann G (2013) In vivo hip joint loading during post-operative physiotherapeutic exercises. PLoS One 8(10):e77807. https://doi.org/10.1371/journal.pone.0077807
13. Seedhom BB, Wallbridge NC (1985) Walking activities and wear of prostheses. Ann Rheum Dis 44(12):838–843
14. Jagatia M, Jalali-Vahid D, Jin ZM (2001) Elastohydrodynamic lubrication analysis of ultra-high molecular weight polyethylene hip joint replacements under squeeze-film motion. Proc Inst Mech Eng H J Eng Med 215(2):141–152. https://doi.org/10.1243/0954411011533706
15. Williams PA, Clarke IC (2009) Understanding polyethylene wear mechanisms by modeling of debris size distributions. Wear 267(1–4):646–652. https://doi.org/10.1016/j.wear.2008.12.069
16. Ingham E, Fisher J (2005) The role of macrophages in osteolysis of total joint replacement. Biomaterials 26(11):1271–1286. https://doi.org/10.1016/j.biomaterials.2004.04.035
17. Austin MS, Higuera CA, Rothman RH (2012) Total hip arthroplasty at the rothman institute. HSS J musculoskelet J Hosp for Spec Surg 8(2):146–150. https://doi.org/10.1007/s11420-012-9268-x
18. Wang L, Liu X, Li D, Liu F, Jin Z (2014) Contact mechanics studies of an ellipsoidal contact bearing surface of metal-on-metal hip prostheses under micro-lateralization. Med Eng Phys 36 (4):419–424. https://doi.org/10.1016/j.medengphy.2013.12.017
19. Saikko V, Calonius O, Keranen J (2004) Effect of slide track shape on the wear of ultra-high molecular weight polyethylene in a pin-on-disk wear simulation of total hip prosthesis. J Biomed Mater Res B Appl Biomater 69(2):141–148. https://doi.org/10.1002/jbm.b.20043
20. Ong KL, Manley MT, Nevelos J, Greene K (2012) Review: biomechanical issues in total hip replacement. Surg Technol Int 22:222–228
21. Heiberg KE, Ekeland A, Bruun-Olsen V, Mengshoel AM (2013) Recovery and prediction of physical functioning outcomes during the first year after total hip arthroplasty. Arch Phys Med Rehabil 94(7):1352–1359. https://doi.org/10.1016/j.apmr.2013.01.017
22. Alberta Hip Improvement P, MacKenzie JR, O'Connor GJ, Marshall DA, Faris PD, Dort LC, Khong H, Parker RD, Werle JR, Beaupre LA, Frank CB (2012) Functional outcomes for 2 years comparing hip resurfacing and total hip arthroplasty. J Arthroplast 27(5):750–757. e752. https://doi.org/10.1016/j.arth.2011.10.004
23. Ewen AM, Stewart S, St Clair Gibson A, Kashyap SN, Caplan N (2012) Post-operative gait analysis in total hip replacement patients-a review of current literature and meta-analysis. Gait Posture 36(1):1–6. https://doi.org/10.1016/j.gaitpost.2011.12.024
24. Jensen C, Aagaard P, Overgaard S (2011) Recovery in mechanical muscle strength following resurfacing vs standard total hip arthroplasty – a randomised clinical trial. Osteoarthr Cartil 19 (9):1108–1116. https://doi.org/10.1016/j.joca.2011.06.011
25. Damsgaard M, Rasmussen J, Christensen ST, Surma E, de Zee M (2006) Analysis of musculoskeletal systems in the anybody modeling system. Simul Model Pract Theory 14 (8):1100–1111. https://doi.org/10.1016/j.simpat.2006.09.001
26. Ren L, Qian Z, Ren L (2014) Biomechanics of musculoskeletal system and its biomimetic implications: a review. J Bionic Eng 11(2):159–175. https://doi.org/10.1016/S1672-6529(14)60033-0

27. Hallaceli H, Uruc V, Uysal HH, Ozden R, Hallaceli C, Soyuer F, Ince Parpucu T, Yengil E, Cavlak U (2014) Normal hip, knee and ankle range of motion in the Turkish population. Acta Orthop Traumatol Turc 48(1):37–42. https://doi.org/10.3944/aott.2014.3113
28. Delp SL, Anderson FC, Arnold AS, Loan P, Habib A, John CT, Guendelman E, Thelen DG (2007) OpenSim: open-source software to create and analyze dynamic simulations of movement. IEEE Trans Biomed Eng 54(11):1940–1950. https://doi.org/10.1109/tbme.2007.901024
29. Manders C, New A, Rasmussen J (2008) Validation of musculoskeletal gait simulation for use in investigation of total hip replacement. J Biomech 41:S488. https://doi.org/10.1016/S0021-9290(08)70487-1
30. Sherman MA, Seth A, Delp SL (2011) Simbody: multibody dynamics for biomedical research. Procedia IUTAM 2:241–261
31. Rydell NW (1966) Forces acting on the femoral head-prosthesis. A study on strain gauge supplied prostheses in living persons. Acta Orthop Scand 37(88):1–132
32. Zhang X, Chen Z, Wang L, Yang W, Li D, Jin Z (2015) Prediction of hip joint load and translation using musculoskeletal modelling with force-dependent kinematics and experimental validation. Proc Inst Mech Eng H J Eng Med 229(7):477–490. https://doi.org/10.1177/0954411915589115
33. Besong AA, Lee R, Farrar R, Jin ZM (2001) Contact mechanics of a novel metal-on-metal total hip replacement. Proc Inst Mech Eng H J Eng Med 215(6):543–548. https://doi.org/10.1243/0954411011536145
34. Wilson DR, Apreleva MV, Eichler MJ, Harrold FR (2003) Accuracy and repeatability of a pressure measurement system in the patellofemoral joint. J Biomech 36(12):1909–1915
35. Muller O, Parak WJ, Wiedemann MG, Martini F (2004) Three-dimensional measurements of the pressure distribution in artificial joints with a capacitive sensor array. J Biomech 37 (10):1623–1625. https://doi.org/10.1016/j.jbiomech.2004.01.024
36. Hua X, Li J, Wang L, Jin Z, Wilcox R, Fisher J (2014) Contact mechanics of modular metal-on-polyethylene total hip replacement under adverse edge loading conditions. J Biomech 47 (13):3303–3309. https://doi.org/10.1016/j.jbiomech.2014.08.015
37. Wang L, Williams S, Udofia I, Isaac G, Fisher J, Jin Z (2012) The effect of cup orientation and coverage on contact mechanics and range of motion of metal-on-metal hip resurfacing arthroplasty. Proc Inst Mech Eng H J Eng Med 226(11):877–886. https://doi.org/10.1177/0954411912456926
38. Hua X, Wroblewski BM, Jin Z, Wang L (2012) The effect of cup inclination and wear on the contact mechanics and cement fixation for ultra high molecular weight polyethylene total hip replacements. Med Eng Phys 34(3):318–325. https://doi.org/10.1016/j.medengphy.2011.07.026
39. Liu F (2005) Contact mechanics and elastohydrodynamic lubrication analysis of metal-on-metal hip implant with a sandwich acetabular cup under transient walking condition. University of Bradford
40. Beaule PE, Amstutz HC (2005) Orientation of the femoral component in surface arthroplasty of the hip. J Bone Joint Surg Am 87(5):1162
41. Brodner W, Grubl A, Jankovsky R, Meisinger V, Lehr S, Gottsauner-Wolf F (2004) Cup inclination and serum concentration of cobalt and chromium after metal-on-metal total hip arthroplasty. J Arthroplast 19(8 Suppl 3):5
42. Isaac GH, Siebel T, Oakeshott RD, McLennan-Smith R, Cobb AG, Schmalzried TP, Vail TP (2009) Changes in whole blood metal ion levels following resurfacing: serial measurements in a multi-centre study. Hip Int J Clin Exp Res Hip Pathol Ther 19(4):330–337
43. Kang L, Galvin AL, Jin ZM, Fisher J (2006) A simple fully integrated contact-coupled wear prediction for ultra-high molecular weight polyethylene hip implants. Proc Inst Mech Eng H J Eng Med 220(1):33–46
44. Liu F, Leslie I, Williams S, Fisher J, Jin Z (2008) Development of computational wear simulation of metal-on-metal hip resurfacing replacements. J Biomech 41(3):686–694. https://doi.org/10.1016/j.jbiomech.2007.09.020

45. Wang L, Yang W, Peng X, Li D, Dong S, Zhang S, Zhu J, Jin Z (2015) Effect of progressive wear on the contact mechanics of hip replacements – does the realistic surface profile matter? J Biomech 48(6):1112–1118. https://doi.org/10.1016/j.jbiomech.2015.01.020
46. Saikko V, Calonius O (2002) Slide track analysis of the relative motion between femoral head and acetabular cup in walking and in hip simulators. J Biomech 35:10
47. Gao Y, Jin Z, Wang L, Wang M (2015) Finite element analysis of sliding distance and contact mechanics of hip implant under dynamic walking conditions. Proc Inst Mech Eng H J Eng Med 229(6):469–474. https://doi.org/10.1177/0954411915585380
48. Craig JJ (1989) Introduction to robotics: mechanics and control. Addison-Welsley, Reading
49. Guyen O, Pibarot V, Vaz G, Chevillotte C, Bejui-Hugues J (2009) Use of a dual mobility socket to manage total hip arthroplasty instability. Clin Orthop Relat Res 467(2):465–472. https://doi.org/10.1007/s11999-008-0476-0
50. Philippot R, Adam P, Farizon F, Fessy MH, Bousquet G (2006) Survival of cementless dual mobility sockets: ten-year follow-up. Revue de chirurgie orthopedique et reparatrice de l'appareil moteur 92(4):326–331
51. Philippot R, Camilleri JP, Boyer B, Adam P, Farizon F (2009) The use of a dual-articulation acetabular cup system to prevent dislocation after primary total hip arthroplasty: analysis of 384 cases at a mean follow-up of 15 years. Int Orthop 33(4):927–932. https://doi.org/10.1007/s00264-008-0589-9
52. Gao Y, Chai W, Wang L, Wang M, Jin Z (2016) Effect of friction and clearance on kinematics and contact mechanics of dual mobility hip implant. Proc Inst Mech Eng H J Eng Med 230(1):39–49. https://doi.org/10.1177/0954411915617198
53. Geringer J, Boyer B, Farizon F (2011) Understanding the dual mobility concept for total hip arthroplasty. Investigations on a multiscale analysis-highlighting the role of arthrofibrosis. Wear 271(9–10):2379–2385. https://doi.org/10.1016/j.wear.2011.02.027
54. Adam P, Farizon F, Fessy MH (2005) Dual articulation retentive acetabular liners and wear: surface analysis of 40 retrieved polyethylene implants. Revue de chirurgie orthopedique et reparatrice de l'appareil moteur 91(7):10
55. Nassutt R, Wimmer MA, Schneider E, Morlock MM (2003) The influence of resting periods on friction in the artificial hip. Clin Orthop Relat Res 407:127–138
56. Bergmann G, Graichen F, Rohlmann A, Verdonschot N, van Lenthe GH (2001) Frictional heating of total hip implants. Part 2: finite element study. J Biomech 34(4):429–435
57. Scholes SC, Unsworth A (2000) Comparison of friction and lubrication of different hip prostheses. Proc Inst Mech Eng H J Eng Med 214(1):49–57. https://doi.org/10.1243/0954411001535237
58. Banchet V, Fridrici V, Abry JC, Kapsa P (2007) Wear and friction characterization of materials for hip prosthesis. Wear 263(7–12):1066–1071. https://doi.org/10.1016/j.wear.2007.01.085
59. Jin ZM, Dowson D, Fisher J (1997) Analysis of fluid film lubrication in artificial hip joint replacements with surfaces of high elastic modulus. Proc Inst Mech Eng H J Eng Med 211(3):247–256. https://doi.org/10.1243/0954411971534359
60. Hamrock BJ, Dowson D (1978) Elastohydrodynamic lubrication of elliptical contacts for materials of low elastic modulus. I: fully flooded conjunction. Trans ASME J Lubric Technol 100(2):236–245
61. Jalali-Vahid D, Jagatia M, Jin ZM, Dowson D (2001) Prediction of lubricating film thickness in UHMWPE hip joint replacements. J Biomech 34(2):261–266
62. Elfick AP (1998) Surface topography of retrieved PCA acetabular liners: proposal for a novel wear mechanism. J Mater Sci Lett 17(13):1085–1088
63. Engh CA Jr, Hopper RH Jr, Huynh C, Ho H, Sritulanondha S, Engh CA Sr (2012) A prospective, randomized study of cross-linked and non-cross-linked polyethylene for total hip arthroplasty at 10-year follow-up. J Arthroplast 27(8 Suppl):2–7. e1. https://doi.org/10.1016/j.arth.2012.03.048
64. Goldring SR, Schiller AL, Roelke M, Rourke CM, O'Neil DA, Harris WH (1983) The synovial-like membrane at the bone-cement interface in loose total hip replacements and its proposed role in bone lysis. J Bone Joint Surg Am 65(5):575–584

65. Mirra JM, Marder RA, Amstutz HC (1982) The pathology of failed total joint arthroplasty. Clin Orthop Relat Res 170:175–183
66. Schmalzried TP, Jasty M, Harris WH (1992) Periprosthetic bone loss in total hip arthroplasty. Polyethylene wear debris and the concept of the effective joint space. J Bone Joint Surg Am 74 (6):849–863
67. Revell PA, al-Saffar N, Kobayashi A (1997) Biological reaction to debris in relation to joint prostheses. Proc Inst Mech Eng H J Eng Med 211(2):187–197. https://doi.org/10.1243/0954411971534304
68. Tipper JL, Ingham E, Hailey JL, Besong AA, Fisher J, Wroblewski BM, Stone MH (2000) Quantitative analysis of polyethylene wear debris, wear rate and head damage in retrieved Charnley hip prostheses. J Mater Sci Mater Med 11(2):117–124
69. Hirakawa K, Bauer TW, Stulberg BN, Wilde AH (1996) Comparison and quantitation of wear debris of failed total hip and total knee arthroplasty. J Biomed Mater Res 31(2):257–263. https://doi.org/10.1002/(SICI)1097-4636(199606)31:2<257::AID-JBM13>3.0.CO;2-I
70. Maloney WJ, Smith RL, Schmalzried TP, Chiba J, Huene D, Rubash H (1995) Isolation and characterization of wear particles generated in patients who have had failure of a hip arthroplasty without cement. J Bone Joint Surg Am 77(9):1301–1310
71. Howling GI, Barnett PI, Tipper JL, Stone MH, Fisher J, Ingham E (2001) Quantitative characterization of polyethylene debris isolated from periprosthetic tissue in early failure knee implants and early and late failure Charnley hip implants. J Biomed Mater Res 58(4):415–420
72. Endo M, Tipper JL, Barton DC, Stone MH, Ingham E, Fisher J (2002) Comparison of wear, wear debris and functional biological activity of moderately crosslinked and non-crosslinked polyethylenes in hip prostheses. Proc Inst Mech Eng H J Eng Med 216(2):111–122. https://doi.org/10.1243/0954411021536333
73. Ingram JH, Stone M, Fisher J, Ingham E (2004) The influence of molecular weight, crosslinking and counterface roughness on TNF-alpha production by macrophages in response to ultra high molecular weight polyethylene particles. Biomaterials 25(17):3511–3522. https://doi.org/10.1016/j.biomaterials.2003.10.054
74. Green TR, Fisher J, Matthews JB, Stone MH, Ingham E (2000) Effect of size and dose on bone resorption activity of macrophages by in vitro clinically relevant ultra high molecular weight polyethylene particles. J Biomed Mater Res 53(5):490–497
75. Matthews JB, Green TR, Stone MH, Wroblewski BM, Fisher J, Ingham E (2000) Comparison of the response of primary human peripheral blood mononuclear phagocytes from different donors to challenge with model polyethylene particles of known size and dose. Biomaterials 21 (20):2033–2044
76. Green TR, Fisher J, Stone M, Wroblewski BM, Ingham E (1998) Polyethylene particles of a 'critical size' are necessary for the induction of cytokines by macrophages in vitro. Biomaterials 19(24):2297–2302
77. Matthews JB, Green TR, Stone MH, Wroblewski BM, Fisher J, Ingham E (2000) Comparison of the response of primary murine peritoneal macrophages and the U937 human histiocytic cell line to challenge with in vitro generated clinically relevant UHMWPE particles. Biomed Mater Eng 10(3–4):229–240
78. Matthews JB, Green TR, Stone MH, Wroblewski BM, Fisher J, Ingham E (2001) Comparison of the response of three human monocytic cell lines to challenge with polyethylene particles of known size and dose. J Mater Sci Mater Med 12(3):249–258
79. Goodman SB, Fornasier VL, Lee J, Kei J (1990) The histological effects of the implantation of different sizes of polyethylene particles in the rabbit tibia. J Biomed Mater Res 24(4):517–524. https://doi.org/10.1002/jbm.820240408
80. Fisher J, Bell J, Barbour PS, Tipper JL, Matthews JB, Besong AA, Stone MH, Ingham E (2001) A novel method for the prediction of functional biological activity of polyethylene wear debris. Proc Inst Mech Eng H J Eng Med 215(2):127–132. https://doi.org/10.1243/0954411011533599
81. Maxian TA, Brown TD, Pedersen DR, Callaghan JJ (1996) A sliding-distance-coupled finite element formulation for polyethylene wear in total hip arthroplasty. J Biomech 29(5):687–692. https://doi.org/10.1016/0021-9290(95)00125-5

82. Teoh SH, Chan WH, Thampuran R (2002) An elasto-plastic finite element model for polyethylene wear in total hip arthroplasty. J Biomech 35:8
83. Buford A, Goswami T (2004) Review of wear mechanisms in hip implants: paper I – general. Mater Des 25(5):385–393. https://doi.org/10.1016/j.matdes.2003.11.010
84. Essner A, Schmidig G, Wang A (2005) The clinical relevance of hip joint simulator testing: in vitro and in vivo comparisons. Wear 259(7–12):882–886. https://doi.org/10.1016/j.wear.2005.02.105
85. Oral E, Christensen SD, Malhi AS, Wannomae KK, Muratoglu OK (2006) Wear resistance and mechanical properties of highly cross-linked, ultrahigh-molecular weight polyethylene doped with vitamin E. J Arthroplast 21(4):580–591. https://doi.org/10.1016/j.arth.2005.07.009
86. Kang L, Galvin AL, Brown TD, Jin Z, Fisher J (2008) Quantification of the effect of cross-shear on the wear of conventional and highly cross-linked UHMWPE. J Biomech 41(2):340–346. https://doi.org/10.1016/j.jbiomech.2007.09.005
87. Kang L, Galvin AL, Fisher J, Jin Z (2009) Enhanced computational prediction of polyethylene wear in hip joints by incorporating cross-shear and contact pressure in additional to load and sliding distance: effect of head diameter. J Biomech 42(7):912–918. https://doi.org/10.1016/j.jbiomech.2009.01.005
88. Jin ZM, Fisher J, Brown TD, Galvin AL, Kang L (2008) Wear simulation of ultra-high molecular weight polyethylene hip implants by incorporating the effects of cross-shear and contact pressure. Proc Inst Mech Eng H J Eng Med 222(7):1049–1064. https://doi.org/10.1243/09544119jeim431
89. Mattei L, Di Puccio F, Ciulli E (2013) A comparative study of wear laws for soft-on-hard hip implants using a mathematical wear model. Tribol Int 63:66–77. https://doi.org/10.1016/j.triboint.2012.03.002
90. Liu F, Fisher J, Jin Z (2012) Computational modelling of polyethylene wear and creep in total hip joint replacements: effect of the bearing clearance and diameter. Proc Inst Mech Eng H J Eng Tribol 226(6):552–563. https://doi.org/10.1177/1350650112441908

Chapter 9
Clinical Application of Roentgen Stereophotogrammetric Analysis Technique to Detect Hip Aseptic Loosening

Jin-Wu Wang, Yuan-Jing Xu, Han Yang, and Ke-Rong Dai

Abstract Artificial joint replacement is a common treatment for joints affected by trauma, osteoarthritis, or rheumatoid arthritis. Prosthetic aseptic loosening is a major clinical problem and one of the most important reasons for revisions after prosthetic replacement. Routine clinical testing to detect aseptic loosening involves physical and imaging examinations; however, they are not sufficiently accurate. The gold standard of orthodontic kinematics measurement is the Roentgen stereophotogrammetric analysis (RSA) technique, which can predict aseptic loosening of implants by measuring micromovements during the first 2 years after joint replacement. The accuracy of RSA technique is in the range 0.05–0.5 mm in translation and 0.15–1.15° in rotation. The RSA technique plays an important role in evaluating implant fixations with high precision and three-dimensional (3D) migration measurements. This chapter explains the basic principles and development of the RSA technique as well as its clinical application to detect hip artificial joint aseptic loosening.

Keywords Artificial joint prostheses · Aseptic loosening · Blocking · Dynamic quantitative analysis · Modeling · 2D–3D matching

J.-W. Wang (✉) · K.-R. Dai
Shanghai Key Laboratory of Orthopaedic Implants, Department of Orthopaedic Surgery, Shanghai Ninth People's Hospital, Shanghai Jiaotong University School of Medicine, Shanghai, China
e-mail: wangjw@sjtu.edu.cn

Y.-J. Xu · H. Yang
Clinical Translational Engineering Research Center of Digital Medicine, Ministry of Education, Shanghai Jiaotong University, Shanghai, China

J. Fu et al. (eds.), *UHMWPE Biomaterials for Joint Implants*, Springer Series in Biomaterials Science and Engineering 13,
https://doi.org/10.1007/978-981-13-6924-7_9

9.1 Introduction

Artificial joint replacement is a common treatment for joints that have been affected by trauma, osteoarthritis, or rheumatoid arthritis. Worldwide, 1.5 million total hip replacements and 500,000 total knee replacements are performed every year. The average lifespan of the prosthesis is approximately 15–20 years. A prosthesis has to be revised when the prosthetic liner has worn out or when the prosthesis has loosened with respect to the surrounding bone [1].

Artificial hip clinical application has been used for decades. It has an excellent medical and rehabilitation effect to restore the patient's joint function and relieve the patient's suffering. In the Netherlands, 20,000 total hip replacements and 7000 total knee replacements are performed annually. Prostheses may be secured in bones by using cement or by bony ingrowth onto the implant's surface. Reports state that there are about 800,000 total hip replacements annually in China, with a rapid increase. However, clinical applications have witnessed a series of problems that seriously limit the in vivo performance and lifespan of the artificial hip implants. Scientists and engineers have been struggling to improve the performance of the implants. Unfortunately, some prostheses have to be revised earlier than the average lifespan. Major causes to short-term revisions include septic loosening (infection), fracture of the prosthesis, or destruction of the polyethylene articular surface (delamination). Aseptic prosthetic loosening is a major clinical problem and is one of the most important reasons for revision. Statistics from Norwegian National Knee Ligament Registry [2] and Swedish Hip Arthroplasty Register [3] show that loosening failure account for approximately 50–60% of the total joint replacement failure. Although this situation gradually improved since the 1990s, aseptic loosening is still the main reason for artificial joint revisional surgery. It is noted that aseptic loosening is not prevailing in early implantation but becomes a major failure cause in the middle and late stages. In 2008, Swedish official data show that dislocation and infection become the major failure modes, which take up 38.5% and 25.9% of the total joint replacement failures, respectively, while aseptic loosening accounts for 18.4%. The loosening increases during 2 to 6 years after implantation and becomes the main reason for replacement failure after 6 years, accounting for approximately 85% of the total number of failures.

Some researchers reported statistical analysis on clinical data. From 1996 to 2003, Clohisy et al. [4] collected 439 cases of revisions and found that aseptic loosening accounted for 55% of the failures through retrospective analysis. From 2000 to 2012, Dalury et al. [5] analyzed 693 total knee arthroplasty renovations and found that aseptic loosening accounted for the highest place (23.1%). Ulrich et al. [6] analyzed 237 hip replacement failures and found aseptic loosening in 123 cases, accounting for 51.9%. Schroer et al. [7] analyzed 844 failed primary total knee arthroplasties that presented for revision surgery, and found that aseptic loosening was the predominant mechanism of failure (31.2%).

Prosthesis aseptic loosening may result from the following: (1) a surrounding osteolysis caused by abrasive particles leading to a decrease in support structure strength of the prosthesis; (2) the prosthesis is unstable or load-excessive due to an improper surgical procedure; (3) a surrounding tissue necrosis and osteolysis caused by an intermittent or continuous high fluid pressure [8].

Early diagnosis of aseptic loosening and osteolysis allows for active prevention, which can delay or avoid catastrophic consequences of severe osteolysis. Therefore, a research on the diagnosis method on artificial joint aseptic loosening can deepen the understanding of pathomechanisms and help guide clinical operations and evaluate postoperative effects.

9.2 Clinical Diagnosis of Artificial Joint Prosthesis Aseptic Loosening

In recent years, with the development of imaging technique and prosthetic interface biology research, there are many methods and indicators for the diagnosis of aseptic loosening. Radiation biopsy [9], bone mineral density measurement [10], biological marker detection [11], radionuclide diagnosis [12], and other methods are applied to detect loosening in the literature. However, most present diagnosis is still based on imaging examinations combined with clinical symptoms, which is far from a precise and early diagnostics of aseptic loosening.

9.2.1 Clinical Physical Examination

Pain reproduction of artificial joint replacement is an important clue to the clinical evaluation of aseptic loosening. Physicians can determine whether artificial joint replacement prosthesis is loose through a detailed physical examination. Lavernia and Alcerro [13] conducted a 1-year follow-up study of total hip arthroplasty (THA) in 98 patients, which showed that the clinical examination indicators of acetabular prosthesis loosening in sequence included femoral axial percussion pain, hip joint pain caused by external rotation, and hip pain. The indicators that can predict femoral stem prosthesis loosening in sequence are femoral axial percussion pain, hip joint pain caused by external rotation, thigh pain, hip joint with internal rotation, hip pain, and knee pain. The largest difference compared with infectious pain is that the pain caused by prosthesis loosening is closely related to weight [14].

9.2.2 Imaging Examination

Imaging examination can help with the diagnosis of prosthesis aseptic loosening and osteolysis. With the improvement of imaging and image processing techniques, imaging examination becomes more important to artificial joint loosening diagnosis.

9.2.2.1 Conventional X-ray Photography Technique

X-ray is the general radiographic method used to diagnose the loosening and blockage of artificial joints in clinic. The apparent displacement of prosthesis and the extent of osteolysis around the prosthesis can be clearly shown on an X-ray image. A translucent line around a prosthesis shows that prosthesis–bone interface integration is bad, and it is an important indication of aseptic loosening. The International Documentation and Evaluation System proposed a radiological standard of prostheses loosening after artificial hip replacement [15], namely, a clinical femoral sinking of more than 3 mm or a translucent band between bone cement and shank larger than 2 mm.

However, X-rays are essentially based on a single plane, static radiography, and are significantly affected by the angle of the radiograph, position of the patient, and experience of the physician. In addition, the variance between imaging markers and clinical performance makes precise diagnosis difficult. Engh et al. [16] asked four surgeons to read the X-ray films of 60 patients with total hip prosthesis replacement to assess osteolysis. Various evaluation results were provided with poor reproducibility. A set of continuously shot X-ray images are better than single ones shot separately for analysis.

9.2.2.2 Computerized Tomography (CT)

CT scanning is more sensitive for assessing osteolysis compared to X-ray, but metal artifacts and high radiation dose limit its application for detecting loosening.

Metals with high attenuation coefficient come in contact with soft tissues with low attenuation coefficient when metal prosthesis is implanted into the human body. Radiation through the metal with rapid attenuation causes beam hardening artifacts, which seriously affects the evaluation of the bone structure around the prosthesis. A method of reducing artifacts involves the use of multiprobe helical CT while increasing the exposure peak (kVp) and time (mAs); the radiation penetration and image quality are improved. The disadvantage is an increase in the ionizing radiation dose. The use of postprocessing software can further improve the image resolution, and applying improved image processing algorithms, such as an edge enhancement algorithm, also helps in imaging diagnosis. Garcia-Cimbrelo et al. [17] used multilayer scanning combined with artifact suppression techniques to evaluate the osteolysis of 60 hip prostheses and found that the sensitivity was superior to that of X-ray films. However, CT radiation dose is high; thus, it is impossible to replace X-ray examination as a routine detection means.

9.2.2.3 Magnetic Resonance Imaging (MRI)

MRI can detect pseudo-joint capsule biological reactions induced by wear particles before the occurrence of osteoclasts. It can also observe the oppressive performance in the local nerve and vascular structure caused by polyethylene foreign body granuloma or pseudo-joint capsule expansion, making up for the deficiency of X-ray and CT on the limited resolution of early synovial changes of granulosis in early soft tissues. Weiland et al. [18] compared MRI to plain film analysis in the assessment of periacetabular bone loss using a cadaver model. MRI was 95% sensitive in the detection of lesions, with 98% specificity and 96% accuracy. The mean absolute deviation in determining the lesion size was 0.8 ± 2.2 cm^3. Using conventional radiographic analysis, the overall sensitivity of lesion detection was 52%, and the specificity was 96%. Using plain film analysis, the identification of true lesions depended on the location with 83% of ilial lesions, 64% of pubic lesions, 55% of chial lesions, and 0% of posterior wall lesions was correctly identified.

In addition, even if there is no abnormal X-ray indication, MRI can also show abnormal soft tissue reactions of early patients after joint replacement, including prosthesis peripheral fluid, bone marrow edema, muscle edema, gluteus and buttocks tendon tear, piriformis and obturator muscle atrophy, medial femoral fracture, and others [19]. As the ferromagnetic metallic prosthesis is adjacent to the weak magnetic or diamagnetic soft tissues and the magnetic field strength is proportional to the magnetic sensitive constant volume of specific materials, MRI has a limitation of magnetic susceptibility artifacts, resulting in image distortion and frequency shift effect and affecting the diagnosis of MRI to a certain extent.

9.2.3 Detection Method of Artificial Joint Prosthesis Aseptic Loosening Based on Imaging

Although the measurement of bone mineral density, biological markers, and radionuclide diagnosis methods are applied to the detection of loosening in laboratory studies, they are not used as the gold standard to detect such diseases in the clinic. In addition, more effective and intuitive methods are desirable in practice. Therefore, a three-dimensional, visual, real-time, and accurate detection method is urgently required in the clinic.

With the development of computing techniques implemented in the medical field, some image-related detecting methods, like finite element (FE) and dual fluoroscopic imaging system, were gradually introduced into this field and played a critical role in the prediction of aseptic loosening of prostheses.

9.2.3.1 Finite Element

Prosthesis chronic aseptic loosening is accompanied by a progressive loss of bone mass, which is mainly due to the effect of bone dissolution and mechanical stress on the hip joint [20]. Stress-induced osteolysis is a major cause of aseptic loosening in artificial joints. The FE technique is widely used in the orthopedic field for preoperative implant surgery and has achieved remarkable results [21].

Lennon et al. [22] applied FE to the prediction of revision of the femoral head aseptic loosening after THA. The study performed predictions of 17 patients with FE modeling and maintained more than a decade of follow-up tracing (except for renovators). The patients' femur and prosthetic models were constructed using reverse engineering techniques, and corresponding parameters were obtained from the patients' postoperative X-ray. The model was established after combination of the postoperative implanted prosthesis status and personalized model grid. The value of the wear was then calculated from the age and sex of the patient. Statistics showed that the revision time of five patients with the highest wear amount was earlier than that of the other patients. The author's hypothesis demonstrated that an individualized FE simulation of patients could predict the displacement and damage of the prosthesis in advance. Moreover, through a clinical follow-up comparison, the method does have some advance predictability.

9.2.3.2 Roentgen Stereophotogrammetric Analysis (RSA)

In 1974, Selvik developed RSA, which could assess a three-dimensional migration of prostheses with accuracy between 0.05 and 0.5 mm for translation and between 0.15° and 1.15° for rotation (95% confidence interval) [23, 24]. By using special tantalum markers inserted into the bone during surgery, one can measure the position and orientation of the bone with RSA.

The position and orientation of prosthesis is measured by fitting a computer-generated projection of a 3D model of the prosthesis to its projection in the X-ray images or by attaching markers to the prosthesis. A submillimeter accuracy can be achieved by using a specially calibrated stereo X-ray setup [25, 26].

Measurements of implant micromovements during the first 2 years after surgery have been proved valuable in predicting subsequent clinical failures because of aseptic loosening and revision. In many previous prosthetic design studies, subsidence of the femoral stem or proximal migration of the acetabular cup between 1 and 2 mm indicates an increased risk of early or intermediate-term revision. As a first step in the clinical evaluation of new implants or surgical techniques, it can be used to lower the potential risk of clinical failure by predicting the value of radiostereometric analysis measurements [11].

9.2.3.3 Dual Fluoroscopic Imaging System (DFIS)

The team of Guoan Li from the Department of Orthopedics and Biomedical Engineering Laboratory of Harvard University Medical School has focused on DFIS in recent years. DFIS is a test of knee joint in imaging of body motion [27]. Detecting the precise motion of six degrees of freedom in the body is a technical challenge. The three-dimensional motion of a joint is characterized by marked points on the surface of the joint, and the markers are directly fixed to the bone and then located through a dual x-slice. With the development of imaging techniques and postprocessing algorithms, joint motion tracking techniques are expanding. In 2006, Zihlmann et al. [28] used a single fluoroscopic imaging device with gait tracking to record joint motion. Based on a previous work, Li proposed a DFIS recording and analysis device using a double projection technique with high accuracy and noninvasive safety. An accurate measurement of the in vivo knee joint kinematics in six degrees of freedom (6DOF) remains a challenge in biomedical engineering. The team has adopted DFIS to investigate various in vivo dynamic knee joint motions. They provided a thorough validation of the accuracy and repeatability of the DFIS system when used to measure 6DOF dynamic knee kinematics. First, the validation utilized standard geometric spheres made from different materials to demonstrate the capability of the DFIS technique to determine the object positions under changing speeds. The translational pose of the spheres could be recreated to less than 0.15 ± 0.09 mm for velocities below 300 mm/s. Next, tantalum beads were inserted into the femur and tibia of two fresh frozen cadaver knees to compare the dynamic kinematics measured by matching knee models to the kinematics using tantalum bead matching – a technique similar to RSA. Each cadaveric knee was attached to the crosshead of a tensile testing machine and vertically translated at a rate of 16.66 mm/s while images were captured by DFIS. Subsequently, the tibia was held fixed and the femur manually flexed from full extension to 90° of flexion, as the DFIS obtained images. In vitro translation of the cadaveric knee using a tensile testing machine deviated from predicted values by 0.08 ± 0.14 mm for the matched knee models. The difference between knee matching and tantalum bead models during the dynamic flexion–extension motion of the knee was 0.1 ± 0.65°/s in flexion speed, 0.24 ± 0.16 mm in posterior femoral translation, and $0.16 \pm 0.61°$ in internal–external tibial rotation. Finally, they applied the method to investigate the knee kinematics of a living subject during a step ascent and treadmill gait. High repeatability was demonstrated for the in vivo application. Thus, DFIS provides an easy and powerful tool to accurately determine 6DOF positions of the knee when performing daily functional activities.

The key is a custom dual fluorescence fluoroscopic imaging projection device, which is used to record motion sequence images. An MRI data reconstruction model matches the two images, and then rebuilds the space state to reproduce the motion process. Compared with the gold standard RSA technique, the deviation of this method in the knee femoral translation is less than 0.24 ± 0.16 mm through a validation test, tibial internal and external rotation is below $0.16 \pm 0.61°$, and other degrees of freedom are highly similar.

RSA is an accurate measurement technique to assess micromotions of joint replacement implants with respect to the surrounding bone [23]. With RSA, the three-dimensional position and orientation of objects is determined by the reconstruction of the three-dimensional position of well-defined markers. However, the implantation of markers poses an extra risk to the patient. Model-based RSA (MBRSA) has been developed with the aim of assessing the three-dimensional position and orientation of complex-shaped prostheses without the need for attaching radio-opaque markers [29]. The method is based on matching the detected contours of an implant with the calculated projected contour of a three-dimensional model of the same implant, achieving a comparable accuracy to a marker-based RSA [30]. Similar techniques have been used for other applications, such as the determination of the position of vertebrae and the assessment of the position and orientation of total knee prostheses from single-focus fluoroscopic images [10–12]. The fluoroscopic stereophotogrammetric analysis (FSA) technique is a type of MBRSA that uses single-focus but dynamic fluoroscopic images. In this method, the matching theory is the same as with other methods. Moreover, the process of joint movement will be obtained through continuous multiframe reconstruction. The method has been discussed in-depth in laboratory studies of knee motion and knee prosthesis in vivo movements [10, 11, 26–28, 31].

9.3 RSA Technique

As previously mentioned, the RSA technique was proposed by Selvik in 1974 to evaluate three-dimensional movements of prosthesis [23]. The RSA technique has been applied to a large number of scientific researches, involving prosthetic fixation, joint stability, joint kinematics, fracture stability, bone growth, vertebral motion, and spinal fusion [32, 33]. The technique is considered to be the gold standard method for the detection of micromotions of implanted human joint prostheses [34].

9.3.1 Principles

Figure 9.1 shows the standard set-up for RSA acquisition. The position and orientation of the prosthesis is calculated by matching the projections of a computer model of the prosthesis to the projections in the X-ray images, or by attaching markers to the prosthesis. The change in the relative position of the prosthesis with respect to the bone over time is called prosthesis migration and can be calculated with submillimeter accuracy (micromotion). By applying automated image analysis techniques, the data processing is automated to a great extent.

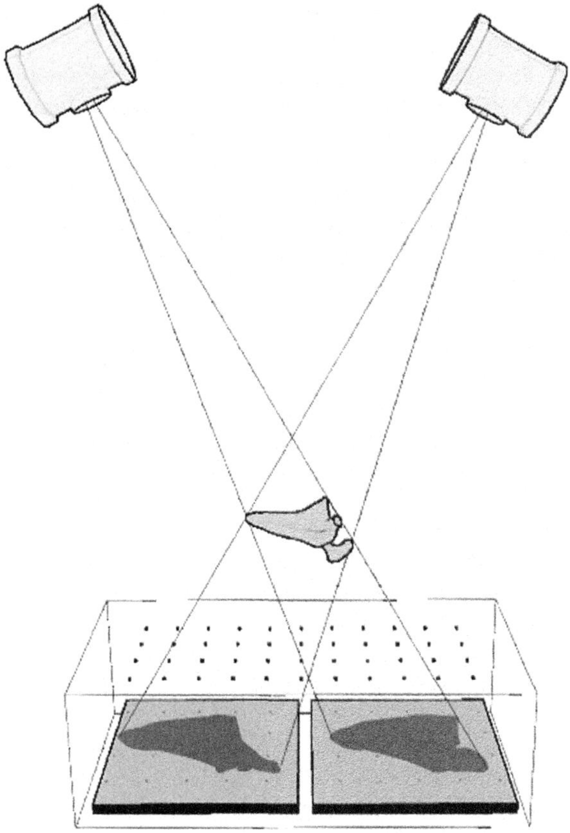

Fig. 9.1 Standard set-up for shoulder RSA acquisition with two Roentgen tubes and the calibration box with the film cassettes under the patient. (Reproduced from Ref. [26] with permission from Elsevier)

RSA is a highly accurate measurement method for the assessment of three-dimensional micromotions of orthopedic implants. The technical details of RSA are as follows.

9.3.1.1 Marking with Tantalum Beads

9.3.1.1.1 Marking Bone

An accurate measurement of migration in RSA radiographs is difficult because the bony landmarks are not sufficiently distinctive. To obtain well-defined measurement points, tantalum beads are inserted into the bone with a special insertion instrument. These beads have a diameter of 0.5 mm, 0.8 mm, or 1 mm. Owing to their small size and spherical shape, their projection will not be influenced by changes in patient position or Roentgen focus position. Therefore, the position of these markers can be measured with high accuracy. The tantalum bead material has a high atomic number that facilitates identification on radiographs; it is biocompatible and is resistant to corrosion [35, 36].

To define a three-dimensional reference coordinate system, at least three nonlinear markers are required. Any change in the prosthesis position and orientation is calculated relative to this static reference coordinate system. The use of additional markers will increase the accuracy of the procedure and will reduce the chance of overprojection of markers by the prosthesis. The stability of the bone markers is verified by the RSA software, which compares the inter-marker distances between consecutive radiographs of a follow-up study. Changes in these distances could be caused by measurement errors or by loosening of a marker. Changes greater than 0.3 mm indicate that a marker is not stable, and it will be excluded from the analysis. In a study of the Interax total knee prosthesis [37], only five of the 300 inserted markers were found to be unstable: the most extreme motion of these single markers relative to the corresponding bone tissue ranged between 0.4 mm and 2 mm.

In addition to the number of markers, the position of markers is also important. The accuracy of the RSA measurements will increase when the markers are appropriately scattered in the bone. Furthermore, to decrease the chance of overprojection, the markers should be positioned at some distance from the prosthesis [38].

9.3.1.1.2 Marking Implants

Most prostheses do not have landmarks that can be measured in a reproducible manner. Therefore, prostheses have to be marked with at least three non-collinear markers. In some knee prostheses and in all-polyethylene cups, the markers are inserted into the polyethylene. It is preferable that the implant manufacturer takes care of this procedure, although it can also be performed intraoperatively by the surgeon performing the arthroplasty. The advantage of the manufacturer's marking is that the position of the markers will be more uniform in all implants. Furthermore, the low-stress areas can be selected using FE model simulations, so that the insertion of markers will not cause stress to rise above acceptable limits. Some hip stems and knee prostheses are marked with beads that are attached to the metal of the implant. The RSA stem of the total hip has a stainless steel marker at its tip and a tantalum marker in a plug at its shoulder. The center of the spherical head of the prosthesis is used as the third marker. To use the center of the head of prosthesis as a marker, it has to be clearly visible. This will not be a problem when all-polyethylene cups are used. However, when tantalum cups or metal-backed cups are paired with ceramic heads, the contrast between the cup and the head will be extremely low for the head to be clearly visible. In such cases, it will be necessary to add a third bead into the stem.

We do not recommend using markers with materials different from the prosthetic base material because galvanic corrosion processes could occur and weaken the prosthesis. For specific stem designs, for example the Mallory-Head prosthesis, prosthetic landmarks are more detectable and it is not necessary to attach beads to define the implant's position and orientation. For the Mallory-Head femoral stem, the center of the head is used as a marker, the projection of the head center on the central axis of the stem is used as a second marker, and the tip is used as a third marker. Moreover, for metal-backed hemispherical cups, such as the Vitalock cup, a markerless technique has been developed to calculate not only the migration of the prosthesis along three axes but also the change in anteversion and inclination angles.

9.3.1.2 The Roentgen Setup

9.3.1.2.1 Two Synchronized Roentgen Tubes

The term stereo in stereophotogrammetry refers to the stereo image of the patient obtained by two synchronized Roentgen tubes. By using two projections of the area of interest, it is possible to reconstruct the three-dimensional positions of markers in that area. The two Roentgen tubes are positioned at approximately 1.60 m above the Roentgen cassette at a 20° angle with the vertical. In clinical practice, one static Roentgen tube and one mobile Roentgen tube are used. To prevent motion artifacts, it is important that both Roentgen tubes are fired simultaneously. This is accomplished by simultaneously pushing both exposure buttons. However, when performing dynamic studies, synchronization is relatively more critical and should be accomplished by using an electronic synchronization device.

9.3.1.2.2 Calibration Box

A calibration box is required to define a three-dimensional laboratory or fiducial coordinate system and to accurately assess the foci positions. For this purpose, tantalum markers are attached to a plane close to the radiograph (fiducial markers) and to a plane that is distant from the radiograph (control markers). These markers are positioned with a computer-controlled device and their positions are known within a few micrometers.

In the past, plexiglass was used as the base material for the calibration box. However, the dimensions of this material are sensitive to changes in temperature and humidity. For this reason, some groups used glass that is less sensitive to these changes. However, glass has the following disadvantages: its high density causes the planes to bend under their own weight and it has Roentgen opacity. Therefore, the new calibration box is made of carbon fiber sandwich panels. These plates are lightweight, stable, and radiolucent.

The carbon cage used for hip and spine studies measures 700 × 460 × 240 mm and holds two 35 cm × 43 cm film cassettes at the bottom. The markers at the lower plane of the cage define the three-dimensional fiducial coordinate system. After calibration, the measured points in the radiograph are transformed to this coordinate system. The markers in the upper plane are designated control markers that are used to determine the spatial position of both Roentgen foci.

9.3.1.2.3 Scatter Grids

When X-rays strike the patient and the prosthesis, they are partially absorbed but they continue to travel in a straight line. This so-called primary bundle will produce the image of the patient. A portion of the X-rays is dispersed by the patient's soft tissue. These scattered X-rays cause a grayish mist that reduces the quality of the radiograph. Normally, a moving scatter grid (bucky) is built into the Roentgen table, and is used to filter out these scattered X-rays. However, in RSA studies, two synchronized Roentgen sources are used, and, thus, this bucky cannot be applied as it would filter out not only the scattered X-rays but also the primary X-rays of one of the two foci.

Therefore, stationary scatter grids that hold lead lamellas that allow the primary X-rays to pass while absorbing the scattered X-rays are used. Scatter grids that are not focused have smaller loss of primary rays than focused scatter grids when the Roentgen beams are improperly aligned with the direction of the lamellas. Hence, non-focused grids have less stringent requirements for the positioning of the Roentgen tubes, which increases the practicality of the system. With these grids, the quality of the images increases considerably, particularly in obese patients. The grids need to be used in hip, spine, and knee studies. The amount of scattered X-rays in smaller joints, like the elbow or the wrist, is relatively low and scatter grids are not required.

9.3.1.2.4 Setting of Roentgen Tubes

As the main purpose of RSA radiographs is to visualize markers and not bony structures, the setting of the Roentgen tubes (kV and mAs) is not the same as in conventional radiography. There are three groups of markers that need to be visualized: calibration box markers, bone markers, and prosthesis markers. The Roentgen technique will have to provide a good contrast between the patient's soft tissue, bone markers, and prosthesis markers, but should not overexpose the calibration box markers (Table 2.1). Sometimes, bags filled with water are positioned next to the patient to provide an evenly distributed exposure of the film.

In all clinical and experimental RSA studies, Kodak Lanex regular film was used. For all studies, except for clinical RSA studies on total elbow arthroplasty, scatter grids were used.

9.3.1.2.5 Radiation Dose

RSA radiographs cannot be used as a substitute for conventional clinical radiographs, as the quality of the images is generally extremely poor to assess regular clinical parameters. In other words, when a patient is included in an RSA study, the RSA radiographs are an additional radiation exposure above that included in a regular clinical follow-up. This additional radiation dose should be kept as low as possible, without compromising the quality of the radiographs.

The amount of exposure was expressed in equivalent dose. In equivalent dose calculations, weighting factors are used for organs and tissues. During the first postoperative year, a patient with THA in an RSA study will receive a total radiation dose (conventional radiographs and RSA radiographs) comparable to the yearly natural background radiation dose that one receives during daily living.

9.3.1.2.6 Measurement of Two-Dimensional Marker Coordinates

Using a measuring table, manual digitization of marker positions yields the coordinates of the projected markers. At the Department of Physiology in Leiden University, a measuring table with a measuring range of 30 × 40 cm was developed. The radiograph was fixed under a glass plate, and it can be moved along two axes. The radiograph was viewed through a microscope and the image is picked up by a CCD (charge-coupled device) camera. On top of the image displayed on a black and white monitor, a crosshair and congruent circles were displayed. The crosshair was positioned at the center of a projected marker, and the two-dimensional coordinates were transferred to a personal computer.

9.3.1.3 Calibration of Roentgen Set-up

9.3.1.3.1 Transformation from Radiographic Coordinates to Fiducial Coordinates

As stated above, the calibration box has two planes that hold tantalum markers for which the three-dimensional position is accurately known. The markers in the plane closest to the radiographic film are the fiducial markers, and these markers define a three-dimensional fiducial coordinate system.

To calculate the three-dimensional positions of the bone markers and prosthesis markers, the coordinates that have been measured using the measuring table have to be transformed to the lower plane of the calibration box. This transformation is accomplished by matching the projected fiducial markers with the actual fiducial markers on the box. The quality of this transformation is expressed by the difference in position of the fiducial markers and the projections of these markers in fiducial coordinates.

9.3.1.3.2 Calculation of Foci Positions

The next calibration step is to determine the positions of both Roentgen foci. For this assessment, the markers in the upper plane of the calibration box (control markers) are used. Through these markers and their transformed projections, projection lines are determined. In an ideal situation, these lines will intersect in one point: the focus of the Roentgen tube. However, measurement errors can occur and the projection lines do not intersect at the same point but cross each other at short distances. Therefore, the position of the focus has to be determined by solving a least squares problem.

9.3.1.3.3 Calculation of Three-Dimensional Marker Positions

The determination of the three-dimensional position of tantalum markers is similar to the determination of the focus position. The three-dimensional position of a marker is the position where its two projection lines intersect. However, because of measurement errors, the lines will not intersect at the same point but will cross each other at a short distance. The three-dimensional position of the marker is located in the middle of the shortest line connecting the two projection lines. This shortest line is denoted as the crossing line error. The three-dimensional marker position is the solution to a least squares problem.

9.3.1.3.4 Calculation of Micromotion

After the three-dimensional positions of the bone markers and the prosthesis markers have been calculated, the relative motion of the bone and the prosthesis can be assessed. The bone markers function as a reference rigid body relative to which the motion of the second rigid body, the prosthesis, is calculated. As patients cannot be positioned in exactly the same position and orientation between follow-ups, the position and orientation of this reference rigid body changes between follow-ups. To obtain the same position and orientation of the bone, the bone markers in the first and in follow-up radiographs are matched onto each other; thereafter, the relative motion of the prosthesis with respect to the bone can be calculated. The results from the motion calculations are a rotation matrix and a translation vector. These parameters are difficult to interpret for the orthopedic surgeon. Therefore, translation is normally expressed as the difference in position of the geometric center of the prosthesis between the first radiograph and the follow-up radiograph. Rotations are expressed as rotation angles about three coordinate axes, normally the transverse axis, longitudinal axis, and sagittal axis. In some cases, for example, when rotational movements are studied, presentation of micromotion in helical axis parameters may be preferred. The accuracy of translation and rotation parameters in a specific study can be obtained by a double examination of the patient during one of the follow-ups. Two RSA radiographs of the patient are taken within a time interval of approximately 10 minutes. As there is a very short time interval between these two pairs of radiographs, we may assume that the implant did not migrate between these two exposures. By measuring, analyzing, and comparing these two pairs of radiographs, the accuracy of the micromotion parameters can be assessed [39]. This method is used in most RSA studies that are carried out in Sweden. At the Leiden University Medical Center, the accuracy is assessed by repeated measurements and analysis of RSA radiographs.

9.3.2 *Motion Calculation of Rigid Body*

The three-dimensional spatial position of the bone and prosthetic marker point are calculated and determined by the spatial information of the projection point and source. As this aspect is not covered by this study, it is not elaborated here. To evaluate the motion information of a rigid body, it is necessary to calculate the marker point information from different situations.

9.3.2.1 Calculation of Rotation Matrix and Displacement Vector

Suppose we obtain the information of n points on a rigid body set to $a_1 ... a_n$ under scene 1 and $b_1 ... b_n$ under scene 2. To calculate the rotation matrix M and the displacement vector d, the following equation needs to be solved:

$$\min_{M,d} \sum_{i=1}^{n} \|Ma_i + d - b_i\|^2; M \text{ is an orthogonal matrix.} \tag{9.1}$$

Soderkvist [41] proposed the best way to solve the equation, which can be expressed by the following equation:

$$d = \frac{1}{n}\sum_{i=1}^{n}(b_i - Ma_i) = \bar{b} - M\bar{a} \tag{9.2}$$

Substituting (9.2) into (9.1), we can obtain the equation containing only one unknown quantity, M.

$$\min_{M} \sum_{\mathrm{i}=1}^{\mathrm{n}} \|M(a_i - \bar{a}) - (b_i - \bar{b})\|^2 \tag{9.3}$$

Defining, $A = [a_1 - \bar{a}, \ldots, a_n - \bar{a}]$, $B = [b_1 - \bar{b}, \ldots, b_n - \bar{b}]$, then (9.3) can be expressed as

$$\min_{M} \|MA - B\|; M \text{ is an orthogonal matrix.} \tag{9.4}$$

The solution of the rotation matrix is

$$M = UV^t, \; BA^t = U \sum V^t \text{ is the singular value decomposition.} \tag{9.5}$$

By substituting (9.5) into (9.2), we can obtain the solution of d.

9.3.2.2 Relative Motion

In RSA clinical studies, people often focus on the movement of the prosthesis relative to the surrounding bones, that is, the relative movement between two rigid bodies. The two rigid bodies are defined as A and B, and the relative motion from time *t0* to *t1* is calculated as follows:

$$A_{t_1} \approx M_A A_{t_0} + d_A e^t,\ \ B_{t_1} \approx M_B B_{t_0} + d_B e^t,\ \ e^t = [1, \ldots, 1] \tag{9.6}$$

The relative motion can be expressed as

$$M_{\text{rel}} = M_A^t M_B. \tag{9.7}$$

Setting the origin at the geometric midpoint

$$\text{Origin} = \frac{1}{n_A} \sum_{i=1}^{n_A} A_{t_0,i}, \tag{9.8}$$

Then, the relative motion can be expressed as

$$d_{\text{rel}} = M_A^t (d_b - d_A) + (M_{\text{rel}} - I_3)\text{Origin}, \tag{9.9}$$

where I_3 is a unit matrix.

9.3.3 *Limitations of Traditional RSA Techniques*

RSA is the most accurate imaging technique in the three-dimensional micromotion evaluation of orthopedic implants. In small-scale clinical trials, it is of great value in the design, coatings, and evaluation of new fixation techniques for new prostheses. However, in clinical practice, the conventional RSA technique still has the following limitations:

1. The conventional RSA is a labor-intensive technique that requires professional researchers to conduct the analysis. The manual matching of the identification points on the radioactive image can take a large amount of time. If this process can be automated, the overall efficiency of the analysis will be improved.
2. In the conventional RSA technique, the location and angle of orthopedic implants can only be identified by implanted or attached points. Another problem arising in clinical research is that these markers are obscured by metal prostheses in the captured image so that the image cannot contain sufficient information to position the prosthesis. At the same time, the addition of markers to the prosthesis and bone is an invasive lesion surgery, and some countries and regions require consent documents that are often difficult for patients to accept.

9.3.4 Development and Application of RSA Technique

Compared to conventional RSA techniques, the computer-based new RSA technique is more time-saving and easy to use. It can achieve spatial orientation information of implants in the same precision without sign points. To avoid injecting tantalum markings, Valstar et al. [40] applied and verified the use of prosthetic boundary projection matching methods in body precision and in submillimeter level. An MBRSA technique was proposed and validated in 2001; however, because of the difference in dimensional tolerances between the implanted prosthesis and model, its accuracy was slightly inferior to that of the RSA method that implanted a tantalum labeling point [33]. In 2003, Kaptein et al. [41] investigated the influence of the accuracy of the implant models that were used for MBRSA by studying both computer aided design (CAD) models as well as models obtained by reverse engineering (RE). The results demonstrated that the RE models provide more accurate results than the CAD models. If these RE models are derived from the very same implant, it is possible to achieve a maximum standard deviation of the error in the migration calculation of 0.06 mm for translations in x- and y-direction and 0.14 mm for the out-of-plane z-direction. For rotations about the y-axis, the standard deviation was approximately 0.1° and for rotations about the x- and z-axis, 0.05°. In 2006, Hurschler et al. [42] compared the model analysis results of the conventional RSA method and MBRSA method for the same axial motion. In the prosthesis and bone model (ZRM), the mean differences of the measured migration using RSA is 0.006 mm (SD $\leq$ 0.048 mm) and MBRSA is less than 0.009 mm (SD $\leq$ 0.049 mm). In prosthesis mode (RM), the mean differences of the measured migration using RSA is less than 0.027 mm (SD $\leq$ 0.047 mm) and MBRSA is less than 0.048 mm (SD $\leq$ 0.037 mm). In both modes, the rotational deviations are less than 0.220° (SD $\leq$ 0.187°). The MBRSA technique functions without prosthesis markers, which presents new opportunities for measuring implant migration. In recent years, the RSA technique has gained new progress and innovation. In 2007, an image-based RSA (IBRSA) proposed by de Bruin et al. [26] reconstructed bone and prosthesis with 3D CT data and generated digitally reconstructed radiographs (DRRs) (Fig. 9.2). Through iterative calculation and 2D–3D image registration, three-dimensional CT DDRs and RSA's two-dimensional perspective constantly match to determine the orientation of the prosthesis and bone. The precision σ in x-, y-, and z-direction was 0.090, 0.077, and 0.220 mm, respectively, for translations and 0.155°, 0.243°, and 0.074°, respectively, for rotations. The results are lower than those of in vitro standard RSA but higher than those of in vivo standard RSA.

9.4 Clinical FSA Method Based on RSA Technique

Based on the RSA technique, Wang [34] et al. took actual situation of Chinese hospitals into account and proposed that FSA technique can be used to clinically detect the aseptic loosening after total hip replacement. At present, the invasive RSA

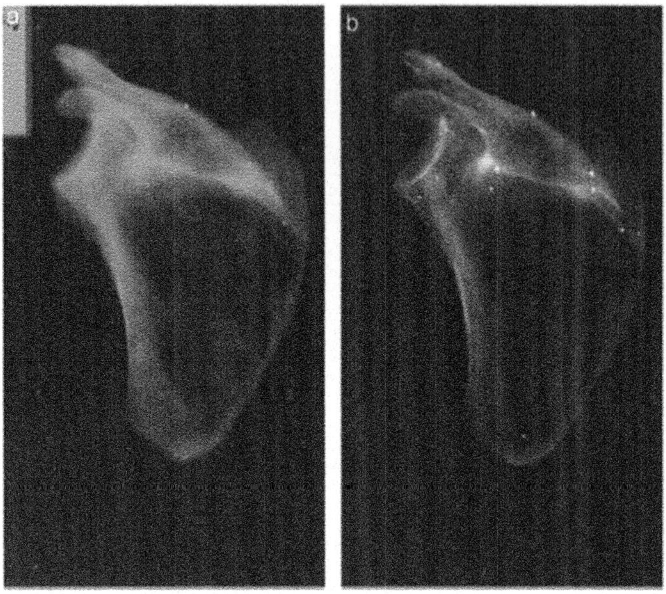

Fig. 9.2 Example of an actual and a virtual RSA radiograph: (**a**) RSA image and (**b**) DRR image. (Reproduced from Ref. [26] with permission from Elsevier)

method is difficult to promote in clinical practice, and most Chinese hospitals do not have the dual light perspective equipment. Wang et al. used the existing CT and DSA equipment and applied the FSA technique into the diagnosis of aseptic loosening of artificial joints for the first time in China.

9.4.1 Methods

9.4.1.1 Patients

Five patients (2 males and 3 females) who underwent total hip revision surgery were assessed in this study. Before the revision surgery, these patients were clinically identified with pain, limited range of motion, and no infection. There were two patients with obvious loosening in the X-ray film before revisional surgeries, and three patients were confirmed to have loosening during surgery by senior doctors. This study was approved by the Institution Board and with all patients' consents.

9.4.1.2 Clinical Data

The first step was to record the patients' X-rays and other imaging diagnostic information, and collect medical records and surgery information. The medical

Table 9.1 Medical records of the patients

Patient	Age	Gender	Prosthesis type	Months since the first surgery	Pain intensity	Surgery record	
						Had surgery	Loosening
Huang	83	Male	Hip	8	Sometimes	N[a]	–
Yin	64	Female	Hip	3.5	Pain when walking	Y	Y
Zhang	76	Female	Hip	11	Pain when walking	Y	Y
Pang	71	Female	Hip	12	Pain when walking	Y	Y
Wang	63	Male	Hip	7	Sometimes	N	–

[a]N – no, Y – yes

records include the patients' age, gender, and prosthesis type. The diagnosis, which is used to verify this study, is confirmed during the surgery by a senior doctor. The details are presented in Table 9.1.

The raw experimental data include dynamic fluoroscopic images, obtained from clinical DSA instruments of patients' hip joint motion and CT images of the affected area, including the hip joint bone and prosthesis. Each patient was asked to perform sequential hip adduction and abduction with a supine position under fluoroscopic surveillance in the coronal plane.

9.4.1.3 Experimental Settings

Two types of data were collected by the hardware: one is the CT data (BrightSpeed Elite Select, GE Healthcare) for the reconstruction of the affected part joint, collected when patients were in a supine position, and another is a perspective sequence diagram of the patients' affected joints for the record of a movement process acquired by the DSA equipment (Innova 3100IQ, GE Healthcare).

According to the clinical hardware device details, the image capturing device settings are as follows:

CT scan parameters: scan resolution (in-plane resolution) = 0.5–1 mm; thickness (scan spacing and slice thickness) = 0.5–2 mm; high signal-to-noise ratio; data is outputted in DICOM format.

DSA acquisition parameters: distance between the focus and receiver is set at maximum image resolution of 1024 × 1024; sampling frequency = 10 Hz; pulse width = 1 ms; bit depth > 10; small focal spot size.

9.4.1.4 Analysis Methods

In the laboratory analysis, each part of the patient's CT data is separately modeled; this forms the STL (stereolithography) files. Then, using the MBRSA software 3.20

FSA module combined with the dynamic fluoroscopic data, each surface model is matched with the X-ray multiframe image. As a result, a reconstruction of the three-dimensional motion process of the patient's affected joints is achieved under normal load. From this, we gain access to the space position information of each model during movements over time. Furthermore, the above data were analyzed to determine if loosening had occurred, and the location and extent of loosening information were recorded.

They have proposed two methods to evaluate the motion. Method 1 is measuring the displacement between the bone and prosthesis in the same frame, called relative displacement (RD). In this method, the key to diagnosing the aseptic loosening of the hip joint is the consistency between the prosthesis and bone during movements, which implies that changes in RD indicate the extent of loosening. This change often occurs in the state of motion changes; thus, two sets of image data are collected–the abduction and adduction of the hip during the course campaign, focusing on the alternating action.

Method 2 is measuring the displacement between the same prosthesis in different frames, called absolute displacement (AD). In this method, three frames are selected. One is in the initial state used as the standard frame, second is in the abduction state, and third is in the adduction state. The angles (between prosthesis and ilium, confirmed by senior doctors) of these three frames were measured. The expected result should show that the latter two are the same or similar. The AD means the displacements between the non-initial state and the initial state. If loosening exists, the ADs will be different.

9.4.2 *Results*

All of the five patients have the same analysis processing. The study shows typical results, first as an example; the complete set of results will follow.

9.4.2.1 Relative Displacement

The study reconstructed the three-dimensional motion process of the patients' affected joints using the models and 2D data in Fig. 9.3. Thus, the spatial data of bones and prostheses are obtained, and then, the RD will be calculated as shown in Fig. 9.4.

In Fig. 9.4, the RDs between the bone and prosthesis in different frames of patient Yin are clearly shown. It is obvious that the RDs in some frames are relatively different from others.

(a)

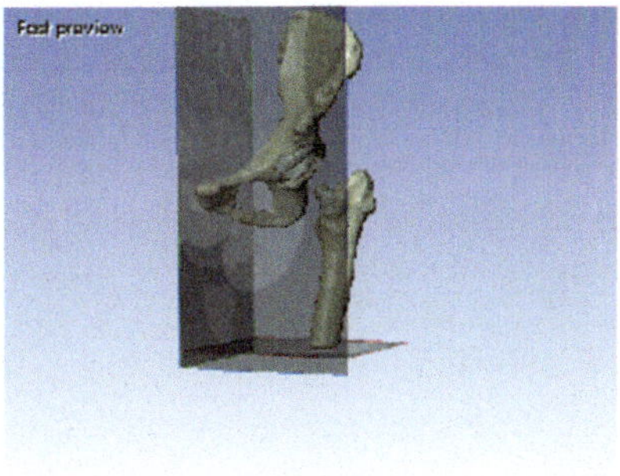

(b)

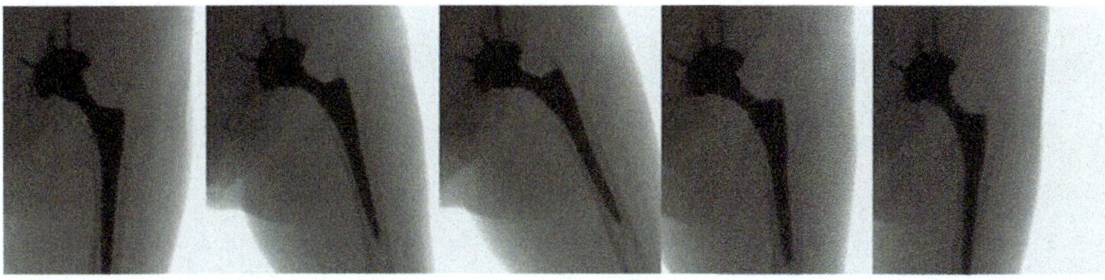

Fig. 9.3 Basic data of patient Yin: (**a**) affected part models of the joint bone and prosthesis reconstructed from CT DICOM data; this is the overall model; (**b**) some interval sequence diagram from abduction to adduction of patient's left leg

9.4.2.2 Absolute Displacement

This method selects three frames as shown in Fig. 9.5. No two frames in abduction and adduction states have exactly the same angle; thus, the closest pair was chosen. Then, the three-dimensional spatial status of the patient's affected joint is reconstructed as shown in Fig. 9.6.

The ADs of the prosthesis in abduction and adduction status with reference to the initial position are shown in Fig. 9.6. This study obtained the spatial data of bones and prostheses, and then, the ADs were calculated as shown in Fig. 9.7. There is a significant difference between these two statuses.

9.4.2.3 Results Obtained from All Patients

The analysis figures of all the five patients are not shown here. Table 9.2 has been constructed to record their analysis and clinical results.

In Table 9.2, the surgery record has two components: one record shows whether the patient has had a revision surgery; the other record shows the doctor's judgments on loosening during surgery. All of the patients who had the surgery were confirmed to experience loosening.

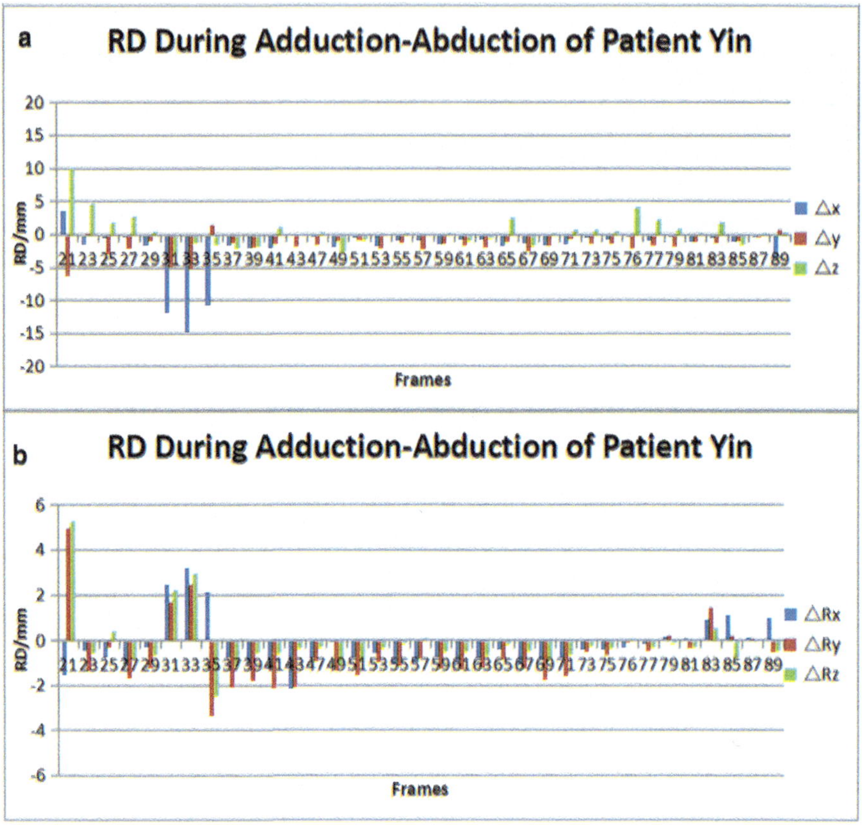

Fig. 9.4 RD during adduction–abduction of patient Yin: (**a**) displacements of DOF x, y, and z; (**b**) displacements of DOF Rx, Ry, and Rz

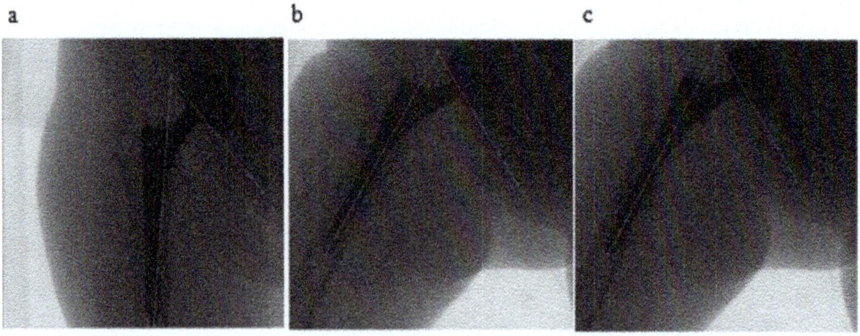

Fig. 9.5 The selected frames: (**a**) initial frame with 43.4°, (**b**) abduction frame with 61.0°, and (**c**) adduction frame with 59.3°

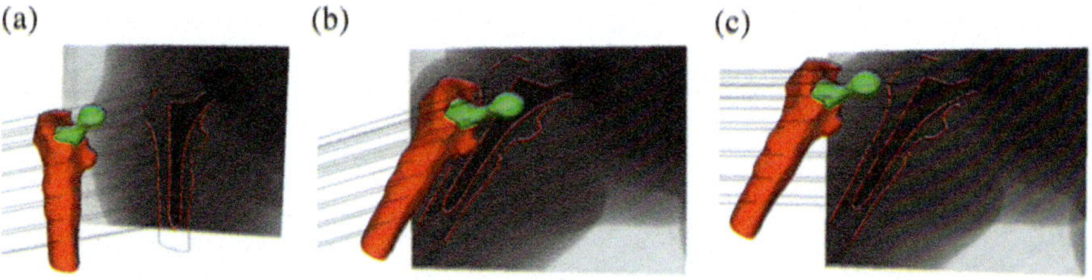

Fig. 9.6 Reconstructed three-dimensional spatial status: (**a**) initial status with 43.4°, (**b**) abduction status with 61.0°, and (**c**) adduction status with 59.3°

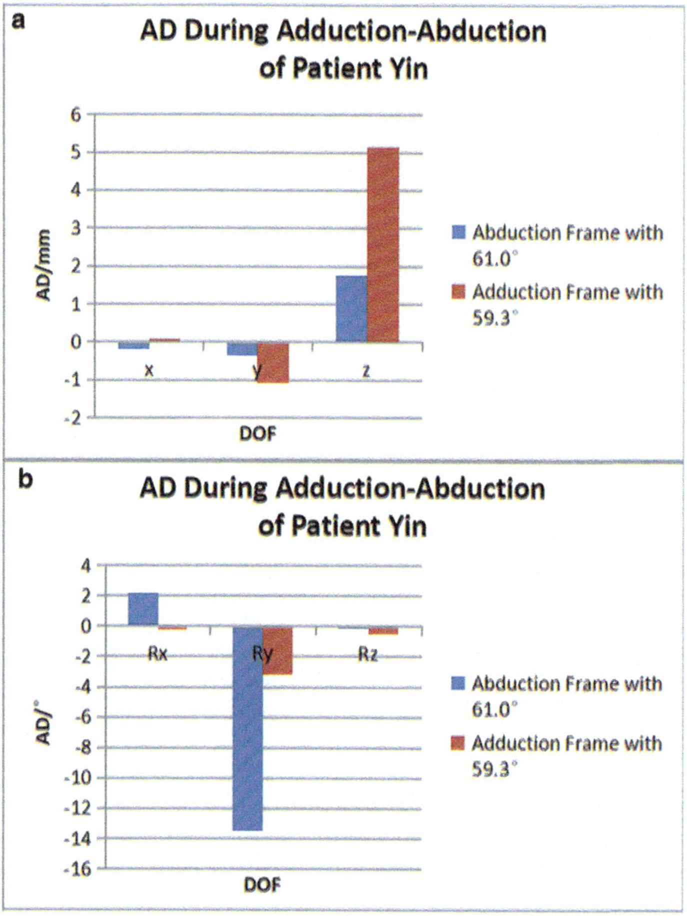

Fig. 9.7 ADs during adduction–abduction of patient Yin: (**a**) displacements of DOF x, y, and z; (**b**) displacements of DOF Rx, Ry, and Rz

Table 9.2 Analysis figures of the five patients

		RD Analysis				AD analysis					
		Upper value of larger cluster		Largest fluctuation		Upper value of larger cluster		The largest difference		Surgery record	
Patient	Gender	mm	°	mm	°	mm	°	mm	°	Had surgery	Loosening
Huang	Male	0.31	0.48	0.67	1.59	0.45	0.60	2.13	3.59	N[a]	–
Yin	Female	1.98	3.38	9.62	10.16	0.72	2.35	3.37	10.29	Y	Y
Zhang	Female	2.33	1.71	11.81	5.32	1.56	4.30	2.87	9.78	Y	Y
Pang	Female	1.67	2.56	8.34	9.21	1.33	3.10	3.01	8.85	Y	Y
Wang	Male	0.81	1.13	1.52	2.97	0.71	0.89	1.98	2.27	N	–

[a]N – no, Y – yes

In RD and AD analysis columns, the upper value of the larger cluster and the largest fluctuation (or difference) are recorded. In both analyses, the data are separated into two groups: a normal one and an abnormal one. A larger cluster means the same as the normal group; thus, the upper value is the upper boundary of the normal case, which is the critical value to access if loosening exists. The largest fluctuation (or difference) is the value in the abnormal group, which may present the extent of loosening.

Among the three patients with surgery verification, the data in both methods show a large difference, which matches the doctor's judgments.

9.4.3 *Discussion*

From the analysis and clinical results, the method can directly reflect the state of motion changes of the affected articular bone and prosthesis over time, showing the relative and absolute motion of the bone and prosthesis. It is validated through the analysis of clinical examples that the data model can show the occurrence of aseptic prosthesis loosening, position of models, and angular information, which possess some reference values. However, this research method is only in its early stages; there are a lot of issues that should be addressed concerning the model's exact reconstruction, matching time, and data analysis.

9.4.3.1 Model's Exact Reconstruction

In this study, the models are reconstructed from the CT data, which bears certain errors caused by metal artifacts. Although we took extreme care in reconstructing the models to minimize errors, the results appeared unsatisfactory. The next step is to use a faster method for metal artifacts reduction and use CAD to reconstruct the prosthesis.

9.4.3.2 Matching Time

During the data processing segment, the matching time costs a lot, resulting in efficiency reduction. This is because the iteration algorithm uses the projected lines as its boundary condition. Future studies will try to use some important points to replace the lines; this may increase the processing speed. The data exporting section may be improved to increase its proclivity.

9.4.3.3 Data Analysis

In the analysis segment, this study proposed two methods, RD and AD. RD measures the displacement between the bone and prosthesis in the same frame, which requires processing all the frames to achieve the result. Although the work involved is large, it can show the entire spatial status and possesses relatively acceptable accuracy. On the other hand, AD measures the displacement between the same prosthesis in different frames; each comparison only requires three frames, which is more convenient. However, in the examples described in the results, it is apparent that the same angle pair is difficult to find, which may increase the error.

In Table 9.1, it can be observed that the results of both methods are different for different patients. Therefore, we cannot propose a common value to define the existing loosening. The differences in the results may be due to the sample size; thus, more data should be collected to determine this standard value.

The FSA technique was used for checking the postoperative aseptic loosening of hip prosthesis, and we provided several examples to verify the effectiveness of the method in clinical applications through a dynamic three-dimensional quantitative image analysis. It was validated through the analysis of clinical examples to determine if the data model can show the occurrence of aseptic prosthesis loosening, position of the models, and angular information, which possesses some reference values. The innovation of this method lies in adding the fourth dimension of time into the quiescent state of three-dimensional space, which can make up the clinical examination of prosthesis loosening, providing reliable data for patients and doctors. It can be forecasted that the method can be used for checking if the joints have loosened by clinical joint surgeons. The combination of clinical symptoms and other laboratory examinations (routine blood test, CT, MRI, CD44, etc.) can be used to confirm its accuracy. This research method is only in its early stages for the exact reconstruction of the model, and matching and establishing the data model. Cases of clinical application will accumulate in the future. More studies have to be conducted to improve accuracy and precision of the method.

9.5 Conclusions

Aseptic loosening is the main reason for artificial hip joint failure in recent years, and in total, its statistical rate has remained the highest [43]. It is also a major problem in clinical judgment and diagnosis. At present, the methods commonly used to detect joint prosthesis failure are clinical physical and general imaging examination, which cannot completely determine the aseptic loosening of joint prosthesis. The gold standard for the detection of joint kinematics in vivo is RSA, but because of the invasive nature of the conventional RSA technique, it is rarely used in the clinic. With the development of the conventional RSA technique, the MBRSA technique based on the model, the FSA technique based on the model and the image, and the

SFIS and DFIS technique, etc. appeared. These new technologies have their own advantages in terms of accuracy, equipment requirements, clinical applications, and others aspects. They have been applied in the clinical diagnosis of aseptic loosening of artificial joint prosthesis. However, the practical application of these new technologies requires further improvement in terms of convenience and hardware requirements.

In the past few years, the quality of mobile Roentgen units has increased considerably. A combination of a single wall-mounted Roentgen tube and a mobile Roentgen tube can deliver high-quality RSA radiographs. Therefore, single-focus techniques are now obsolete for clinical RSA studies because high-quality mobile Roentgen tubes are now commonly available in most hospitals and the accuracy of single-focus techniques is demonstrated to be significantly lower than that of RSA. However, these techniques may still be of interest for research groups that operate outside hospitals and have only one Roentgen source at their disposal.

For a clinical RSA study of the total hip prosthesis, radiographs were taken using a modern wall-mounted Roentgen tube and a new mobile Roentgen tube. In addition, appropriate scatter grids were used. The quality of the radiographs increased considerably and a greater number of radiographs could be used for analysis. The placement of tantalum markers was facilitated by a new insertion instrument. Improved planning in the outpatient clinic resulted in fewer patients being lost to follow-up. With these improvements, the clinical RSA study of the total hip prosthesis was carried out in a more successful manner, and the results of this study were published.

In some RSA studies, marker coordinates are manually digitized. The procedures are started by manually labeling the markers with ink. Thereafter, each marker is measured on a measuring table. The coordinates of the markers are then transferred to the RSA software for subsequent analysis. This procedure takes up to 60 minutes per radiograph. During this study, the requirement for a lesser labor-intensive automated RSA method became apparent. The ultimate goal for the development of a clinical measurement technique is to actually use it in clinical studies. Several clinical RSA studies have been performed using the RSA system developed by our group. In this thesis, two of these clinical studies are described.

The aim of performing THA is to reduce pain and to restore elbow function. However, the fixation of the implant to the bone should also be satisfactory. In THA, this fixation is of concern and high rates of revision caused by aseptic loosening have been reported. Changes in prosthesis design are considered to improve the fixation of the hip components. As the micromotion of implants detected in the present study cannot be detected in conventional radiographs, the effect of this change in design should be studied by means of a randomized RSA study in which the old and the new design would be compared. Only then can improvement in fixation be clearly assessed in a reasonably short follow-up period. Otherwise, long-term follow-up studies with a much larger patient population will be required to obtain the same insight in the eventual improvement of fixation of the new hip component design. We used the RSA system also in a number of clinical and experimental RSA studies that have not been described in this thesis or are not yet completed.

Currently, several clinical RSA studies are being conducted in our hospital, and several others will start in the near future. Ongoing studies include micromotion of cups after revision arthroplasty with bone impaction grafting, the effect on micromotion following the use of a new cement in THA, and the stability of a new cup in THA.

According to Murray et al. [44], a total of 62 different primary total hip prostheses were in the market in the United Kingdom for which only 30% had published results in peer-reviewed journals. The authors advise the use of implants with good results in peer-reviewed long-term clinical trials with at least 10 years follow-up. However, the results of these long-term studies may differ considerably as a result of subjective measurements.

Furthermore, the design of many implants is altered during the time that they are in production. This could mean that at the time that long-term results of an implant become available and the implant has been altered in the meantime, the results cannot be extrapolated to predict the fixation of the newer design. As there is a correlation between short-term micromotion and long term aseptic loosening, it is our firm belief that small-scale clinical RSA studies should be performed for all implants that are introduced into the market. Objective results on fixation will then be available after 2 years follow-up.

References

1. Kaptein BL, GarlingEH VER et al (2008) Roentgen stereophotogrammetric analysis: technique and clinical relevance. J Med Biomech 23:337–346
2. The Norwegian Arthroplasty Register (2010) Annual report 2010. The Norwegian Arthroplasty Register. Available from http://nrlweb.ihelse.net/eng/Report_2010.pdf
3. Garellick G, Karrholm J, Rogmark C, Herberts P (2010) Goteborg, Sweden: Swedish Hip Arthroplasty Register Annual Report. Available at: www.shpr.se/Libraries/Documents/AnnualReport-2010-2-eng.sflb.ashx. Accessed 28 May 2013
4. Clohisy JC, Calvert G, Tull F et al (2004) Reasons for revision hip surgery: a retrospective review. Clin Orthop Relat Res 429:188–1 92
5. Dalury DF, Pomeroy DL, Gorab RS et al (2013) Why are total knee arthroplasties being revised? J Arthroplast 28:120–121
6. Ulrich SD, Seyler TM, Bennett D et al (2008) Total hip arthroplasties: what are the reasons for revision? Int Orthop 32:597–604
7. Schroer WC, Berend KR, Lombardi AV et al (2013) Why are total knees failing today? Etiology of total knee revision in 2010 and 2011. J Arthroplast 28:116–119
8. van der Vis HM (1997) Wear, particles and physical factors in loosening of the hip protheses. Dissertation, University of Amsterdam
9. Hurschler C, Seehaus F, Emmerich et al (2008) Accuracy of model-based RSA contour reduction in a typical clinical application. Clin Orthop Relat Res 466:1978–1986
10. Banks S, Hodge W (1996) Accurate measurements of three-dimensional knee replacement kinematics using single-plane fluoroscopy. IEEE Trans Biomed Eng 43:638–649
11. Zuffi S, Leardini A, Catani F et al (1999) A model-based method for the reconstruction of total knee replacement kinematics. IEEE Trans Med Imaging 18:981–991
12. Walker SA, Hoff W, Komistek R et al (1996) 'In vivo' pose estimation of artificial knee implants using computer vision. Biomed Sci Instrum 32:143–150

13. Lavernia CJ, Alcerro JC (2011) Quality of life and cost-effectiveness 1 year after total hip arthroplasty. J Arthroplast 26:705–709
14. Yang Q (2009) Hypothyroidism after total hip arthroplasty. Chin J Surg Electron 3:4
15. Yamaguchi M, Bauer TW, Hashimoto Y (1997) Three-dimensional analysis of multiple wear vectors in retrieved acetabular cups. J Bone Joint Surg Am 79(10):1539–1544
16. Engh CA, Sychterz CJ, Young AM et al (2002) Interobserver and intraobserver variability in radiographic assessment of osteolysis. J Arthroplast 17:752–759
17. Garcia-Cimbrelo E, Tapia M, Martin-Hervas C (2007) Multislice computed tomography for evaluating acetabular defects in revision THA. Clin Orthop Relat R:138–143
18. Weiland DE, Walde TA, Leung SB et al (2005) Magnetic resonance imaging in the evaluation of periprosthetic acetabular osteolysis: a cadaveric study. J Orthop Res 23:713–719
19. Toms AP, Marshall TJ, Cahir J et al (2008) MRI of early symptomatic metal-on-metal total hip arthroplasty: a retrospective review of radiological findings in 20 hips. Clin Radiol 63:49–58
20. Pang Q (2011) Imaging diagnosis of aseptic loosening of prosthesis after total hip arthroplasty. Mod Pract Med 23:3
21. Viceconti M, Brusi G, Pancanti A et al (2006) Primary stability of an anatomical cementless hip stem: a statistical analysis. J Biomech 39:1169–1179
22. Lennon AB, Britton JR, MacNiocaill R et al (2007) Predicting revision risk for aseptic loosening of femoral components in total hip arthroplasty in individual patients – a finite element study. J Orthop Res 25:779–788
23. Selvik G (1989) Roentgen stereophotogrammetry: a method for the study of the kinematics of the skeletal system. Acta Orthop Scand 232:1–51
24. Kärrholm J (1989) Roentgen stereophotogrammetry: review of orthopedic applications. Acta Orthop Scand 60:491–503
25. Hoff W, Komistek R, Dennis D et al (1996) Pose estimation of artificial knee implants in fluoroscopy images using a template matching technique. In: Applications of computer vision, pp 181–186
26. de Bruin PWKB, Stoel BC, Reiber JH et al (2008) Image-based RSA: Roentgen stereophotogrammetric analysis based on 2D-3D image registration. J Biomech 41:10
27. Li G, VandeVelde S, Bingham J (2008) Validation of a non-invasive fluoroscopic imaging technique for the measurement of dynamic knee joint motion. J Biomech 41:1616–1622
28. Zihlmann MS, Gerber H, Stacoff A et al (2006) Three-dimensional kinematics and kinetics of total knee arthroplasty during level walking using single plane video-fluoroscopy and force plates: a pilot study. Gait Posture 24:475–481
29. Valstar ER, de Jong FW, Vrooman HA et al (2001) Model-based Roentgen stereophotogrammetry of orthopaedic implants. J Biomech 34:715–722
30. Kaptein BL, Valstar ER, Spoor CW et al (2006) Model-based RSA of a femoral hip stem using surface and geometricalshape models. Clin Orthop Relat R 448:92–97
31. Tashman S, Anderst W (2003) In-vivo measurement of dynamic joint motion using high speed biplane radiography and CT: application to canine ACL deficiency. J Biomech Eng 125:238–245
32. Kr J (1989) Roentgen stereophotogrammetry: review of orthopedic application. Acta Orthop Scand 60:13
33. Valstar ER (2001) Digital Roentgen stereophotogrammetry: development, validation, and clinical application. Dissertation, Leiden University
34. Wang (2009) Research progress of aseptic loosening of artificial joint prosthesis. Chin J Exp Surg 25:133–134
35. Alberius P (1983) Bone reactions to tantalum markers. Cells Tissues Organs 115(4):310–318
36. Aronson AS, Jonsson N, Alberius P (1985) Tantalum markers in radiography. Skelet Radiol 14 (3):207–211
37. Nelissen RG, Valstar ER, Rozing PM (1998) The effect of hydroxyapatite on the micromotion of total knee prostheses. A prospective, randomized, double-blind study. J Bone Joint Surg Am 80(11):1665–1672

38. Ranstam J, Ryd L, Onsten I (2000) Accurate accuracy assessment: review of basic principles. Acta Orthop Scand 71:106–108
39. Söderkvist I, Wedin PA (1993) Determining the movements of the skeleton using well-configuredmarkers. J Biomech 26(12):1473–1477
40. Valstar E, Spoor C, Nelissen R et al (1997) Roentgen stereophotogrammetric analysis of metal-backed hemispherical cups without attached markers. J Orthop Res 15:869–873
41. Kaptein B, Valstar E, Stoel B et al (2003) A new model-based RSA methodvalidated using CAD models and models from reversed engineering. J Biomech 36:873–882
42. Hurschler C, Seehaus F, Emmerich J et al (2006) Accuracy and precision of a model-based RSA technique for measuring implant migration. J Biomech 39:129–130
43. Garling E, Kaptein B, Nelissen R et al (2007) Limited rotation of the mobile-bearing in a rotating platform total knee prosthesis. J Biomech 40:25–30
44. Murray CB, Kagan CR, Bawendi MG (1995) Self-organization of CdSe nanocrystallites into three-dimensional quantum dot superlattices. Science 270(5240):1335

Chapter 10
Wear and Diagnostic Analysis of Clinical Failures of Artificial Hip Joints

Jin-Wu Wang, Han Yang, Cheng-Tao Wang, Zhong-Min Jin, and Ke-Rong Dai

Abstract Wear is an important factor in the life of artificial joint prosthesis. This chapter mainly describes the wear mechanism of artificial hip joints, the definition of wear life criteria, causes of abnormal wear, and clinical manifestations of wear failure and establishes an analytical procedure for assessing failure incidents. The clinical manifestations of the wear failure are classified through a wear-osteolysis morphological matrix. The procedure for assessing failure causes is established through a clinical investigation. Nine types of clinical manifestations of wear failure were found. Primary wear processes in artificial hip joints include boundary and mixed friction, adhesive, ploughing, and third-body (abrasive) wear. Surface quality, fit clearance between the acetabular cup and the femoral head, and roundness greatly affect the early abnormal wear. The diversity of clinical manifestations of wear failure is the morphological result of mechanical wear and osteolysis, which is helpful for assessing failure incidents.

Keywords Artificial hip joint · Wear · Failure · Mechanics · Biology

---

J.-W. Wang (✉) · K.-R. Dai
Shanghai Key Laboratory of Orthopaedic Implants, Department of Orthopaedic Surgery, Shanghai Ninth People's Hospital, Shanghai Jiaotong University School of Medicine, Shanghai, China
e-mail: wangjw@sjtu.edu.cn

H. Yang · C.-T. Wang
Clinical Translational Engineering Research Center of Digital Medicine, Ministry of Education, Shanghai Jiaotong University, Shanghai, China

Z.-M. Jin
Tribology Research Institute, School of Mechanical Engineering, Southwest Jiaotong University, Chengdu, Sichuan, China

State Key Laboratory of Manufacturing System Engineering, School of Mechanical Engineering, Xi'an Jiaotong University, Xi'an, Shaanxi, China

Institute of Medical and Biological Engineering, School of Mechanical Engineering, University of Leeds, Leeds, UK

J. Fu et al. (eds.), *UHMWPE Biomaterials for Joint Implants*, Springer Series in Biomaterials Science and Engineering 13,
https://doi.org/10.1007/978-981-13-6924-7_10

10.1 Introduction

Although hip, knee, and other orthopedic implants have proven effective prostheses, the wear of these implants after implantation remains poorly understood. Wear of an articular bearing surface is one of the main causes of failure after 5 years from the date of implant and, very often, wear is accepted as a natural consequence of use. Tribological investigations into the wear of artificial joint implants are critical to their in vivo performance and longevity [1].

Wear is a phenomenon that occurs between two bodies in contact. The term "wear" can be defined as an undesirable progressive loss of material from one or both of the surfaces during relative motion between them [2]. Wear can result from a variety of different processes triggered by the sliding contact between two surfaces and can cause different effects depending on the component in question and its functional requirements. In the wear process, worn material that is expelled from the contact surfaces may cause adverse reactions, such as massive bone loss around the implant and consequent loosening of the fixation [3].

The wear mechanism is very complex, and five wear processes have been clearly recognized based on the physical nature of the wear: abrasion, adhesion, fatigue, erosion, and corrosion [1]. Currently, three different articulating surfaces, including metal/ceramic-on-ultra-high molecular-weight polyethylene (UHMWPE), metal-on-metal, and ceramic-on-ceramic, have been used in patient implants. Because of the different mechanical and tribological properties of these materials, their significant tribological behavior and wear properties have been widely investigated and reported to describe and/or predict the in vivo performance and longevity of the joint implants, and also to guide the design of new materials and geometries of joint implants [4].

As illustrated in Fig. 10.1, modern artificial hip joints contain two types of tribological interfaces: (1) the articulating surface between the femoral head and the acetabulum, which is a major contributor to the artificial joint wear (C in Fig. 10.1), and (2) the interfaces between the connective components or between the components and the host bone (A, B, C, and D in Fig. 10.1). Under pressure, fretting and fretting wear occur between the interfaces because of the difference in the elastic moduli and the structural stiffness, which can also cause wear failure events. The contrast in mechanical properties of articulating pairs significantly influences the wear of each side. The tribological conditions between the articulating interfaces strongly depend on their surface geometry, roughness, lubrication, and stiffness [5]. Therefore, tribological analysis of the articulating interfaces must be performed based on the material and mechanical properties and surface geometries of both materials. Such tribological analyses contribute to defining categories of tribological status and wear mechanisms of different joint implants [1]. Moreover, by combining these analyses with clinical data, we are able to establish a series criteria to enhance our understanding of how wear debris can cause osteolysis. Finally, these criteria can be used to create guidelines for clinical diagnostics of wear-related artificial implant failures based on radio photometry.

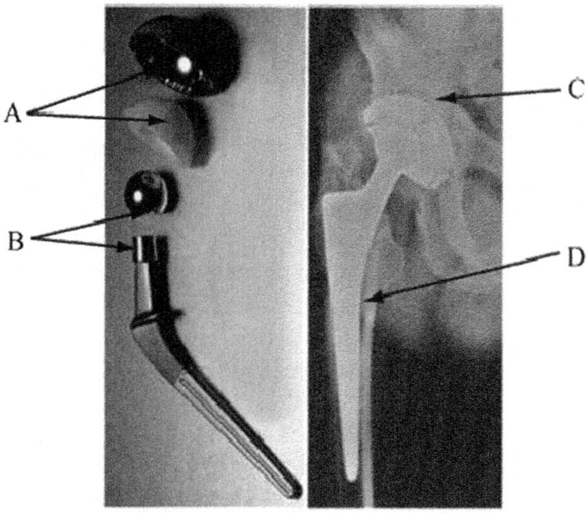

Fig. 10.1 Bearing surface of the artificial hip joint. Reproduced from Ref. [58] with permission from the *Journal of Medical Biomechanics*

10.2 Tribological Status Analyses of Artificial Hip Joints

Tribological analysis is an important tool to determine the long-term success of artificial hip joints. In the artificial joint design process, the lubrication condition determined by friction analysis provides significant information regarding the friction state of the joints' working surface, which can be used to predict wear, evaluate the impact of joint design parameters, and optimize the tribological design of artificial joints [6].

Although the joint capsule is removed during artificial joint replacement, the artificial joint remains filled with body fluids and, a layer of lubricating film remains between the friction surfaces. This membrane can be divided into the boundary lubricating film and hydrodynamic lubricating membrane [6]. The boundary lubricating film is formed by physical adsorption, chemical adsorption, or chemical reaction between the lubricating medium and the friction surface, and the boundary lubricating film's thickness is only of a few molecular layers. This layer is characterized as a border friction state, in which friction and wear are relatively large, and the nature of the materials of the friction surfaces is an important lubrication determinant. In contrast, the hydrodynamic lubricating film is formed by the relative movements of the friction surfaces. When the thickness of this layer is sufficient to separate the two friction surfaces, the friction is converted into the internal friction of the fluid layer, where the friction occurs in the liquid friction state, the friction coefficient and wear are very low, and the nature of the materials of the friction surfaces has little impact on the lubrication condition.

After a total hip arthroplasty, a synovial fluid type of lubricant develops in the joint capsule, which may help to maximize lubricant protection of the implant surfaces. Surface-adherent boundary additives can be placed in the synovial fluid to reduce the amount of direct surface contact, but repeated and frequent replacement of such material might be necessary. Therefore, it may be more practical to develop bearings that promote fluid film lubrication (while avoiding fatigue from subsurface stresses), thus separating the surfaces wholly or partially and reducing the proportion of the total load carried by asperity contacts. Toward this objective, the mechanism of fluid film lubrication in multiform hip implants must be more fully understood [7].

Considering the large deformation of the artificial joint surface, the elastic design of the joint design is usually conducted based on elastic hydrodynamic lubrication (EHL) theory [6]. The Reynolds equation is used to calculate the thickness of the synovial membrane and the hydrodynamic pressure between the femoral head of the hip joint and the acetabulum. The deformation of the acetabulum under the oil film pressure can be calculated by applying the finite element method, and the shape of the lubrication gap between the femoral head and the acetabulum after deformation is set as the boundary condition for the solution of the Reynolds equation. Various iterative calculation methods have been used in the literature [7–9] to calculate the minimum thickness of the synovial membrane $h_{\min}(\mu m)$. Assessments of the lubrication regimes of fluid film have been used to design hip implants in conjunction with conventional engineering methods. These assessments can be divided into two categories: experimental measurements and theoretical predictions. The former typically consists of two aspects: friction measurements that are related to the Stribeck curve, and the detection of separation between the two bearing surfaces with a simple resistivity technique. The theoretical predictions have usually been based on the ratio defined as follows [10, 11]:

$$\Lambda = \frac{h_{\min}}{\sqrt{\delta_1^2 + \delta_2^2}}, \tag{10.1}$$

where δ_1 and δ_2 represent the micro-convex body height of the femoral head and the acetabular surface roughness (μm), respectively. In the hydrodynamic lubrication area, the micro-convex body of the friction surface is completely separated by the lubricating membrane, and the friction coefficient is approximately 0.001–0.015. In the boundary lubrication area, the micro-convex body of the friction surface is in contact, although it can be separated by the boundary lubricating film, and the friction coefficient is approximately 0.08–0.40. Table 10.1 summarizes the calculation results for different types of artificial hip joints.

As shown in Table 10.1, the UHMWPE–CoCrMo joint surface is mainly in the boundary friction state, where adhesive, ploughing, and three-body wear (abrasive wear) occur. The finite element analysis of the prosthesis shows that the maximum contact stress is 10 MPa, which is considerably lower than the yield strength of UHMWPE (20–25 MPa) [12]. Therefore, boundary friction will dominate over the entire hip joint, without a high risk of yield damage of UHMWPE during articulation. Under normal circumstances, the joint pair will continue to work in the aforementioned wear conditions without a high risk of fatigue damage of the material. Vassiliou et al. [13] found that the wear rate of the UHMWPE–CoCrMo artificial joint is small, with in vitro tests showing wear of 5–10 μm per million cycles.

Table 10.1 Tribological and geometric properties for artificial hip joints [7, 8]

Type of artificial hip joint	Friction coefficient	Lubrication film thickness/nm	Surface roughness/nm	Friction state
CoCrMo–UHMWPE	0.06–0.1	100	$\delta_1 = 10 - 25$ $\delta_2 = 50 - 1000$	$0.1 \leq \Lambda < 2$ Boundary friction
CoCrMo–CoCrMo	0.1–0.3	25–50	$\delta_1 = \delta_2 = 10 - 25$	$0.7 \leq \Lambda < 3.5$ Boundary and liquid mixed friction
Ceramics–ceramics	0.002–0.07	25	$\delta_1 = \delta_2 = 5$	$\Lambda = 3.5$ Mixed or liquid friction

In contrast, analyses and calculations based on the elastic hydrodynamic lubrication of the artificial knee joint yield an $h_{\min}$ of approximately 60 nm [14, 15] and $\Lambda = 0.06 - 1.18$, indicating a boundary friction state. Therein, the contact stress is approximately 20 MPa, which is close to the yield strength of UHMWPE. Adhesive wear occurs at an early stage in artificial knee joints; subsequently, fatigue wear initiates and grows gradually until it dominates at later stages, particularly when the UHMWPE oxidizes. The UHMWPE tibial plateau may have a relatively higher risk of fatigue failure than hip liners.

These analyses suggest that the tribology and wear of artificial joint implants are strongly related to the mechanical properties and surface geometries of the articulating surfaces. These issues have been extensively and intensively investigated from both biomechanical and biotribological perspectives [5, 6, 8, 11, 12]. In this chapter, tribology analysis will be performed to investigate both the mechanical conditions and biological effects of joint implants, particularly for those based on UHMWPE. The results help elucidate the tribological status, friction condition, and wear patterns of joint implants. Based on the results presented in literature, clinical images are demonstrated to diagnose and predict in vivo performance of the artificial joints.

10.3 Mechanical Criterion for Evaluating Joint Implant Wear

One of the main causes to hip prosthesis failure is the wear of the contacting materials [16]. Although hip implants are lubricated by synovial fluid, the lubricant film is often insufficiently thick to separate the solid surfaces completely. Therefore, solid–solid contact occurs during a substantial portion of the implant's life in vivo, resulting in gradual wear of one or both of the contacting surfaces, eventually resulting in the failure of the device [5].

10.3.1 Mechanical Considerations for Joint Implant Wear

Two load sources act on a hip joint during normal activities: the ground reaction force during the stance period when the foot contacts the ground, and the muscles that apply forces and moments to the joint. From a tribological perspective on the failure of the implant surfaces, the predominant force is the resultant contact force acting at the interface between the contacting components of the prosthesis [5]. Researchers have determined the forces acting on knee and hip joints using both experimental and analytical methods. It is relatively easy to measure the ground reaction force using an instrumented force plate; however, in vivo contact forces are much more difficult to measure. Experimental methods that utilize instrumented implants that telemetrically transmit load, strain, or pressure data have been successfully used in the hip joint [17, 18].

Analytical studies of joint forces require analysis of the free body diagrams of the lower leg as the leg moves during the gait cycle. This is a difficult task because the human leg contains over 45 muscles and each joint is acted upon by three resultant force components and three resultant moment components [5]. In addition, the system of forces acting on a model of the human leg is statically indeterminate, and solutions must be determined for many orientations of the leg during a gait cycle or any other activity.

Several researchers have studied the contact forces that occur in hip joints during a walking cycle. Bergmann et al. [17] used instrumented hip prostheses to measure hip joint forces in a series of individuals during walking and other activities. Their results show that all three components of the hip joint force, along with the resultant hip contact force, vary considerably during a typical gait cycle. The orientation of the resultant force may also change slightly during the walking cycle as the femur goes through flexion–extension, abduction–adduction, and internal–external rotation motions; however, the dominant force component is the vertical component (F_Z), and therefore the resultant contact force is generally oriented close to vertically down.

Metal-on-polyethylene hip prostheses often fail because of the wear of the polyethylene acetabular liner, which leads to the production of millions of small wear particles [19]. These small wear particles can be responsible for osteolysis that eventually loosens the prosthesis, which is the main reason for the failure of these types of hip implants [20].

From a mechanical perspective, the linear or volume wear rate of joint implants is the most important measure of the extent of wear and of the performance of the articulating materials. In vitro wear tests have been widely performed using joint simulators that run in patterns similar to human gaits with bovine serum as a lubricant. Numerous investigators around the world have used hip joint simulators to obtain an in vitro measurement of the amount of wear that might occur in vivo with a given hip implant design or with a given combination of hip implant materials [21]. However, such tests are often complex and expensive. As a complement to wear testing, or as an alternative, it is often useful to model the wear process to predict the amount of wear analytically to determine the effects of different materials or operating variables. The methodologies that have been used to model wear of hip prostheses have been discussed in the review paper by Mattei et al. [22], along with some of the wear predictions that have been developed by various wear models. A typical contact-coupled wear model methodology for metal-on-polyethylene (MoP) hip implants is described by Kang et al. [23]: the contact pressure is determined at each time interval, usually using a numerical method such as a finite-element method as described by Mattei et al. [22], and the sliding distance in an interval is then calculated.

The results by Kang et al. [23] show that the maximum contact pressure in the MoP hip implant during a gait cycle generally follows the transient hip joint contact force. However, the magnitude of the contact pressure decreases as wear of the polyethylene acetabular cup occurs, and the worn surface of the acetabular cup becomes more conformal with the surface of the metallic femoral head. Both linear and volumetric wear predicted by the model have been found to be in good agreement with experimental data for UHMWPE acetabular cups, and the relationships between wear and geometric variables such as the femoral head diameter and diametric clearance were also found to agree well with experimental results. Therefore, wear modeling is useful considering the effect of implant design or material selection on the prediction of wear performance of hip joint implants.

Metal-on-metal (MoM) hip prostheses have been implanted in large numbers in the past decade in an attempt to avoid the wear problems encountered with MoP hip implants. However, the in vivo performance of MoM hip implants has been disappointing for many patients [24, 25]. The thickness of the lubricant film in an MoM prosthesis is very small (<50 nm), and because the surface roughness of the metallic components is generally on the order of 15–70 nm [22], mixed lubrication is expected to prevail in the contact, and some solid–solid contact is expected to occur. This inevitably results in small amounts of metallic wear debris liberated within the contact region. In an attempt to gather more information about the wear of MoM hip implants, many MoM implants that failed in service have been retrieved and analyzed [26]. Numerous in vitro experimental tests, generally using hip simulators, have been run [27]. The results of simulations have shown that the wear rates of MoM prostheses are generally quite low, and much lower than for MoP hip implants [27]. However, wear rate information by itself does not necessarily indicate an excellent performance of the implant in vivo. In fact, the biological reaction of an individual patient to metallic wear debris could dictate which could not be determined by in vitro wear tests. Mattei et al. [22] have developed wear models for MoM hip implants that have been found useful in predicting the amount of wear (volumetric or linear) that occurs in typical walking cycles, and the resulting effects of various design changes on wear and contact pressure.

Ceramic-on-ceramic (CoC) hip prostheses show even lower wear rates than MoM implants in hip simulation tests, and much lower wear rates than MoP designs [28]. This is because of the high hardness and low surface roughness of the ceramic materials, enabling them to maintain a hydrodynamic lubricant film better than rougher MoM or MoP bearings [5]. However, the long-term resistance of ceramic hip components to fracture is yet to be determined.

Table 10.2 Basic wear data

Material matching pair	Wear amount per year/mm			
	Average amount of line wear/mm	Average amount of volume wear/mm	Amount of abrasive	Abrasive size/μm
CoCrMo/ UHMWPE	0.10–0.20	40–100	4×10^{10}	2–5 (breaking-in period)
	0.5 (youth)			0.1–1.0 (after the breaking-in period)
CoCrMo/ CoCrMo	0.025–0.035 (breaking-in period)	1.0 (breaking-in period)	$7 - 70 \times 10^{12}$	1.0–3.0 (breaking-in period)
	0.005 (after the breaking-in period)	0.1–0.3 (after the breaking-in period)		0.030 (after the breaking-in period)
Ceramics/ ceramics	0.016	1.0 (breaking-in period)	$9 - 191 \times 10^{13}$	0.2–1.0 (breaking-in period)
		0.05–0.10 (after the breaking-in period)		0.010 (after the breaking-in period)

Representative simulator results are summarized in Table 10.2. Considering the aging process caused by the long-term use of UHMWPE in vivo, Jin et al. [7, 8] conducted an oxidation accelerated wear test. Their results showed that the amount of wear remained within the scope of Table 10.2. The wear life of high-crosslinked UHMWPE was higher, and the linear wear of the CoCrMo/CoCrMo and the CoC artificial hip joints was very low.

10.3.2 Mechanical Factors Affecting Joint Implant Wear

The wear-related failure of joint implants depend on many factors, including the materials, joint design, femoral head diameter, and the ratio between the femoral head and neck.

In terms of materials, conventional UHMWPE sterilized by gamma irradiation is prone to oxidation on the shelf or in vivo. In the mid-1990s, scientists found that sterilization using gamma irradiation renders subsequent oxidation of the material, thereby changing the mechanical properties of the finished product [29, 30]. During radiation sterilization, gamma radiation from a ^{60}Co source disrupts all biological activity. Simultaneously, the radiation causes chain scission and hydrogen abstraction from the carbon backbone of the polymer chain. Free radicals created within the material seek a lower energy state by recombining in their original conformation, combining with each other or remaining in a high-energy state until they combine with other chemical species found in the polymer bulk [31].

Less understood is the oxidation profile in relation to depth. It was originally thought that the radiation dose varied through the depth of the polymer, resulting in an increased free radical population below the surface [32]. However, later studies determined that bulk-irradiated polyethylene could be cut to expose new surfaces, and the same characteristic oxidation profile would appear [33]. Researchers have been further perplexed by retrieved devices that showed non-uniform oxidation around the free surfaces. In some cases, regions near the edge of a tibial insert or the flange of an acetabular cup oxidize faster than the articular surfaces [34]. Oxidation is detectable as more than just double-bonded oxygen species within the polymer bulk.

Because the mechanical properties of polyethylene change with oxidation, any freshly cut surfaces will show the appearance of "white bands" below the exposed surface [31]. The white band is the result of localized brittle failure during sectioning and is often the locus of failure in retrieved devices. The maximum oxidation level of a component exposed to atmospheric oxygen will increase exponentially with time. This degradation will fail an orthopedic bearing in about 5 years because of the loss of toughness [35]. Therefore, to avoid shelf oxidation of orthopedic bearings, all US manufacturers have discontinued radiation sterilization in the presence of oxygen.

Since the 1990s, the development and clinical use of highly crosslinked UHMWPE has significantly reduced the wear rate of joint implants. In vitro studies have demonstrated a promising low wear of liners based on highly crosslinked UHMWPE. Moreover, midterm follow-up studies based on thousands of patients have showed an 87% reduction of wear-related osteolysis and revisions.

The wear life of the artificial joint is defined as the use cycle from prosthesis implantation to the loss of the mechanical function of the prosthesis. The material may be worn out because the femoral head has ground to the acetabular depth, which reduces the scope of the prosthesis activity in the frontal plane (Fig. 10.2a) and the cross section (Fig. 10.2b) from θ_1 to θ_2. This reduction leads to edge collision and dislocation. Figure 10.2c shows a clinical example of such failure.

Despite the development of low wear materials, mechanical factors have been widely recognized critical to the wear and clinical performance of joint implants. Nevertheless, the wear life is the result of the following comprehensive product design factors.

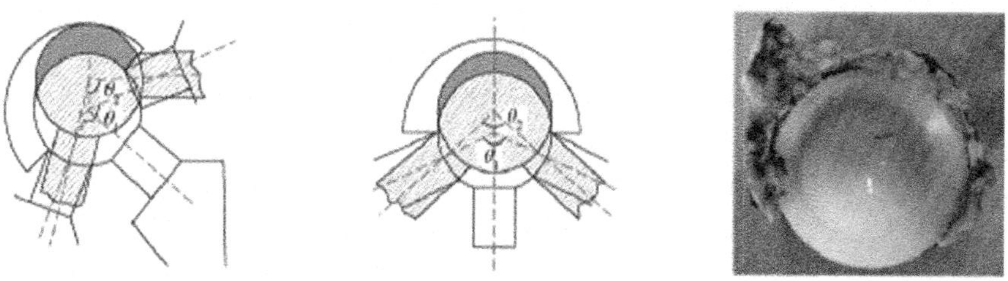

(a) (b) (c)

Fig. 10.2 Edge collision caused by the ball head migration after wear of the artificial hip joint. Reproduced from ref. [58] by permission of Journal of Medical Biomechanics

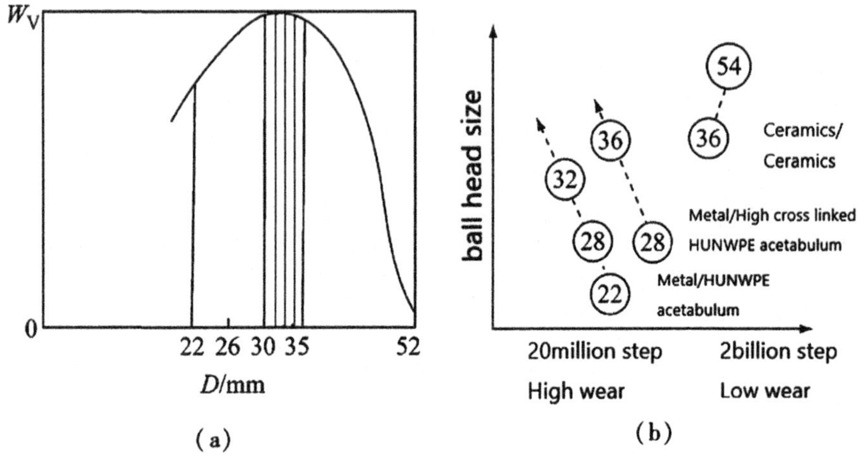

Fig. 10.3 Relationship between ball head diameter and wear. Reproduced from Ref. [58] with permission from the *Journal of Medical Biomechanics*

1. Femoral head diameter: One study [12] shows the relationship between the ball head diameter of the hip joint D and the friction work W_v, which is shown in Fig. 10.3a. The head diameter of 30–35 mm incurs the largest friction work and the largest ball head wear, whereas smaller ball heads incur small amounts of wear. Above this range, even though the ball head diameter increases, the wear also decreases. Figure 10.3b presents the wear test results with different ball head diameters from University of Leeds, which leads to the same conclusion [36].
2. Configuration of the joint implants: The ratio of the head diameter to the neck diameter (head–neck ratio) determines the joint's degree of motion. The range of motion is large when the head–neck ratio is large. Furthermore, the effect of the ball head displacement is small.
3. Surface roughness: The analysis presented in the Sect. 10.3.1 implies that the roughness of the friction surfaces θ_1 and θ_2 determines the surface friction properties. The surface micro-convex contact collision probability increases as the roughness becomes greater, thereby resulting in increased adhesion wear.
4. Geometry of artificial joints: The fit clearance between the hip joint head and the spherical surface of the acetabulum is an important factor in determining the synovial membrane thickness $h_{\min}$. As the lubrication clearance decreases, $h_{\min}$ increases, and the adhesive wear decreases [9]. However, an extremely small lubrication gap will also lead to joint occlusion and early wear of the acetabular joint. The internal spherical surface and the ball head roundness play a decisive role at this point.
5. Positioning of the implants. An incorrect placement of the implant causes the neck to rub against the acetabulum rim, resulting in severe wear. MoM prostheses are particularly sensitive to the angle of the acetabular implant. An in vitro study shows that the wear rate significantly increases when the angle of the acetabular

cup increases from 45° to 55°. The contact between the acetabular rim and the femoral head will concentrate stress, which prevents the lubricant from entering and thus destroys the lubrication effect. These studies are consistent with the recent clinical observations [37–39].

10.4 Biological Considerations for Joint Implant Wear

Wear debris are known to induce osteolysis at the prosthesis periphery, which has been a major problem with UHMWPE-based joint implants from the 1970s to the 1990s. Evidence in support of this theory include the following phenomena: osteolysis has been correlated with higher wear rates [40] and vast numbers of wear particles have been found attached to the periprosthetic interfacial membrane when they are removed during revision surgery [41]. Furthermore, experiments have shown that particulate debris can induce osteolysis in a variety of animal models [42] and inflammatory responses in cultured macrophages [43]. Recent studies have investigated the relevance of the particle size, geometry, and number density to osteolysis. Moreover, the generation, transport, and migration of wear particles have been investigated, illuminating new joint implant designs that minimize the invasion of wear particles into the periphery tissues and reduce the risk of osteolysis.

Biological responses to wear particles are critical concerns for joint arthroplasty. Wear particles include the abrasive UHMWPE debris and particles of metal or ceramics generated during long-term articulation. Such wear particles can migrate within the host tissue, near the prosthesis, or even invade the circulatory system. For example, UHMWPE wear particles usually induce osteolysis [44, 45]. However, metal particles will either cause osteolysis or release metal ions that migrate to and accumulate in many other organs. Although metal ions are not a major initiator of osteolysis, they do play a role in increasing the concentration of key inflammatory cytokines that stimulate osteoclasts and thus contribute to osteolysis and aseptic loosening [46].

The amount of cytotoxic materials produced by wear over time is recognized as a biological evaluation index of the wear performance of the joint pair. Almost 10 years of studies have shown that this kind of reaction is mainly abrasive and caused by osteolysis of the host bone. In recent years, the CoCrMo/CoCrMo artificial hip joint wear process is found to produce a large number of metal ions that impair kidney function and can adversely affect fetal health.

10.4.1 Osteolysis

Osteolysis may be periprosthetic and/or linear [47]. Periprosthetic or focal osteolysis is described as focal bone loss around a component that may or may not exhibit loosening. The etiology of osteolysis around the prosthesis is not fully understood;

however, osteolysis is believed to be an immune response caused by wear debris [48, 49]. Linear osteolysis has been used to refer to aseptic loosening [47], which is a loss of fixation in a previously well-fixed implant due to interfacial bone loss. Osteolysis usually develops before it is identified as a linear radiolucency in radiographs [47]. In addition, mechanical factors in the absence of wear debris or metal ions may also lead to aseptic loosening [50, 51].

The immune response is considered as the main inflammatory initiator secondary to the metal ions. Inflammation secondary to the wear debris is believed to be the most common reason for periprosthetic osteolysis in total hip replacements [52]. Although the inflammation's most important cellular target is the macrophage, typically, the macrophage responds to a particle challenge in two distinct ways, both of which lead to increased bone resorption [53].

Proinflammatory signaling activated by wear debris is well known to result in increased osteoclast recruitment and activation. More recently, it has been reported that the activity of antiosteoclastogenic cytokines such as interferon gamma are also inhibited by wear debris, which in turn leads to improved differentiation of macrophages to bone-resorbing osteoclasts. Moreover, osteoblasts, fibroblasts, and possibly lymphocytes may also be involved in the process. MAP kinase cascades, NF-kB, and other transcription factors are activated at a molecular level, as are the suppressors of cytokine signaling.

The magnitude of the effect of abrasive osteolysis is related to the number of abrasive grains produced per unit time, and is especially related to the abrasive concentration in the tissue lesion site. Therefore, the effect is also related to the material mechanics and wear properties.

The toxicity of the abrasive particles depends on the material itself. Research shows that the toxicity of the UHMWPE abrasive debris triggers significant osteolysis, whereas ceramic abrasive triggers only slight osteolysis. The wear product toxicity is dependent on the size and shape of the abrasive particles. Small circular abrasive particles are more likely to be phagocytosed by macrophage-induced osteolysis than larger and slender abrasives [44, 45].

Jin et al. [7, 8] performed wear tests on a simulated test machine and obtained the average size of the wear particles from three typical distribution pairs. The data used in the present study (Table 10.2) estimates the number of wear particles assuming that the wear particles are spherical.

10.4.2 Effect of Wear Debris Geometry on Osteolysis

At the macroscale, polypropylene is quiescent in the body, and it is therefore used for a wide range of medical applications beyond knee and hip bearings. However, as a piece of polyethylene decreases in size, the cellular self-defense mechanisms of the human body begin to act against the foreign material. Phagocytic cells throughout the body aim to rid the body of old blood cells, cellular debris, and microorganisms. The phagocytic cells operate by binding to, and subsequently engulfing, the invader or dead tissue. Because this recognition and interaction occur at the cellular level, most of the targeted inflammatory particles are cellular or subcellular in size. In early studies, researchers determined that polyethylene particles must be in the "phagocytosable size range of 0.3 to 10 microns to be biologically active" [54]. Later studies have shown that particle sizes in the range of 0.1–1 microns are the most biologically active [55].

If a wear particle is biologically active, it stimulates a cascade of events that will ultimately lead to a loss of bone stock and implant loosening. In addition to phagocytosing the particle, a macrophage will send a variety of signals to other cells in and around the joint space. These signals include proteolytic enzymes, inflammatory mediators, and osteolytic mediators [54, 55]. Osteoclasts are either recruited by or differentiated from the macrophages and begin to break down the bone in the vicinity of the joint device. For the patient, this translates to potential pain, swelling, and implant loosening. In some cases, osteolysis is pain-free, and the loss of bone stock continues asymptomatically. This clinical condition of bone resorption because of biological activity of wear debris is known as osteolysis.

10.4.3 Biological Criteria for Joint Wear Life

One study [7, 8] showed that the threshold of polyethylene abrasives that leads to osteolysis is approximately 500–800 mm^3. The wear rate measured by the joint simulator was 40 mm^3/MC (million cycles). According to calculations from elderly patients who take approximately 10^6 steps per year, the wear life of a polyethylene acetabulum is only 12–20 years or less for young patients, whereas the wear life of the polyethylene acetabulum is only 5–8 years if the wear rate is 100 mm^3/MC. The ceramic implant experiments showed that a ratio of ceramic wear particle size to the number of mononuclear cells higher than 500 μm^3:1 will cause osteolysis. The annual wear of the ceramic materials implied that more than 100 years of grinding grain accumulation were needed to incur osteolysis with these materials [56].

10.4.3.1 Effect of Debris Migration and Accumulation on Osteolysis

In addition to the biological wear properties of the material, the following three conditions are also important components for the wear life of the artificial hip joint.

1. Abrasive flow and abrasive accumulation: Fig. 10.4(a) shows the abrasive flow produced by the moving surface and the micro-surface of the artificial hip joint and assembled at the interface of the joint stem and host bone. This abrasive flow makes the joint stem prone to proximal osteolysis and distal diffusion. The fretting wear particles at the interface of the UHMWPE acetabulum and the metal cup aggregate in the nail holes and lead to multiple osteolysis sites. Artificial knee joints of the same material show more osteolysis because the abrasive flow misses the interface in the artificial hip joint (Fig. 10.4(b)). A

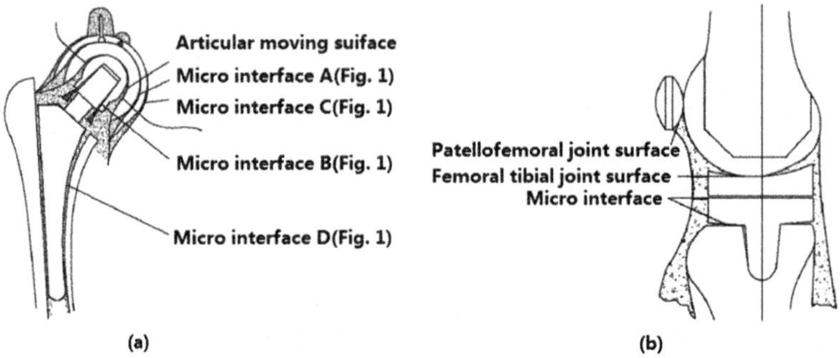

Fig. 10.4 Wear particles and their flow route. Reproduced from Ref. [58] with permission from the *Journal of Medical Biomechanics*

reasonable structural design (e.g., blocking the excess screw holes in the metal cup) can reduce or block the flow of abrasive particles.

2. The quality of the prosthesis–host bone interface: The interface provides a channel for the abrasive particles to enter the host bone, resulting in a contaminated area with abrasive particle contents in the host bone. Very few abrasive particles can enter a securely closed interface. However, early loosening of the interface with the movements of the human body will cause cyclic switching motion at the interface that forms pump suction for the body fluid in the surrounding environment. The abrasive particles will then enter this effective channel of abrasive flow into the medullary cavity, which causes bone resorption of the distal interface. The interface quality will be progressively deteriorated in patients with severe osteoporosis, large weight, and insufficient or excessive activity, or other conditions.
3. Sensitivity of the bone tissue of the contaminated area to the biological response of the abrasive particles: The osteolysis effects of abrasive particles are patient-dependent due to the different immunological and physiological characteristics between individuals. The fundamental reason for this is yet to be studied.

10.4.3.2 Ion Release

Stainless steel, CoCrMo, pure Ti, Ti6A14V, and TiNi alloys are frequently used biomaterials because of their high corrosion resistance and durability. The metal's ability to resist corrosion is associated with the oxidation level at the surface. These surface oxide films are important inhibitors to ion release [46]. When the surface oxide film is disrupted, ions are released according to the electrochemical rule, which stipulates that when ions are released there must be an equal number of electrons transferred during the process. Many retrieval studies of surgical implants have confirmed this disruption by documenting the changes in the composition of surface oxide films. Other electrolytes and biomolecules in the human body surrounding the metal implant, amino acids, proteins, dissolved oxygen, inorganic ions, and cells in the immediate environment of the metal implant all influence the release of metal ions. Changes in the pH level of the human body can even serve as a potential accelerator of ion release [46].

The surface oxide film can be disrupted, allowing ions to be released into the surrounding environment, by anodic dissolution and by a hypersensitivity reaction. Anodic dissolution is a reversible procedure, where metal ions released could be reduced into metals under properly controlled conditions. If the metal ion release dominates, the release of metal ions must occur at a faster rate than the return. The change of electric potential for these biomaterials is usually small; therefore, the metal ion release is normally slow. During a hypersensitivity reaction, macrophages adhere to the metal implant directly because the macrophages recognize the metal implant as a foreign substance and initiate an inflammatory cascade [57].

Even when metal ions are released, not all ions will initiate the cascade leading to osteolysis and aseptic loosening. To become cytotoxic or initiate an allergic reaction, metal ions must combine with biomolecules to begin a more destructive path before combining with water molecules or other anions to form less harmful oxides, hydroxides, and inorganic salts. Ti ions are highly active, often readily combining with hydroxyl radicals and anions, which leaves little time for Ti ions to combine with other biomolecules. Other ions, such as Ni and Cu from different alloys and stainless steel, are less active than Ti ions, which do not immediately react with water molecules or inorganic anions, creating a greater chance to combine with other biomolecules and establish toxicity [57]. The regeneration times of the surface oxide films also contribute to the number of ions released once the surface oxide film is disrupted: until the surface oxide film is regenerated, metal ions can be released. Stainless steel has been shown to have a longer regeneration time; therefore, more ions are released. Ti has been shown to have the shortest regeneration time; therefore, fewer ions are released.

10.5 Wear-Related Clinical Failures of Artificial Hip Joints and Diagnostic Principles

This section summarizes numerous wear-related failure modes of artificial hip joints clinically observed.

10.5.1 Clinical Forms of Artificial Hip Wear Failure

Clinical manifestations of the wear failure of CoCrMo–UHMWPE hip joints have many forms. Based on the wear failure mechanism and the speed of the mechanical wear and osteolysis, Wang et al. [58] use a morphological matrix to define two categories of factors; the failure modes are summarized in Table 10.3 and Fig. 10.5.

Table 10.3 Clinical wear failures of CoCrMo–UHMWPE artificial hip joints

		Early (abnormal) wear	Rapid wear	Normal wear
	Osteolysis sensitivity	Annual wear rate of 1.0–1.5 mm	Left column > annual wear rate > right column	Annual wear rate of 0.1–0.2 mm
Sensitive	Through a large number of abrasive flows	The acetabulum exhibited rapid wear in 2–5 years, and the center of the ball has a significant offset	The acetabulum exhibited wear in 5–15 years, and the center of the ball offset without dislocation	A small amount of acetabular wear, and the ball deviates only a small amount
	Interface loosening channel is open	Significant osteolysis and prosthesis loosening before dislocation	Significant osteolysis and prosthesis loosening	Significant osteolysis and prosthesis loosening
	Bone tissue response is sensitive	Typical failure pattern: Fig. 10.5(a)	Typical failure pattern: Fig. 10.5(b)	Typical failure pattern: Fig. 10.5(c)
General	Through abrasive flow	The center of the ball shifted, even grinding exposed metal backing, and the edge collision dislocation occurred before loosening failure	The acetabulum exhibited wear in 5–15 years, and the center of the ball shifted; even grinding exposed metal backing	After 15 years of acetabular wear, the ball head is offset
	Partial opening near the interface port	Visible osteolysis, without loosening failure	Visible osteolysis and loosening failure	Significant osteolysis and prosthesis loosening
	Bone tissue reaction is general	Typical failure pattern: Fig. 10.5(d)	Typical failure pattern: Fig. 10.5(e)	Typical failure pattern: Fig. 10.5(e)
Low	Abrasive flow isolation or barrier	The center of the ball head is obviously offset, and edge collision and dislocation occur	The acetabulum exhibited wear in 5–15 years, and the center of the ball shifted; even grinding exposed metal backing and caused the dislocation failure	The ball head is obviously offset; even grinding exposed metal backing; and the edge collision and dislocation failure occur
	The interface and channel are closed	Osteolysis is not obvious	Osteolysis is not obvious, without loosening failure	A small amount of osteolysis, without loosening
	Bone tissue reaction is low	Typical failure pattern: Fig. 10.5(d)	Typical failure pattern: Fig. 10.5(f)	Typical failure pattern: Fig. 10.5(d)

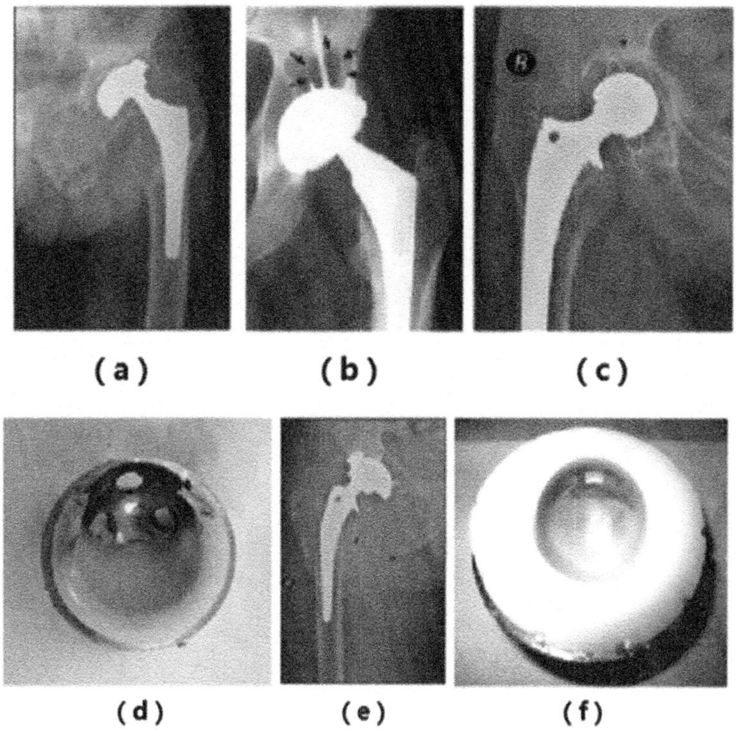

Fig. 10.5 Clinical wear failure of artificial hip joints. Reproduced from Ref. [58] with permission from the *Journal of Medical Biomechanics*

10.5.2 Analytical Procedure for Artificial Hip Joint Failure Events

The proposed wear failure analysis is based on the aforementioned research and the clinical failure events described in Table 10.4. Incidents are primarily assessed as either normal or abnormal wear (Table 10.3), and abnormal wear is assessed based on its cause. The proposed assessment procedure is divided into three levels:

First level: Based on the X-ray images, the ball-head offset is measured which helps to premliminarily identify whether the corrosion is normal or not and to determine the possible reasons.

Second level: If necessary, the retrieval from a joint revision surgery is used to analyze the causes of the abnormal wear.

Third level: If necessary, an in-depth analysis is performed, including finite element analysis, to determine the rationality of the prosthesis size. Parallel implants never implanted will be used to extract critical parameters including size and other design features of the implants for detailed analysis. Such assessment is performed to evaluate the contributions of several factors that may cause failures of implants.

Table 10.4 Wear failure assessment procedure

<table>
<tr><td rowspan="2">Assessment levels</td><td colspan="17">Postoperative time/a</td></tr>
<tr><td>Perioperative period</td><td>1</td><td>2</td><td>3</td><td>4</td><td>5</td><td>6</td><td>7</td><td>8</td><td>9</td><td>10</td><td>11</td><td>12</td><td>13</td><td>14</td><td>15</td><td>>15</td></tr>
<tr><td rowspan="4">First level</td><td rowspan="7">Wear problems should not occur</td><td colspan="16">Annual line wear from x-ray film images (mm/a)</td></tr>
<tr><td colspan="5">≥1.0–1.5
Wear of metal backing in 5 years is considered abnormal wear</td><td colspan="5">≥0.6
Wear of metal backing in 10 years is considered excessive wear</td><td colspan="5">≥0.50
Wear of metal backing in 15 years is considered fast wear</td><td>0.10–0.20
Normal wear</td></tr>
<tr><td colspan="16">Evaluation is taken from x-ray images assuming a known prosthesis size</td></tr>
<tr><td colspan="5" rowspan="2">Implant position improper or loosening of the acetabulum, the occurrence of edge rub causing severe abrasive wear of the articular surface
Observe X ray film
Observe the removed objects and check the acetabular rim</td><td colspan="10">Acetabular loosening and displacement, occurrence of edge collision, and debris-induced wear of articular surface
Observe x-ray film</td><td>—</td></tr>
<tr><td rowspan="3">Second level</td><td colspan="10">Joint surface mixed with a foreign body, resulting in a vicious circle of abrasive wear
Observe the surface wear scratches on the removed objects</td><td rowspan="3">—</td></tr>
<tr><td colspan="5" rowspan="2">Fixed cup screw head is incompletely embedded in the nail hole, leading to articular surface movement disorder and a combined failure of the UHMWPE acetabulum and the metal backing
Observe the removed objects and check the interface</td><td colspan="10">The ball head diameter is in the range of 32–36 mm; the volume wear increases; and line wear is increased by abrasive wear
Observe the articular surface wear scratches on the removed objects</td></tr>
<tr><td colspan="3"></td><td colspan="7">Early products did not produce measurements after γ ray disinfection
Observe the removed objects to understand the production level</td></tr>
</table>

(continued)

Third level		Using COMSOL or other 3d wear modeling software, calculate the volume of the joint volume wear from the x-ray images to determine the amount of annual wear (less than 40–100 mm^3/a is normal wear)		
			Suspected patient weight is too large, or daily activities are too aggressive When necessary, finite-element modeling and analysis are performed; compare the calculated contact stress with the normal contact stress value (10) of the articular surface	—
			Suspect product quality conditions, including surface roughness, the sphericity of the ball head and the ball socket, the clearance of the joint, or the material itself When necessary, the same products from the same period should be analyzed	

10.6 Clinical Diagnosis of Total Hip Arthroplasty: A Case Study

Figure 10.6a shows the X-ray results of a patient with right hip pain, who had received total hip arthroplasty 14 years ago. The implant showed severe ball head deviation, periprosthetic femoral osteolysis, and prosthesis loosening.

According to the medical record, the prosthesis diameter was 32 mm, and the external diameter of the acetabular cup was 50 mm at the time of surgery. Based on the external diameter of the acetabular cup in the X-ray image, the original position of the ball head and the acetabulum and the center position of the ball head should be restored. The offset of the center of the ball head was 4.78 mm (Fig. 10.6(b)). Wear occurred over 14 years, and the annual line wear volume was 0.34 mm. Table 10.3 compares different wear categories, including fast wear.

Figure 10.6c shows the simulated wear path of the ball head in the acetabulum based on the 3D COMSOL software. The analysis based on the 3D model suggested that the severe wear with the ball head diameter of 32 mm was primarily attributed to the poor wear resistance of UHMWPE, instead of the ball head.

Patients receiving total hip arthroplasty revision surgery visibly exhibited significant osteolysis in the greater trochanter and multiple cystic cavities in front of and inside the hip joint, which contained a large number of brown liquid and sediment samples of wear particle tissues. The polyethylene acetabulum was worn through after the femoral prosthesis was removed. The acetabulum bone tissue also showed severe osteolysis after the removal of the metal acetabular cup.

10.7 Conclusion

The elastic hydrodynamic lubrication calculation showed that the boundary friction is major for the CoCrMo–UHMWPE artificial joints, and adhesive wear was the main wear condition. The wear failure is primarily caused by mechanical

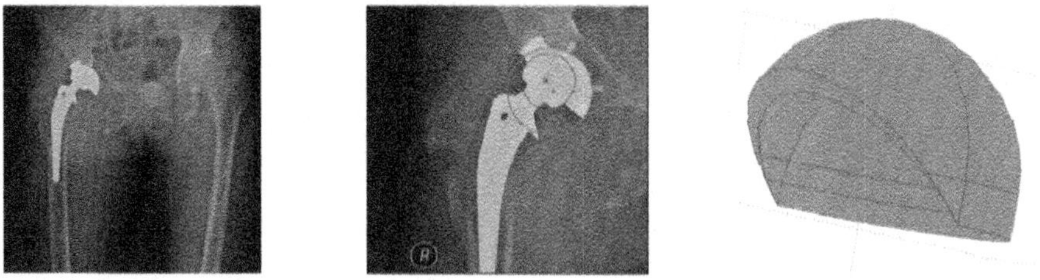

(a) (b) (c)

Fig. 10.6 Prosthesis dislocations caused by wear. Reproduced from Ref. [58] with permission from the *Journal of Medical Biomechanics*

properties and biological responses. There were approximately six to nine clinical failure patterns caused by wear. Restricted by the osteolysis effect, the CoCrMo–UHMWPE joint's biological wear life is approximately 15 years, which agrees with other current studies on prostheses of this type. The failure assessment procedure provided herein is mainly used to determine whether or not the wear is normal and to obtain possible causes of abnormal and excessive wear.

References

1. Affatato S, Brando D (2013) Introduction to wear phenomena of orthopaedic implants. In: Wear of orthopaedic implants & artificial joints, pp 3–26. https://doi.org/10.1533/9780857096128.1.3
2. Bhushan B (1999) Definition and history of tribology. Principles and applications of tribology. Ind Lubr Tribol 98:635
3. Brown SS, Clarke IC (2006) A review of lubricant conditions for wear simulation in artificial hip joint replacements. Tribol Trans 49:72–78
4. Cao SF, Mischler S (2016) Assessment of a recent tribocorrosion model for wear of metal-on-metal hip joints: Comparison between model predictions and simulator results. Wear 362–363:170–178
5. Kennedy FE (2013) Biomechanics of the hip and knee: implant wear. In: Wear of orthopaedic implants & artificial joints, pp 56–92
6. Wang CT (2009) Tribology problems in natural and artificial joint. J Med Biomech 24:317–325
7. Jin ZM, Medley JB, Dowson D (2003) Fluid film lubrication in artificial hip joints. Proc 29th Leeds-Lyon Symp Tribol 211:237–256
8. Jin ZM, Tipper JL, Stone MH et al (2010) Hip Joints: Artificial-biotribology of bearings and biological response of wear debris. In: Webster JG (ed) Encyclopedia of medical devices and instrumentation. John Wiley & Sons Inc, Somerset, pp 514–525
9. Wang CT, Wang YL, Chen QL et al (1990) Calculation of elasto-hydrodynamic lubrication film thickness for hip prostheses during normal walking. Tribol Trans 33:239–245
10. Dowson D (2001) New joints for the Millennium: wear control in total replacement hip joints. Proc Inst Mech Eng H J Eng Med 215:335–358
11. Medley JB, Bobyn JD, Krygier JJ et al (2001) Elastohydrodynamic lubrication and wear of metal-on-metal hip implants. In: Rieker C et al (eds) World tribology forum in arthroplasty. HansHuber, Bern, pp 125–136
12. Wang CT (2008) Human biotribology. Science Press, Beijing
13. Vassiliou K, Elfick A, Scholes S et al (2006) The effect of 'running–in' on the tribology and surface morphology of metal–on–metal BHR device in simulator studies. Proc Inst Mech Eng H J Eng Med 220:269–277
14. Su YL, Yang PR, Fu ZL et al (2010) Elastohydrodynamic lubrication analysis of artificial knee joint line contact under the condition of gait. Tribology 30:80–87
15. Su YL, Yang PR, Wang CT (2010) A full numerical analysis of elastohydrodynamic lubrication in knee prosthesis under walking condition. J Mech Med 10:621–641
16. Bozic KJ, Kurtz SM, Lau E et al (2009) The epidemiology of revision total hip arthroplasty in the United States. J Bone Joint Surg 91:128–133
17. Bergmann G, Deuretzbacher G, Heller M et al (2001) Hip contact forces and gait patterns from routine activities. J Biomech 34:859–871
18. Rydell NW (1966) Forces acting on the femoral head-prosthesis: a study on strain gauge supplied prostheses in living persons. Doctoral thesis, University of Gothenburg, Sweden

19. Ries MD, Scott NK, Jani S (2001) Relationship between gravimetric wear and particle generation in hip simulators: conventional compared with cross-linked polyethylene. J Bone Joint Surg 83A:116–122
20. Ulrich SD, Seyler TM, Bennett D et al (2008) Total hip arthroplasties: what are the reasons for revision? Int Orthopaed (SICOT) 32:597–604
21. Affatato S, Spinelli M, Zavalloni M et al (2008) Tribology and total hip joint replacement: current concepts in mechanical simulation. Med Eng Phys 30:1305–1317
22. Mattei L, Di Puccio F, Piccigallo B et al (2011) Lubrication and wear modelling of artificial hip joints: a review. Tribol Int 44:532–549
23. Kang I, Galvin AL, Jin ZM et al (2006) A simple fully integrated contact-coupled wear prediction for ultra-high molecular weight polyethylene hip implants. Proc Inst Mech Eng H J Eng Med 220:33–46
24. MacDonald SJ, Hanssen AD (2004) Metal-on-metal total hip arthroplasty: the concerns. Clin Orthop Relat Res 429:86–93
25. Mikhael MM, Hanssen SD, Sierra RJ (2009) Failure of metal-on-metal total hip arthroplasty mimicking hip infection: a report of two cases. J Bone Joint Surg 91:443–446
26. Sieber H-P, Rieker CB, Kottig P (1999) Analysis of 118 second-generation metal-on-metal retrieved hip implants. J Bone Joint Surg 81:46–50
27. Fisher J, Jin ZM, Tipper J et al (2006) Tribology of alternative bearings. Clin Orthop Relat Res 453:25–34
28. Williams S, Schepers A, Isaac G et al (2007) Ceramic-on-metal hip arthroplasties: a comparative in vitro and in vivo study. Clin Orthop Relat Res 465:23–32
29. Collier JP, Sutula LC, Currier BH et al (1996) Overview of polyethylene as a bearing material: comparison of sterilization methods. Clin Orthop Relat Res 333:76–86
30. Currier BH, Currier JH, Collier JP et al (1997) Shelf life and in vivo duration. Impacts on performance of tibial bearings. Clin Orthop Relat Res 342:111–122
31. Citters DWV (2012) Failure analysis of orthopaedic implants. In: Wear of orthopaedic implants & artificial joints, pp 377–402
32. Blanchet TA, Burroughs B (2001) Numerical oxidation model for gamma radiation-sterilized UHMWPE: consideration of dose-depth profile. J Biomed Mater Res 58:684–693
33. Van Citters DW (2003) Fatigue failure of UHMWPE: the development and application of novel methods and devices. Master of science thesis, Dartmouth College, NH
34. Currier BH, Currier JH, Collier JP et al (2007) In vivo oxidation of gamma-barrier sterilized UHMWPE bearings. J Arthroplast 22:721–731
35. Sutula LC, Collier JP, Saum K et al (1995) Impact of gamma sterilization on clinical performance of polyethylene in the hip. Clin Orthop Relat Res 319:28–40
36. Fisher J (2009) Tribological and wear problems in joint replacement. Sino-British Symposium on Tribological Performance Test of Artificial Joint in Tianjin, China, 2009
37. Leslie IJ, Williams S, Isaac G et al (2009) High cup angle and microseparation increase the wear of hip surface replacements. Clin Orthop Relat Res 467:2259–2265
38. Williams S, Leslie I, Isaac G et al (2008) Tribology and wear of metal-on-metal hip prostheses: influence of cup angle and head position. J Bone Joint Surg 90:111–117
39. Morlock MM, Bishop N, Zustin J et al (2008) Modes of implant failure after hip resurfacing: morphological and wear analysis of 267 retrieval specimens. J Bone Joint Surg 90:89–95
40. Dumbleton JH, Manley MT, Edidin AA (2002) A literature review of the association between wear rate and osteolysis in total hip arthroplasty. J Arthroplast 17:649–661
41. Schmalzried TP, Jasty M, Harris WH (1992) Periprosthetic bone loss in total hip arthroplasty. Polyethylene wear debris and the concept of the effective joint space. J Bone Joint Surg 74:849–863
42. Schwarz EM, Benz EB, Lu AP et al (2000) Quantitative small-animal surrogate to evaluate drug efficacy in preventing wear debris-induced osteolysis. J Orthop Res 18:849–855
43. Merkel KD, Erdmann JM, McHugh KP et al (1999) Tumor necrosis factor-alpha mediates orthopedic implant osteolysis. Am J Pathol 154:203–210

44. Mao YQ, Zhu ZA, Tang TT et al (2006) The difference of macrophage response to polyethylene and titanium alloy particles. J Shanghai Jiaotong Univ 26:476–479
45. Mao YQ, Zhu ZA, Tang TT et al (2006) Experimental study of the effect of ultra-high molecular weight polyethylene particles on macrophages. Chin J Surg 44:852–855
46. Magone K, Luckenbill D, Goswami T (2015) Metal ions as inflammatory initiators of osteolysis. Arch Orthop Trauma Surg 135:683–695
47. Callaghan JJ, Rosenberg AG, Rubash HE (2007) The adult. Hip 2:1505–1506
48. Gallo J, Goodman SB, Konttinen YT et al (2012) Particle disease: biological mechanisms of periprosthetic osteolysis in total hip arthroplasty. Innate Immun 19:1–12
49. Grosse S, Haugland HK, Lilleng P et al (2014) Wear particles and ions from cemented and uncemented titanium-based hip prostheses-a histological and chemical analysis of retrieval material. J Biomed Mater Res Part B Appl Biomater 103:709–717
50. Ingham E, Fisher J, Stone MH (2003) Wear of historical polyethylenes in total hip prostheses. Biomechanical success and biological failure. Hip Int 13:17–27
51. Cadosch D, Chan E, Gautschi OP et al (2009) Metal is not inert: role of metal ions released by biocorrosion in aseptic loosening-current concepts. J Biomed Mater Res A 91A:1252–1262
52. Baxter RM, Freeman TA, Kurtz SM et al (2011) Do tissues from the revision of highly crosslinked uhmwpe liners contain wear debris and associated inflammation? Clin Orthop Relat Res® 469(8):2308
53. Purdue PE, Koulouvaris NBJ et al (2006) The central role of wear debris in periprosthetic osteolysis. HSS J 2:102
54. Green TR, Fisher J, Stone M et al (1998) Polyethylene particles of a 'critical size' are necessary for the induction of cytokines by macrophages in vitro. Biomaterials 19:2297–2302
55. Ingham E, Fisher J (2005) The role of macrophages in osteolysis of total joint replacement. Biomaterials 26:1271–1286
56. Hatton A, Nevelos JE, Metthews JB et al (2003) Effects of clinically relevant alumina ceramic wear particles on TNF-A production by human peripheral blood mononuclear phagocytes. Biomaterials 24:1193–1204
57. Hanawa T (2004) Metal ion release from metal implants. Mater Sci Eng C 24:745–752
58. Wang CT, Jin ZM, Liao GS et al (2012) Wear analysis and diagnostic reasoning on clinical failure of artificial hip joints. J Med Biomech 27:361–368

The manufacturer's authorised representative in the EU is Springer Nature Customer Service Centre GmbH, Europaplatz 3, 69115 Heidelberg, Germany. If you have any concerns regarding our products, please contact ProductSafety@springernature.com

Printed and bound by CPI Group (UK) Ltd, Croydon, CR0 4YY

30/07/2026

02182034-0001